Composite Materials

Composite Materials

Concurrent Engineering Approach

S.M. Sapuan

ELSEVIER

Butterworth-Heinemann
An imprint of Elsevier

Butterworth-Heinemann is an imprint of Elsevier
The Boulevard, Langford Lane, Kidlington, Oxford OX5 1GB, United Kingdom
50 Hampshire Street, 5th Floor, Cambridge, MA 02139, United States

Library of Congress Cataloging-in-Publication Data
A catalog record for this book is available from the Library of Congress

British Library Cataloguing-in-Publication Data
A catalogue record for this book is available from the British Library

ISBN: 978-0-12-802507-9

For information on all Butterworth-Heinemann publications visit our website at https://www.elsevier.com/books-and-journals

Publisher: Matthew Deans
Acquisition Editor: Christina Gifford
Editorial Project Manager: Anna Valutkevich
Production Project Manager: Susan Li
Designer: Maria Ines Cruz

Typeset by Thomson Digital

To my parents, the Late Haji Nordin (Salit) bin Hasan, and the Late Hajjah Rugayah binti Wagimon and to my beloved wife Nadiah Zainal Abidin and lovely daughter Qurratu Aini Mohd Sapuan.

Contents

ABOUT THE AUTHOR .... **xi**
PREFACE .... **xiii**
ACKNOWLEDGMENTS .... **xv**
NOMENCLATURES .... **xvii**

CHAPTER 1 Introduction 1
Background 1
Design for the environment/design for sustainability 8
Why composites? 9
Concurrent engineering in composite materials development 12
Conceptual design for composites under CE environment 16
Materials selection of composites under CE environment 20
Design for sustainability of composites under CE Environment 22
About this book 24
References 25

CHAPTER 2 Concurrent engineering, product design, and development 29
Introduction 29
The importance of product development 29
Definitions of product design and development 31
Different design process models developed by design experts 31
Description of each stage of total design model 43
The scope of design for CE of composites 49
Conclusions 54
References 54

CHAPTER 3 Composite materials 57
Introduction 57
History of composites 58

Polymer matrix composites 59
Advantages and disadvantages of polymer composites 66
Applications of polymer composites 67
Manufacturing methods of polymer composite materials 75
Metal matrix composites..82
Ceramic matrix composites 84
Nanocomposites..85
Conclusions 88
References..89

CHAPTER 4 Concurrent engineering in design and development of composite products..95
Introduction 95
Previous work on CE in composite product development 98
Some commercially relevant examples of recent Development, trend, and advances in the area of CE for composites 106
CE seems to be an obvious route to any sort of new composite product development: how it is differentiated?...........111
CE have changed the ways industries work: industrial experience of CE for composites at steelcase 111
Previous works by the author and his coworkers in CE for composites 116
Summary 133
References...134

CHAPTER 5 Conceptual design in concurrent engineering for composites...141
Introduction 141
What is conceptual design?...142
Conceptual design in CE environment...144
Design and development of composite products emphasizing on conceptual design 145
Conceptual design methods for composites 149
Conclusions 203
References...204

CHAPTER 6 Materials selection for composites: concurrent engineering perspective...209
Introduction 209
Composite materials selection 211
Concurrent engineering and materials selection.........................213
Attributes of a good materials selection system 214

Materials selection of composite materials using materials data book 215
Procedure-based versus computer-based composite materials selection 215
Pugh concept selection method 216
Materials selection using digital logic method 218
Materials selection using quality function deployment (eQFD) 226
Materials selection of composite materials using materials databases 226
Materials selection of composite materials using Ashby's chart 228
Materials selection of composite materials using knowledge-based system 229
Java-based materials selection 239
Materials selection of composite materials using multicriteria decision making 241
Analytical hierarchy process 246
TOPSIS method 256
Materials selection of composite materials using neural network 258
Materials selection of biocomposites 262
Conclusions 265
References 265

CHAPTER 7 Design for sustainability in composite product development 273
Introduction 273
Design for the environment or ecodesign 273
Evolution of design-for-environment to design-for-sustainability (D4S) 274
Design for sustainability (D4S) 275
Design for sustainability of natural fiber composite materials 281
Summary 291
References 292

APPENDIX A Product design specification (PDS) of composite pedal box system 295
Performance 295
Environment 296
Target product cost 296
Size and weight 297

Materials 297
Ergonomics 297
Safety 297

APPENDIX B Product design specification document for automotive parking brake lever (Mansor, 2014) 299

GLOSSARY 301
INDEX 311

About the Author

S.M. Sapuan (also known as Mohd Sapuan Salit) is a professor of composite materials at the Department of Mechanical and Manufacturing Engineering, Universiti Putra Malaysia and also holds a joint appointment as a head at Laboratory of Biocomposite Technology, Institute of Tropical Forestry and Forest Products (INTROP), Universiti Putra Malaysia (UPM). He is a head of Composite Technology Program at UPM, Head of Engineering Composites Research Group, Faculty of Engineering, UPM and Vice President of Asian Polymer Association. He currently holds the SEARCA Regional Professorial Chair. He is a fellow of Society of Automotive Engineers International, Malaysian Scientific Association, Institute of Materials Malaysia, and Plastics and Rubber Institute, Malaysia. Professor S.M. Sapuan received the BEng, MSc, and PhD degrees from University of Newcastle, Australia; Loughborough University, Leicestershire, UK; and De Montfort University, Leicester, UK. He specializes in the concurrent engineering for composites, natural fiber composites, materials selection for composites, conceptual design for composites, and design for sustainability for composites. He has authored more than 550 journal papers and 80 chapters in books and authored/edited 17 books. He has supervised more than 50 PhD candidates, who had already completed their studies. He has received ISESCO Science Prize, Khwarizmi International Award, University of Newcastle, Australia Alumni Award, Vice Chancellor Fellowship Award, UPM, and Rotary Research Award. He is the author of *Tropical Natural Fibre Composites*, and *Materials Selection and Design*. He is also the editor of *Composite Materials Technology: Neural Network Applications*, *Manufacturing of Natural Fibre Reinforced Polymer Composites*, and *Green Composites: Manufacturing and Properties*.

Preface

The essence of concurrent engineering (CE) is the consideration of manufacturing and other related issues early in the design stage of product development. Composite material is considered an important class of material used in many industries such as automotive, aerospace, defense, marine, and construction industries. A number of conference and journal papers were recently written in the area of CE (design for manufacture) for composites but it is very difficult to find any book either authored or edited dealing with the topic. Some books are written or edited with the title "design and manufacture" of composites but to the author's knowledge, book with the title "Design for Manufacture" (CE) for composites is not available in the market in the recent years.

In this book, important aspects of CE for composites are presented such as CE activities in composite product development, design of products from composites emphasizing on conceptual design for composites, materials selection under the umbrella of CE for composites, and design for sustainability for composites. Some of the information and knowledge presented in the book are reviews of previous work on this topic while the works of the author and his coworkers related to this topic are also included.

In addition, together with such topics mentioned earlier, the readers have the opportunity to read chapters on engineering design process, engineering design models, and basic knowledge of composite materials. Design methods used to generate design concepts for products from composites include brainstorming, biomimetics, cross-industry innovation, exploitation of existing system, gallery method, Why?Why?Why?, morphological chart, blue ocean strategy (BOS), mind mapping, and TRIZ. Many case studies and examples given in this book are related to the design and development of products of composites with the fiber reinforcements made from natural fibers ("green" materials).

S.M. Sapuan
Serdang, Selangor, Malaysia
2017

Acknowledgments

Alhamdulillah—all praises to almighty Allah who has made it possible for the author to complete this book.

Research work related to this book received financial supports in the forms of research grants from various agencies such as Universiti Putra Malaysia, Ministry of Science, Technology and Innovation Malaysia, Ministry of Higher Education, Malaysia, Ministry of Agriculture and Agro-Based Industry Malaysia, and SEARCA Regional Professorial Chair Grant.

Finally the contribution of the following individuals is also acknowledged: Nadiah Zainal Abidin, Qurratu Aini Mohd Sapuan, Hamim Izwan Mohd Hamdan, Ahmed Ali Basher Ahmed, Lee Ho Boon, Anne Hashim, Salihuddin Hamzah, Majid Davoodi Makinejad, Janti Mohamadeen, Hamdan Mohd Ali, Adanan Nipah, and Azli Mohd Aridi.

Nomenclatures

2D	Two dimensional
3D	Three dimensional
ABS	Acrylonitrile-butadienestyrene
AFM	Atomic force microscopy
AFP	Automated fiber placement
AHP	Analytical hierarchy process
AISI	American Iron and Steel Institute
ANN	Artificial neural network
ANP	Analytical network process
ARB	Antiroll bar
ASTM	American Standards for Testing of Materials
ATF	Advanced Tactical Fighter
BMC	Bulk molding compound
BMI	Bismaleimide
BOS	Blue ocean strategy
BS	British Standards
CA	Concurrent approach
CAD	Computer aided design
CAE	Computer aided engineering
CAM	Computer aided manufacturing
CAPEX	Capital expenditure
CAS	Chemical Abstracts Service
CC	Composite characteristics
CDSS	Conceptual design support system
CE	Concurrent engineering
CES	Cambridge Engineering Selector
CESAR	Cost effective small aircraft
CFRC	Carbon fiber reinforced composites
CI	Consistency index
CIE	Concurrent/integrated engineering
CIM	Computer integrated manufacturing
CMC	Ceramic matrix composites
CMNC	Ceramic matrix nanocomposites
CNC	Computer numerical control
CP	Composite performance
CP	Conceptual parameters

CPD	Concurrent product development
CR	Consistency ratio
CTE	Coefficient of thermal expansion
CVD	Chemical vapor deposition
CVI	Chemical vapor infiltration
D4S	Design for sustainability
DFA	Design for assembly
DFC	Design for cost
DFE	Design for the environment; design for environment
DfE	Design for the environment; design for environment
DFM	Design for manufacture; design for manufacturing; design for manufacturability
DFMI	Design for metal inserts
DFMW	Design for minimal weight
DFQ	Design for quality
DFR	Design for recycling; design for recyclability
DFS	Design for sustainability
DFX	Design for "X"
DLM	Digital logic method
DMC	Dough molding compound
DMU	Digital mock-up
DoD	US Department of Defense
DoE	Design of experiments
DPF	Date palm fibers
DSC	Differential scanning calorimetry
DSP	Decision support problems
DTC	Design to cost
EC	European Commission
ECD	Environmentally conscious design
ECE	Economic Commission for Europe
EFB	Empty fruit bunch
ELECTRE	ELimination and Choice Expressing REality
EN	European Standards
EPP	Expanded polypropylene
EU	European Union
FEA	Finite element analysis
FEM	Finite element method
FMEA	Failure mode and effect analysis
FMVSS	US Federal Motor Vehicle Safety Standards
FRP	Fiber reinforced polymer
FSW	Friction stir welding
FTIR	Fourier transformed infrared spectroscopy
GD	Green design
GF	Glass fiber
GFRP	Glass fiber reinforced polymer
GMT	Glass mat thermoplastics
GPD	Global product development
GUI	Graphical user interfaces
GVWR	Gross vehicle weight rating
HDPE	High density polyethylene

HMS	High mountain syndrome
HoQ	House of quality
HSE	Health Safety and Environment
ICS	Integrated conceptual selection
ICT	Information communication technology
IDA	Institute for Defense Analyses
IEC	International Electrotechnical Commission
IMC	Injection molding compound
IPD	Integrated product development
ISCIE	Information system for concurrent/integrated engineering
ISO	International Organization for Standardization
IT	Information technology
JIS	Japanese Industrial Standards
KBE	Knowledge-based engineering
KBS	Knowledge-based system
KEE	Knowledge engineering environment
LCD	Life-cycle design
LCP	Liquid crystal polymer
LDPE	Low density polyethylene
MADM	Multiple attribute decision making; multiattribute decision marking
MC	Moisture content
MCA	Multicriteria analysis
MCDM	Multicriteria decision making: multiple criteria decision making
MEKP	Methyl ethyl ketone peroxide
MMC	Metal matrix composites
MMNC	Metal matrix nanocomposites
MODM	Multiple objective decision making; multiobjective decision making
MPI	Moldflow insight
NBOS	National Blue Ocean Strategy
NDA	Nondisclosure agreement
NFC	Natural fiber composites
NFP	Natural fiber properties
NFRC	Natural fiber reinforced composites
NHTSA	US National Highway Traffic Safety Administration
NMR	Nuclear magnetic resonance
NN	Neural network
OOA	Out of autoclave
OPEX	Operational expenditure
PA	Polyamide
PACKS	Parametric composite knowledge system
PBP	Polymer-based properties
PBT	Polybutylene terepthalate
PDC	Product development center
PDM	Product data management
PDS	Product design specification
PE	Polyethylene
PEEK	Polyether etherketone
PEI	Polyetherimide
PEK	Polyether ketone

PES	Polyether sulfone
PET	Polyethylene terephthalate [sometimes written poly(ethylene terephthalate)],
PHA	Polyhydroxyalkanoate
PLA	Poly(lactic acid) or polylactic acid or polylactide
PMC	Polymer matrix composites
PMNC	Polymer matrix nanocomposites
PP	Polypropylene
PPE	Polyphenylene ether
PPO	Polyphenylene oxide
PPS	Polyphenylene sulfide
PSO	Particle swarm optimization
PSU	Polysulphones
PU	Polyurethane
PVC	Poly(vinyl chloride)
PVD	Physical vapor deposition
QC	Quality control
QFD	Quality function deployment
QFDE	Quality function deployment for environment
RI	Random index
RTM	Resin transfer molding
SAN	Styrene acrylonitrile
SCAMPER	Substitute, Combine, Adapt, Magnify, Put to Other Uses, Eliminate, Rearrange
SE	Simultaneous engineering
SEM	Scanning electron microscopy
SHS	Self-propagating high temperature synthesis
SLA	Stereolithography
SMA	Styrene maleic anhydride
SMC	Sheet molding compound
SPF	Sugar Palm Fibre
SPHC	It is a symbol used in the Japanese Industrial Standards for steel sheets.
SPS	Sugar Palm Starch
STM	Scanning tunneling microscopy
TAPPI	Technical Association of the Pulp and Paper Industry
TBL	Triple bottom line
TEM	Transmission electron microscopy
TOPSIS	Technique for order of preference by similarity to ideal solution
TRIZ	*Teoriya resheniya izobretatelskikh zadach* (the theory of inventive problem solving)
UPM	Universiti Putra Malaysia
VARTM	Vacuum assisted resin transfer molding
VIKOR	VIse Kriterijumska Optimizacija Kompromisno Resenje
VOC	Volatile organic compounds
WCED	World Commission for Environment and Development
WPG	Weight percentage gain
WPM	Weighted property method
www	World Wide Web
X	"X" ability
XPS	X Ray photoelectron spectroscopy

CHAPTER 1

Introduction

BACKGROUND

In the modern business world, manufacturing companies are facing many challenges and fierce competition from their competitors and they have to steadfast in their pursuits in order to remain relevant and acceptable in the marketplace. In the turbulent business and economic environment, companies can no longer be complacent with their current practices if they were to stay relevant in the marketplace. In the knowledge-based era, strong dependence on knowledge and information would ensure the production of more successful products. New design and manufacturing approaches should be sought in product development. In traditional sequential product development approach, the product is designed by the design engineers, and then the design and drawings are sent to the manufacturing engineers for the manufacture. This traditional approach in product design and development suffers from the absence or limited knowledge and information regarding the later stage of product development, such as fabrication, packaging, sales and marketing, service, maintenance, disposal, assembly, sustainability, and recycling. The sequential nature of the activities had caused longer development time, incurred more cost, and led to poor quality products, where further activities cannot be commenced until the earlier activities are completed, that is, all stages have to be performed consecutively. It is particularly true, for instance, in the case of automobile development, where packaging is very important as many components have to be installed in a very tight space. Without early consideration of packaging, redesign and delays could be expected. It leads to many engineering design changes, production problems, such as scrap, defects, and rework, and a product will be less competitive in the marketplace. This approach is termed serial engineering or sequential engineering (Fig. 1.1A) and it is also known as "the waterfall model" (Fig. 1.2). This model is taken from software development process where the sequence of events in product development is flowing downward to resemble a waterfall.

Design engineers can no longer afford to work in isolation and the communication and information barriers need to be broken. Therefore concurrent engineering (CE) approach (Fig. 1.1B) provides the solution for this problem. In

CONTENTS

Background 1

Beyond Manufacturing Competition 4

Concurrent Engineering Definition 5

Approaches to CE Implementation 5

Aliases of CE 6

Benefits of CE 7

Design for the Environment/Design for Sustainability 8

Why Composites? 9

Concurrent Engineering in Composite Materials Development 12

Previous Studies of Concurrent Engineering for Composites 14

Conceptual Design for Composites Under CE Environment 16

Materials Selection of Composites Under CE Environment 20

Importance of Materials Selection 21

Composite Materials Selection 22

(Continued)

Composite Materials. http://dx.doi.org/10.1016/B978-0-12-802507-9.00001-5

Design for Sustainability of Composites Under CE Environment..........22
About This Book..........24
References..................25

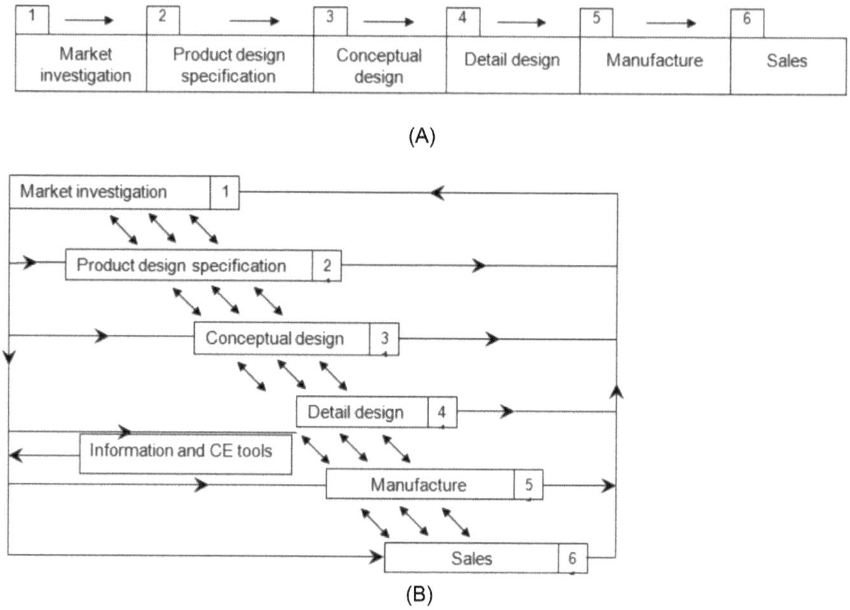

FIGURE 1.1
(A) Serial engineering. (B) Concurrent engineering. *Courtesy of Dr. Hambali Ariff, UTeM, Melaka, Malaysia.*

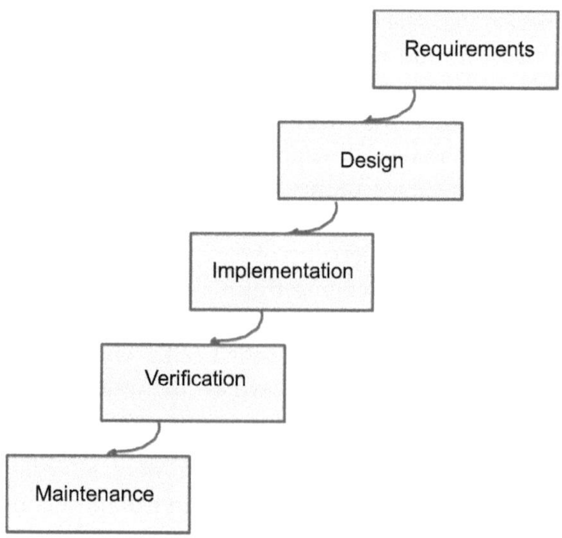

FIGURE 1.2 The waterfall model.

product development process, the physical or virtual departmental barriers between design and manufacture should be removed in order to achieve products that are produced at lower cost, shorter time, and higher quality. The "throw over the wall syndrome" (Fig. 1.3A) should be replaced with concurrent engineering (Fig. 1.3B) where the CE team carried out coordinated activities to

FIGURE 1.3
(A)"Throw over the wall syndrome" (sequential engineering). (B) Concurrent engineering.

ensure better communication among the team members and design reworks, defects, and scraps are omitted or reduced by virtue of the strong emphasis on developing a product right the first time. CE is a powerful business blueprint that is focusing on long-term measures and benefits.

CE seeks to consider all life-cycle product development activities including manufacture, sale, disposal, maintenance, assembly, and recycling in the earlier stage of the design process in order to achieve principally cost reduction, time compression, and higher quality product (Sapuan and Mansor, 2014; Sapuan et al., 2006) in view of fulfilling customers' satisfaction. CE requires that the manufacturing and other downstream activities be considered at the early stage of design process so that many unnecessary problems in the later stages of product development can be overcome. CE is the way to reduce the lead-time between the start of the design and the manufacture of a product by ensuring that design for production consideration is commenced from the inception of product design. Traditional CE was normally focusing on the activities within the organization but beyond that norm, the suppliers' involvement in product development was later emphasized, so that the development becomes more holistic in nature. It is because vendors and suppliers are, in reality, the major contributors toward successful product development.

Through the adoption of CE, Japanese, in the past, had the ability to develop products with high-technology specifications and ability to meet the customers' need. CE relies on team work and the use of disciplined techniques. Japanese were regarded as more technically competent, spending less money to produce things compared other leading nations and they paid more attention to the early stage of design (Wiendahl and Stritzke, 1998). These practices should be followed by other nations for the successful implementation of CE.

Beyond Manufacturing Competition

CE had successfully helped various industries to remain competitive at marketplace. However, Mastura et al. (2016) explored the possibility of implementing CE by making the competition irrelevant as generally promoted in *Blue Ocean Strategy (BOS)* (Kim and Mauborgne, 2005). The major frameworks of BOS are value innovation, the strategic canvas, the four-action framework, and six paths framework. In their work, Mastura et al. (2016) implemented two of the frameworks of BOS, that is, the four-action framework and strategy canvas. It helps to craft a new value curve through four main questions in the product development in the design of automotive antiroll bar from composites. These four actions are: eliminate, reduce, raise, and create.

- *Eliminate*: Which parts of the design are not necessary, and should be eliminated?
- *Raise*: Which parts should be raised well above the design's standard?

- *Reduce*: Which parts should be reduced well below the design's standard?
- *Create*: Which new parts should be created?

A subsection in Chapter 5 of this book on the National Blue Ocean Strategy (NBOS) implemented by the Malaysian government is also discussed. The objective of this initiative is to organize programs or to provide services to the public, which can be implemented in short time, with low cost and high quality, that is, impactful projects. In fact, these three attributes (cost, time, and quality) are the major benefits in CE as emphasized by CE experts. The project involves utilization of agricultural wastes and to transform them into valuable products. This value creation initiative focuses on taking full benefits of "unwanted" and abandoned sugar palm trees in a village in Malaysia. NBOS supports researchers financially and technically in the development of products from sugar palm trees. The products being developed include sugar palm starch (for biopolymer) and sugar palm fibers; two important constituents of composites. The project also involved developing fully degradable biopolymer composites, which is an important element of design for sustainability (DFS), one of the "X" abilities of CE.

Concurrent Engineering Definition

One of the popular definitions of CE was given by the United States Institute for Defense Analyses (IDA) Report R-338 (1986) and it is:

> *A systematic approach to the integrated, concurrent design of products and their related processes including manufacture and support. This approach is to cause the developers, from the outset, to consider all elements of the product life cycle from concept through disposal, including quality, cost, schedule, and user requirement.*

Approaches to CE Implementation

CE is approached in different manners by different researchers. Kitto (1995) pointed out that CE and computer integrated manufacturing (CIM), such as computer numerical control (CNC), finite element analysis (FEA), robotic, rapid prototyping, computer aided design (CAD) and computer aided manufacturing (CAM) are regarded as two separate entities. He integrated both concepts in the teaching of manufacturing engineering program and believed that CE is actually a team work. Kulak and Plaskacz (1996) carried out research on numerical method for studying structural integrity of polymer composite automotive components. The system has the capability to display engineering analysis models at several geographical locations to reduce the time from design to manufacture, which is one important element of CE. In their work, CE and numerical tools are considered as one single entity and not separate concepts.

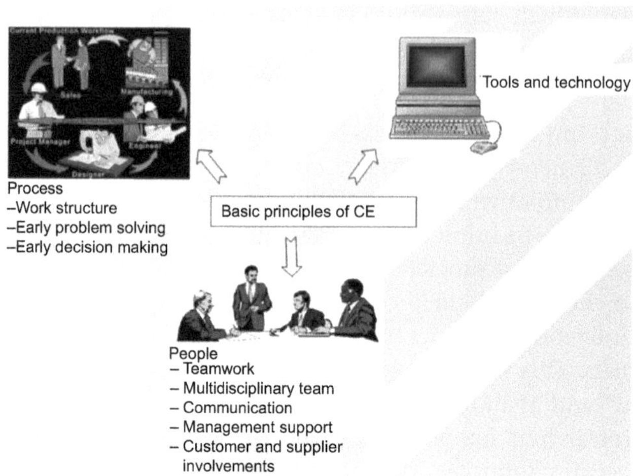

FIGURE 1.4 Basic principles of CE.
Courtesy of Dr. Hambali Ariff, UTeM, Melaka, Malaysia.

According to Balamuralikrishna et al. (2000) the driving force behind the increased practice of CE is modern computer technology but they believed that in CE, managing the modern production organization is more important than advanced technology. Poolton and Barclay (1998) divided CE into two broad categories namely soft and hard CE. Soft CE is further divided into people (like team leadership) and process (like project management) while hard CE includes tools and techniques like CAD and FEA, as well as formal methods like design for assembly (DFA) and failure mode and effect analysis (FMEA). Jo et al. (1993, p. 7) reported that there are two main approaches in the implementation of CE: team-based and computer-based approaches. Basic principles of CE can be summarized into several categories, such as tools and technology, process, and people as shown in Fig. 1.4.

Aliases of CE

Experts have called CE with many different terms but two most common terms are CE and simultaneous engineering (SE). Many terms are more specific to the application for instance, if the emphasis is more related to the environment, CE is termed design for the environment (DFE). In general, CE is known with many other terms, such as: design for manufacture (DFM), design for manufacturability (DFM), design for manufacturing (DFM), design for assembly (DFA), design for quality (DFQ), concurrent product development (CPD), parallel engineering, concurrent product and process design, integrated product and process development, multidisciplinary team approach, design to cost (DTC), design for recyclability (DFR), design for reliability (DFR), design for the environment (DFE, DfE), design for test, design for sustainability (DFS and D4S), design fusion, concurrent/integrated engineering (CIE), synchronized engineering, integrated product development (IPD), design for "X" (DFX), and design for "X" ability (Prabhakaran et al., 2006). In design for "X," "X" is referred to as different attributes, such as assembly, quality, reliability, minimal weight, environment, and safety. The main essence of CE is to consider manufacturing issues early in the design process.

Benefits of CE

The following quote from Andrew Burton, General Manager, Structural Science Composites Ltd. (SSC), Barrow-in-Furness, UK (Concurrent Engineering Ltd., 2016) is an example of a real industrial application of CE.

> *We've reduced the time it takes for product design to prototyping from 6-9 months to just over 8 weeks with the help that Concurrent Engineering has given us in the successful implementation and use of the Creo 3D product design tool.*

Obviously time compression is the major benefit of CE and it leads to other benefits to the organization. The following are general advantages of CE implementation:

- Improved overall product quality
- Reduced time to market
- Decreased product design and development time
- Optimized engineering design cycles
- Enhanced productivity
- Early solving of downstream issues
- Reduced cost in the long run
- Reduced scrap
- Reduced design rework
- Early detection/discoveries of design problems
- Integrated product development team
- Integrated project management
- Increased sale of the products
- Encouraged information sharing
- Promoted team communication
- Reduced production problems
- Prolonged service life
- Decreased engineering changes
- Reduced product defects
- Increased customer satisfaction
- High assets return on investment
- High productivity of white-collar workers
- Facilitated ease of solving conflict

- Facilitated product weight reduction
- Reduced field failure rate
- Increased confident level in manufacturing
- Environment and sustainability consideration

However, CE also suffers from the following disadvantages and challenges:

- Huge investment in terms of computer software and hardware, networks, manpower, and other resources
- Design reviews have to be done early in product development process
- Communication among the CE team members may not be effective
- Data and information management may be difficult
- Resistance to organizational change
- The perception of CE as a mere manufacturing jargon
- The perception of CE as only relevant in the nineties

DESIGN FOR THE ENVIRONMENT/DESIGN FOR SUSTAINABILITY

The traditional approach in CE is to focus on design, manufacture, and maintenance of a product. Although, it is clearly stated in CE definition that, product life cycle, from inception to disposal is taken into consideration, CE practitioners normally limit themselves within the design, manufacture, and maintenance of a product and less emphasis on the environment. The environmental sustainability issues have become paramount in the recent years. Environmental and sustainability issues in product development have become important because the consumers have more awareness of the environmentally conscious products, as well as the implementation of global environmental legislation.

In fact, nowadays automotive and aerospace industries are looking for "green" materials in their attempts to replace some traditional materials with these materials in selected components. In Malaysia, the effort toward making the country "green" has been initiated by the government. For instance, the state government of Selangor has declared Saturday as "no plastic bags" day in the hypermarkets. At Universiti Putra Malaysia (UPM) the practice of giving away plastic bags to patients in pharmacy department has been discontinued as an attempt to build a "green campus."

It is the duty of an engineering designer to design and develop a product that fulfils the customer's needs. But he must remember that in doing so, due consideration should be given to natural and man-made resources. Carrying capacity of ecosystems should not be sacrificed and the option of resources for future generation should be restricted (Fuad-Luke, 2009). It is important to design a product by considering the environment early in the design process. It is done with the concern of safeguarding the planet for future generations.

The terms like green manufacturing, sustainable design, sustainable development, green design, DFS, and design for the environment (DFE) have become important topics of discussion in product development process. DFE and DFS are new terms being coined in relation to current product design and CE. Therefore, the study of CE must also consider DFS and DFE. Vallero and Brasier (2008) emphasized that CE is found to be in line with the term "green design." In fact DFE and DFE, which are the subsets of CE, have the aim to achieve "green product."

DFE and DFS are two closely related concepts and Ashby (2005) clarified that DFE as:

> *The effort to adjust our present design methods to correct known, measurable, environmental degradation; the time-scale of this thinking is 10 years or so, an average product's expected life.*

Ashby (2005) further stated that DFS is the extension of DFE:

> *For longer view: that of adaptation to a lifestyle that meets present need without compromising the needs of future generation (decades or centuries).*

Szczerbicki and Drinkwater (2004) considered DFE to include design for recyclability, reusability, durability, and maintainability of a product. Several environmental attributes, such as ease of recycling, ease of disassembly, ease of reusability, and ease of remanufacturing were considered and these attributes were applied in product development in some selected manufacturing companies. For instance, for the attribute of ease of recycling, which is related to design for recycling, criteria, such as the use of recycled materials, and the minimum use of materials should be considered in the design process. At the end of the study, among others, the use of modular parts, and the reuse of packaging have been suggested for the products. In recent studies by Eastwood and Haapala (2015) on DFS, they have divided sustainability into several categories, such as economic, environmental, and social sustainability. Models were developed to quantify and aggregate sustainability performance matrices for an engineering product, with different design concepts, in terms of economic, environmental, social aspects of sustainability (Eastwood and Haapala, 2015).

WHY COMPOSITES?

Composite materials or simply composites are engineering materials with unique properties. Composites are a combination of at least two materials, with distinct performance compared to the constituent materials acting alone.

The constituent elements, mainly reinforcement fibers and matrices retain their respective properties in composites. These materials have nowadays been found in many different applications ranging from consumer goods like kitchen utensils to aircraft bodies. The major drive in using composites in industries is due to exceptional lightweight performance in addition to excellent strength and stiffness properties. The main reinforcement fibers used in composites are glass, carbon, and aramid fibers. The matrices, which hold the fibers together comprise thermoplastics and thermosetting materials. However, in the recent years, metal and ceramics have been increasingly used as matrices in advanced composites. Recent advances in composite technology include the development and utilization of high-performance composites, such as nanocomposites including composite nanotubes, and graphene-based composites. These relatively new classes of composites demonstrate exceptional performances in service. It is not a surprise to know that graphene can hold an elephant.

The use of advanced composites in high-performance applications, such as in Formula One racing cars, aircraft composites, and sport cars are made possible by means of advanced manufacturing processes, such as vacuum bagging, automated tape layering, and out-of-autoclave (OOA) methods. However, quite surprisingly, the traditional composite fabrication technique, hand lay-up is still dominant in producing such high-performance vehicles and products. Everything is done manually by highly skilful operators and it is a labor-intensive process.

Composites are not only "friends" of high-performance products. They are increasingly used in mundane application, such as making food packaging to replace polystyrene-based food packaging. Composites can address environmental issues if the correct choices of matrices and reinforcements are made. Imagine how "green" the world would be if the fibers are taken from oil palm trunk and the polymer is derived from corn starch. Fully degradable biopolymer composites are now replacing many important consumer products, such as shopping bags, utensils, cutleries, and food packaging used in the form of film for wrapping foodstuff or for replacing polystyrene-based packaging. It should be noted that Henry Ford used to produce a car made from natural fibers mixed with soy-based polymers (Hogg, 2004). So it is evident that fully biodegradable biopolymer composites can be used in simple application but they are also suitable candidates for more advanced applications.

Composites are also being used in general engineering applications, such as in boat building, building and construction, automotive components, furniture, sporting goods, railway industry, energy sector like making electric poles, medical devices, and telecommunication industry.

The composite manufacturing processes are outranging; pultrusion, filament winding, hand lay-up, spray-up, calendaring, pulforming, compression molding, injection molding, resin transfer molding (RTM), extrusion, vacuum bagging, blow molding, autoclave molding, out-of-autoclave (OOA) processing, and automated tape layering. They can cater for one-off products, such as an aircraft body, mass produced automotive components, a small size plate to a huge catamaran hull, a very flexible film for food packaging to a very strong carbon fiber composite bridges. It must also be noted that metal matrix composites are produced from different processes, such as diffusion bonding, powder metallurgy, melt stirring, compocasting/rheocasting, squeeze casting, liquid infiltration under gas pressure, spray codeposition and in situ processes and ceramic matrix composites are made from conventional mixing and pressing, techniques involving slurries, matrix transfer molding, sol–gel processing, chemical vapor deposition (CVD), chemical vapor infiltration (CVI), Lanxide process and in situ technique (Matthews and Rawlings, 1994). Nanocomposites can be developed using chemical vapor deposition, melt intercalation, in situ intercalative polymerization, and exfoliation-adsorption (Hussain et al., 2006).

In composites, fibers are prepared in different arrangements, orientations, forms, and length. Continuous fibers can be laid in unidirectional, bidirectional, and multidirectional ways. Fiber strands can be chopped and made in the form of random mats after applying polymer binder (chopped strand mat) and they can also be used in the form of woven and nonwoven mats. Fibers can also be crushed or be prepared into particulate form and can act as fillers in polymer composites. Fibers bonded with polymer matrices can be made in the form of lamina or ply and those plies can be stacked together to form laminated composites. Fibers in the form of continuous yarns or roving can be wrapped around a mandrel to make filament wound pipes or tubes and they can be preimpreganted with polymer resin to produce pultruded rods or beams.

Composites have numerous excellent properties, such as high strength and stiffness, corrosion resistance, good fatigue properties, good impact properties, low density, good damping properties, which make them attractive and suitable for different industries.

Another important class of fibers are natural fibers or plant-or animal-based fibers. These fibers are attracting the attention of many researchers and companies and they can replace conventional fibers especially glass fibers as reinforcements for polymer composites in less demanding applications. Fig. 1.5 shows kenaf fiber, a natural fiber with excellent mechanical properties suitable to be used as reinforcement for composite in engineering components.

With all these achievements and performances, composites should become very a important class of materials and thorough study of their design and development should be made and reported. Since CE is an important concept in

FIGURE 1.5 Kenaf fiber extracted from kenaf plant.

product development, with major advantages, such as high-quality products, short development time, and low cost investments in mind, it is the right time to focus on CE for composites.

CONCURRENT ENGINEERING IN COMPOSITE MATERIALS DEVELOPMENT

Design of composites required the balance among different attributes, such as light weight, high performance and low materials and fabrication cost. In CE for composites, design, analysis, and fabrication should be carried out differently from other materials where design and analysis should be performed by keeping fabrication in mind. The simultaneous consideration of design, material and manufacturing process, is the key factor in the optimization of composite components. Otherwise major design change is required and it involves a huge amount of investment. For instance, in designing an automotive component using glass fiber reinforced polyamide composites, design engineers proposed to include in the design some additional features, such as ribs and bosses, and then they passed the design to manufacturing engineers. Upon receiving the design, the manufacturing engineers studied the possibilities of fabrication of the component using injection molding process. However, the manufacturing engineers felt that adding such features may incur extra cost in fabrication of the mold, and that the fabrication process would be difficult to carry out, the manufacturing engineers decided to reject the design. The design had to be returned to the design engineers for them to make the necessary changes.

The implementation of CE is very important in composite product development because majority of composite products are developed based on tailor-made design, which means that while designing the products and selecting the materials, manufacturing process issues are studied at earlier stage of design. The design and manufacture of composite products cannot be treated like metals. They require not only an understanding of the material properties but also knowledge of how they are made.

The properties of composites depend very much on how they are processed. Because of this, the manufacturing process must be considered early in the design process. The engineers responsible for fabricating the product must be involved in the design of the components. Such consideration is in fact the essence of CE. The nature of composite product development requires the design of component as well as manufacturing process are considered at the early stage of design process and this consideration fits within the definition of CE. This approach is very crucial in the composite manufacturing industry in order to avoid making costly changes in the final stage of the product development.

The development of composite components using CE technique requires the involvement of all elements of the component life cycle from market investigation through sales including performance, cost, and user requirements. The implementation of CE in the development of composite components will result in the reduction of manufacturing cost and improvement of performance.

The nature of composite product development requires the design of component, as well as manufacturing process at the early stage of design process makes it fits within the definition of CE. While in the recent years, the emphasis in product development is the concurrent design of product and manufacturing process, composite industry has already implemented such ideas knowingly or unknowingly. It is very difficult to design a composite product without designing the materials and without knowing the manufacturing process at the early stage of design.

In the case of polymer composites, whatever the manufacturing technique will be, whether it is injection molding, RTM, filament winding, hand lay-up, spray up, pulforming, autoclave molding, extrusion, vacuum bag molding, vacuum infusion molding, calendaring, or pultrusion, the knowledge of composite materials and their constituent materials are required at the early stage of design process. It is equally applicable whether the fiber system is continuous or random, or in unidirectional, bidirectional woven, multidirectional woven form, or particulate form. In all cases, they have to be decided early in the design stage.

However, as stated earlier, the term CE itself is understood differently by different individuals. Many people believed that CE is computer. According to Prasad (1996), CE can be divided into team-oriented CE and IT-oriented CE, particularly knowledge-based engineering (KBE) oriented CE. In composite, both approaches of CE can be adopted very successfully.

Previous Studies of Concurrent Engineering for Composites

It was pointed out by Mayer (1993) that CE is essential in composite design, whereby material properties are only fully realized on manufacture. Polymer composite products are mainly made through tailor-made design, which means that while designing the product and selecting the materials, manufacturing process is studied at the very early stage of the development process and it is in fact the essence of CE.

Schwartz (1997) predicted that small software would design both the products and the production method. He added that CE of composite materials is the technique needed to meet and keep a generation ahead of competition and stay focused on manufacturability. The interest in design of composite components had been caused by the advent of computer simulation packed with a multitude of analysis programs. Engineers no longer do the task of product analysis and product design separately but the task has been integrated and was performed concurrently by a sole engineer. Furthermore these user-friendly analysis programs permit a designer to evaluate not only the form and function of a design but also the manufacturability of that design very early in the design cycle. This concurrent design approach is essential for polymer composite since processing affects fiber orientation, which in turn greatly affects the mechanical properties of the composites. Nowadays, the prediction of Schwartz (1997) came into reality because such software system are available in the market. In this book, software system and other tools, developed and studied by various workers within the CE environment for composite development are reviewed.

Koudelam (2006) reported that Advanced Research Laboratory (ARL), University of Pennsylvania, USA employs CE in the development of composite structural products. The integration between all departments including acoustics, materials, structures, manufacturing, and quality control was ensured and they also interact continuously among each other. This is carried out with the aim to achieve optimum performance, cost, structural integrity, and reproducibility in the design, fabrication, and testing of full-scale composite structures from the customers. In line with many other works in this research area, integration has been the major strength of this project.

Kulak and Plaskacz (1996) developed virtual reality tools to display computational mechanics results used for enhancing numerical work and CE applications. The numerical methods were used to evaluate the structural integrity of laminated polymer composite automotive components. This resulted in economical design and manufacturing of lightweight, crashworthy automobiles.

Prabhakaran et al. (2006) used one of the important composite manufacturing techniques called RTM in their study for design for "X"-abilities [design for minimal weight (DFMW), DFM, design for metal inserts (DFMI), DFQ, design for reliability, and DFE] based on an approach called a graph theory. It combines all the design aspects to generate solutions by a mathematical model, in particular the matrix algebra. This work is very unique because it attempted to integrate the very traditional method of graph theory with modern CE approach in advanced composite manufacturing technique.

Sobey (2010) studied the development of leisure boat using composite material using CE approach. He paid particular attention to automated communication. The work was supported by means of various tools and techniques, such as multiobjective optimization tool (genetic algorithm), reliability tool like Monte Carlo simulation and design history tool based on artificial neural networks (ANN).

In general, CE for composites, is highly dependent on individual researchers' expertise and strength. For IT experts, CE for composites is carried out using various software tools, which at the end of the day comprise products with high quality, low cost, and shorter time, fulfilling customer expectation are developed. Similar results would be expected for the cases of the composite experts who are specializing in design, fabrication, costing, and management.

The author has been involved in CE for composites since 1994 (Sapuan et al., 1995). The activities performed include implementing CE in the development of automotive components with composites. Different components, such as pedal box system, bumper system, parking brake lever, and antiroll bar were developed. Tools and techniques used include mold flow analysis, finite element analysis, expert system, analytical hierarchy process (AHP) (Hambali et al., 2010), neural network, fuzzy logic, quality function deployment for environment, life cycle analysis, rapid prototyping, and Taguchi method. Experimental work involving materials characterization, and crashworthiness (Fig. 1.6) (Alkbir et al., 2016) analysis have been carried out extensively. The works were carried out in collaboration with composite and automotive industries. However, within these two decades, he found the work on CE in composites is still very limited at a global level. Similarly very limited number of publications were written related to this topic.

A major international furniture manufacturing company, Steelcase, which has production plants in Mexico, France, China, India, and Malaysia, invested on development of furniture components from injection molded glass fiber reinforced polyamide (nylon) composites and the company adopted CE in their product development approach. An important and unique activity related to this topic is that the company implemented global communication in the development of their product. Integration and communication in developing a product is in fact an important agenda of CE. A detailed implementation of CE in composite product development by Steelcase is presented in the form of a case study in Chapter 4 of this book.

FIGURE 1.6 Crashworthiness test of kenaf fiber composites.
Courtesy of Mohamed Al Kateb, Universiti Putra Malaysia.

CONCEPTUAL DESIGN FOR COMPOSITES UNDER CE ENVIRONMENT

As a general rule, design with composites differs from design with metals because of the nature of the materials themselves. However, as far as design methodologies are concerned, traditional design methodologies, such as market investigation, concept generation techniques, product design specification, and IT tools in design are still similar.

Experiences proved that successful fiber reinforced polymer composite product development relied on implementing proper total design approach, such as carrying out proper market investigation, preparing comprehensive product design specification, generating and evaluating design concepts, carrying out detail design and fabrication of products. The author placed great emphasis on the conceptual design stage of product development, as well as materials selection of fiber composites to select the most appropriate materials for a particular component. During this product development stage of design creativity and innovation in design is important. Many concept generation techniques, such as function analysis, morphology chart, value analysis/value engineering, and concept evaluation have been proposed by the experts to come up with the most suitable design solution.

Problem formulation is an important activity in conceptual design within CE environment. When designing a product, a design engineer must be very clear on the design need. Client should be consulted so that design brief is fully understood. Similarly for consumer products, customer requirements should be thoroughly studied and identified. Then he should define the design problem in terms of goal, objectives, and constraints. Problem formulation for composites may be different from conventional design and materials, as many issues related to composites are strangers to other industries.

Information gathering is another important step in conceptual design. A design engineer should gather as much information as possible, to help him in the design process. Sources of information related to the product, to be developed from composites, should be explored and compiled. Examples are books, patents, standards, reports, conference and journal papers, materials databases, and indexes and abstracts.

The product design specification (PDS) should be prepared because this document would help the design engineer to know the limitations and boundaries of his design. PDS should be regarded as a control document and it is a form of a dynamic document, that is, it can be updated as the design progresses from conceptual to detail design stages.

Design with composites is different from design with metals. Many features required in metal components are not required in composites and vice versa. Design is also related to manufacturing. Different manufacturing processes have different design approaches. For instance, injection molded parts may require ribs and bosses, and in hand lay-up, the approach is different. It is normally referred to as CE.

Innovation is important in designing with fiber reinforced polymer composites. Aesthetic value presents materials in different contexts and forms. It is possible to achieve these attributes because design of fiber composites offer some form of innovation and strive toward aesthetically pleasing products. It is for these reasons, among others, a chapter is devoted to innovation and creativity in the design of products from composites (Chapter 5).

Aesthetic value is already embedded in composite products, at least in terms of color and shape. Some manufacturing processes can produce attractive surface finish with the addition of pigment in gel coats. This eliminates the need of secondary painting. Natural fiber composites normally retain their natural or "woody" color. However, study of the aesthetic aspect of composites is beyond the scope of this book.

Conceptual design requires the use of design methods and idea generation techniques. TRIZ, AHP, brainstorming, bimimetics, BOS, gallery method, morphological chart, among others, are needed to generate design concepts. With the help of these methods many unexpected brilliant ideas can be generated and developed.

Brainstorming looks like a simple meeting and discussion but when properly implemented, it can produce remarkably interesting results and findings. Therefore design engineers should try to solve their design problems using this method. One crazy idea may lead to a very important idea. In brainstorming, judgment is suspended and every idea is a good idea. At the end of the session, the members of the design team should decide selected concepts to be developed further.

TRIZ is a design approach to inventions developed by Genrich Altshuller from Russia in 1940s. Altshuller (1998) made significant contribution in the area of idea generation by studying hundreds of thousands of patents and made the important conclusion that the solutions for majority of the problems are quite similar, so they can be grouped into several categories. Problems, pattern of evolution and solutions are repeated in different industries and innovations used scientific effects in other fields of study. He proposed 40 inventive solutions for all the technical problems. Worsening and improving situations are suggested and they are compared with contradiction matrix. Then the solutions are proposed. In composites, what we normally want to achieve is a lightweight product. But according to Altshuller (1998), weight reduction (improving) will be obtained at some prices; that is, ease of manufacture, reliability, and strength are reduced. But fortunately, through contradiction matrix, solutions are proposed. But we need not accept all that is being proposed.

Morphological chart is an established concept generation technique normally used to generate design concepts from a wide range of concepts by breaking down the main function into subfunctions. When we observe around us, very limited inventions or products have a single function, that is, a table comprises table top and stand, car involves body, chassis, engine, and power system and the list goes on. We need a morphological chart to help generate as many solutions as possible. The objective of this method is that all possible solutions of product subfunctions are being considered. Many products made from composites have different subfunctions and morphological chart can play a role. For instance, in the design of automotive accelerator pedal from composites, subfunctions such as means of controlling pedal movement, pedal connection to carburetor, pedal attachment to cable or cylinder, pedal lever profile, ribbing pattern, and pivot shaft location had been considered..

Police incident team in the United Kingdom carried out crime investigation and found information about crime suspects by performing the gallery method (Wright, 1998). Similarly professional dealing with composites should be involved in a gallery method session. In this method contribution from individual and group are equally significant. The individual design concepts are sketched in a paper (one concept per paper) and all the concepts are taped on to the wall (wall mounted). All members walk past the concepts as if they are in an art gallery. They will discuss and suggest improvement on the concepts and a final concept is decided as a group work.

One day, the author was cracking a joke by telling his colleague that his 1.5-year-old child was the author's teacher. The colleague was puzzled with the remark concerning his daughter. Then the author told him, if he wants to learn creativity, he should learn from children as they have a lot of interesting ideas to present. That is why, Why?Why?Why? idea generation technique has been documented for a long period of time and it is being compared to the acts of a persistent child, who has an inquisitive mind. In designing a product from composites, more ideas are welcomed as this is a very new area of study.

Mansor et al. (2014) carried out conceptual design of kenaf fiber polymer composite automotive parking brake lever using integrated TRIZ-morphological chart-AHP method. In the past, efforts to integrate several concept generation techniques had been practiced. It includes integrating AHP with QFD, AHP with TRIZ, TRIZ and fuzzy AHP, TRIZ and axiomatic design, TRIZ, FMEA and EcoDesign.

Davoodi et al. (2011) studied conceptual design of hybrid glass-kenaf polymer composites for automotive bumper beam. They performed concept evaluation using a method called TOPSIS method and this approach is similar to Pugh selection method (Sapuan, 2010) and they came up with the best design concept. Concept selection and evaluation are important because they are related to decision making and making inappropriate decision can be very costly. Pugh selection method has been accepted worldwide an important tool in concept selection for decades.

Magic number of 70% is always quoted by product development experts when they want to highlight that 70% of the product cost is determined during the early stage of the design. This supports the assertion that conceptual design is an important activity in CE. It leads to the need to write the Chapter 5 of this book, that is, concerning conceptual design as an important activity in CE for composites.

Some technological breakthroughs were achieved by observing and mimicking the nature. Leonardo Da Vinci, in an attempt to develop a "flying machine," studied the flight of birds. Wright brothers, who flew the first aircraft had done thorough study and made observations on the flight on pigeons. Biomimetic concept was applied in inventing Velcro tape, which mimicked multiple hooked structures like burs. As far as development of composites is concerned, biomimetics have also made significant contribution in the area. Perhaps one major breakthrough in composites as far as biomimetics is concerned is on honeycomb structure. Honeycomb structure derives its name and idea from honeycomb in a beehive (Pahl et al., 2007). It is light, but when it is compressed it is very strong (Dorworth et al., 2009). Therefore, in thinking about idea generation in composites, biomimetics offer a good tool for it.

Design engineers dealing with composites must also pay attention to the existing technology or analogous products. Designing a composite clutch pedal obviously requires the design engineer to study closely metal clutch pedal. It is a natural choice (Sapuan and Abdalla, 1998). But design engineer should study the function and characteristics of a traditional chair when he is assigned a task to design a car seat. He actually has to study an analogous product that helps him with design in terms of subfunctions like seat and backrest.

Beyond conceptual design, embodiment, and detail designs are very important. However they are not treated in this book because it is outside the scope of the book. It does not mean they are not important. Readers are advised to read books related to embodiment and detail design of composites in order to extend the knowledge gained in this book.

MATERIALS SELECTION OF COMPOSITES UNDER CE ENVIRONMENT

Materials selection is a process to determine the most suitable material for a new product or new design and is normally carried out by materials or design engineers. Materials selection is one of the most important activities in the design process (Hambali et al., 2010). The selection of the most suitable material for a particular application is a crucial task in the design of a product. Fig. 1.7 illustrates the process of materials selection for a new product. The process is generally started by considering various design constraints, such as properties, function, process, and shape. Then, the most suitable material is determined

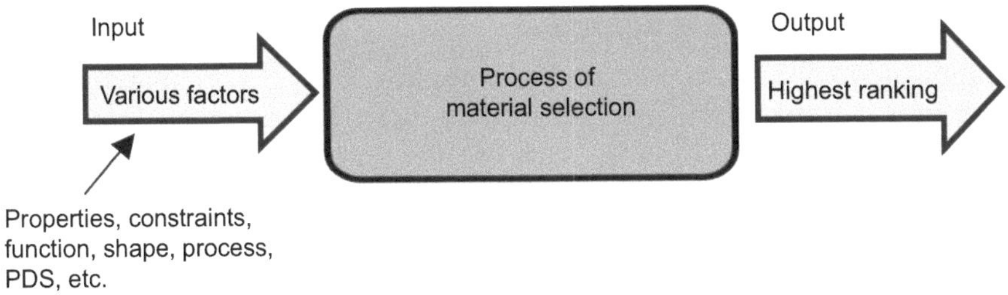

FIGURE 1.7 Material selection process.
Courtesy of Dr. Hambali Ariff, UTeM, Melaka, Malaysia.

by employing materials selection tools or techniques, such as expert system (Sapuan, 2001), artificial neural network (Sapuan and Mujtaba, 2010, p. 317), Pugh selection method (Dieter and Schmidt, 2009), AHP (Ahmed Ali et al., 2015) and Cambridge Engineering Selector (CES) software. Finally, the material that has the highest ranking among all candidates, which satisfied the design constraints, is chosen as the most suitable material.

Ashby (2005) who is successful in his work of materials selection has come up with software CES. It is a commercialized software and has been accepted worldwide. However, materials selection is not just about commercializing software. In pursuit of the knowledge on materials and their selection, one will gain tremendous amount of knowledge and knowledge itself is a great wealth.

Importance of Materials Selection

The selection of an appropriate material for a product design is a difficult and complex task. It is the responsibility of the designers or materials engineers, not only to design a product that meets the specific need, but also to select the most suitable material for the product. A product's success relies on an appropriate materials selection, which reduces product failure and therefore costs, time to market, and returns. It is estimated that there are between 40,000 and 80,000 materials and at least 1,000 different ways to process them and they are continuously being developed. This reveals the need of a systematic approach in selecting the best material for a given application. Materials selection process is often performed in the industry as engineers are always searching for better materials in order to improve the performance of their products. Although other product design considerations matter, a product's overall performance is mainly affected and determined by material selection. Material selection becomes vital for civil and mechanical structures where material cost is almost 50% of the total product cost.

Composite Materials Selection

Composite materials are formed by combining two or more materials that have quite different properties. Composite materials have been applied in various industrial applications, such as automotive, aerospace, civil infrastructure, marine, etc. Recently, the use of composite has been rapidly increased due to improved awareness regarding product performance and increased competition in the global market for lightweight components. Determining the right composite material for a product is a difficult task. It is because many factors have to be considered in ensuring that the selected composite material can produce a component with the desired performance, quality, and cost. Since the composite material can be combined by various kinds of fibers and matrices, the selection of suitable composite materials for a product becomes a difficult task for inexperienced designers. Thus, various material selection tools or methods have been developed in assisting materials and design engineers to select the right composite materials for a product. Although this endeavor takes longer time to come to fruition, and only a small group of experts seemed to benefit greatly from this study, it is strongly believed that the research findings will have significant impacts on the community. It is suggested that effort should be intensified to develop user-friendly software packages that can be used by individuals who have interests on the materials. Simple materials selection system is required to assist design engineers to select the most suitable composite materials in completing their task.

DESIGN FOR SUSTAINABILITY OF COMPOSITES UNDER CE ENVIRONMENT

The importance of sustainability-oriented design and manufacture among materials scientists and engineering community has started to be realized. DFS has manifold targets and among them is the use of biodegradable materials (Kucuksayrac, 2015). For composite materials, the issues of environment may be difficult to address as composite is traditionally using synthetic polymers and fibers. Despite many claims made by researchers that natural fiber composites are biodegradable materials, the claim is not totally correct. It is because, biodegradable composites should have natural fibers as reinforcements (that are biodegradable) as well polymer matrices that are also biodegradable. But in practice, the researchers normally use biodegradable natural fibers as fiber reinforcements but they are still using synthetic polymers as their matrices, which are nondegradable. Therefore claiming these composites as biodegradable composites can be misleading. Perhaps it is partially biodegradable. Synthetic polymers will take a long time to degrade in the environment. However, 100% biobased polymer composites fulfil the requirement of biodegradability as it caters for well-being of future generation. In addition, the raw materials from plants can be renewable and therefore sustainable. Therefore, one of the focuses of this book is on discussion of 100% biobased polymer composites, when the chapter on DFE and DFS is presented. Despite the advantages that they offer, especially when dealing with DFS, the major setbacks with biopolymer composites is that they are costly and difficult for making the products.

Henry Ford practiced the philosophy of CE when he first built a car; he was involved in the design, as well as in the manufacture of the car (Jo et al., 1993, p 5). Fully biodegradable composites had been used since the days of Henry Ford, where it is reported that he was also involved in the testing of the car made from 100% biodegradable materials in 1938; the polymer matrix was made from soy-based resin and was reinforced with straw fibers. Poly(lactic acid) (PLA), one of the commercially available biobased polymers is normally used as a matrix in the study of natural fiber composites. Marques et al. (2012) developed cork-poly(lactic acid) (PLA) composites and used them as materials for the body of city buses in Portugal. A recent research is conducted on CE/DFS of natural fiber reinforced biopolymer composites in food packaging (Sapuan, 2015) and the use of locally available plant, sugar palm, where the sources of starch to form biopolymer and fibers are mainly focused (Figs. 1.8 and 1.9). Many interesting research findings have been obtained from extensive research in this area, which show great potential these materials to be developed as materials for biodegradable packaging (Sahari et al., 2013a,b; Sanyang et al., 2016a,b,c). This topic will also be discussed in Chapter 5.

FIGURE 1.8 Sugar palm fiber and sugar palm starch to form 100% biopolymer composites.
Courtesy of Dr. Sahari Japar, Universiti Malaysia Sabah, Malaysia.

FIGURE 1.9 Sugar palm fibers in woven form.

The life cycle analysis is another area of research related to DFS and the topic has been considered when DFS is studied.

ABOUT THIS BOOK

This book is about the implementation of CE in composite material development. The case for publication of the book is to fill a gap in CE and composites. In many cases, there is a missing link between these two topics. The task can be difficult because CE is normally regarded as "soft" engineering or engineering management subject while "composite" is understood to be "hard" engineering or materials engineering subject. These contrasting subject matters pose a big challenge to the author to come up with the best information. It is not possible to discuss the implementation of CE in all composite manufacturing processes as composite processing is a technology established decades ago. In this book, as much as possible any work on the implementation of CE in composite development either obtained from published literature or from the author's experience is presented. This book is intended mainly for research reference in CE for composite materials. However, the readers from different fields can benefit from the knowledge and information presented, such as in the fields of design, technology management, agriculture and forestry, manufacturing engineering, and materials engineering.

A detailed description of the design process for composite product development is presented in Chapter 2. Many design models developed by design experts, such as from French, Pugh, Pahl and Beitz, Ashby, Jones and Ertas, and Haik, are reviewed. Chapter 3 is devoted to detailed information on composite materials comprising the history of composites, types of composites, types of fibers and matrices, composite fabrication processes, applications of composites, advanced topics in composites like nanocomposites and other topics related to composites. Chapter 4 discusses the various CE models for composite design and development, developed by various CE and composite experts. The chapters also review a number of studies carried out by several experts from major companies worldwide and a case study of CE implementation by an international company producing furniture components using composites is also presented. The chapter also covers the work of the author and his coresearchers on CE implementation in the development of products from composites. A study of conceptual design of composite materials design is presented in detail in Chapter 5. It includes product design specifications (PDS), various methods of concept generation, and concept evaluation techniques that can be applied and related to composite products. The techniques using concept generation include brainstorming, extending search space method also known as WhyWhyWhy?, biomimicry, cross-industry innovation, gallery method, exploitation of existing system, TRIZ, blue ocean strategy, mind mapping, and morphological chart. For concept evaluation process, methods, such as weighted objectives method, Pugh selection method, AHP, direct comparison method, and TOPSIS were studied. Materials selection for composites is presented in Chapter 6. DFS in composite product development is presented in Chapter 7. Glossary of terms related to the topics of CE and composite materials is finally presented at the end of the book.

References

Ahmed Ali, B.A., Sapuan, S.M., Zainudin, E.S., Othman, M., 2015. Implementation of expert decision system for environmental assessment in composite material selection. J. Cleaner Prod. 107, 557–567.

Alkbir, M.F.M., Sapuan, S.M., Nuraini, A.A., Ishak, M.R., 2016. Fibre properties and crashworthiness parameters of natural fibre-reinforced composite structure: a literature review. Compos. Struct. 148, 59–73.

Altshuller, G., 1998. 40 Principles: TRIZ Keys to InnovationTRIZTools, Worchester, MA.

Ashby, M.F., 2005. Materials Selection in Mechanical Design, third ed. Elsevier Butterworth Heinemann, Amsterdam.

Balamuralikrishna, R., Athinarayanan, R., Song, X., 2000. The relevance of concurrent engineering in industrial technology programs. J. Ind. Technol. 16, 1–5.

Concurrent Engineering Ltd., 2016. Composites manufacturer sees 30% year-on-year growth with Reduced Lead Times. Solihull, West Midlands, UK. Available from: http://www.concurrent-engineering.co.uk/creo-case-study-structural-science-composites

Davoodi, M.M., Sapuan, S.M., Ahmad, D., Aidy, A., Khalina, A., Jonoobi, M., 2011. Concept selection of car bumper beam with developed hybrid bio-composite material. Mater. Des. 32, 4857–4865.

Dieter, G.E., Schmidt, L., 2009. Engineering Design, fourth ed. McGraw-Hill Companies, Inc, Boston.

Dorworth, L.C., Gardiner, G.L., Mellema, G.M., 2009. Essentials of Advanced Composite Fabrication & Repair. Aviation Supplies & Academics, Inc, Newcastle, Washington.

Eastwood, M.D., Haapala, K.R., 2015. A unit process model based methodology to assist product sustainability assessment during design for manufacturing. J. Cleaner Prod. 108, 54–64.

Fuad-Luke, A., 2009. EcoDesign, The Sourcebook, third ed. Chronicle Book, San Francisco.

Hambali, A., Sapuan, S.M., Ismail, N., Nukman, Y., 2010. Materials selection of polymer based composite bumper beam using analytical hierarchy process. J. Cent. South Univ. Technol. 17, 244–256.

Hogg, P.J., 2004. The roles of materials in creating a sustainable energy. Presented at the Fourth International Materials Technology Conference and Exhibition 2004, Kuala Lumpur, 23–25 March.

Hussain, F., Hojjati, M., Akomoto, M., Gorga, R.E., 2006. Polymer-matrix nanocomposites, processing, manufacturing, and application: an overview. J. Compos. Mater. 40, 1511–1565.

Institute for Defense Analyses (IDA), 1986. Report R-338. Available from: http://www.dtic.mil/dtic/tr/fulltext/u2/a222714.pdf

Jo, H.H., Parsaei, H.R., Sullivan, W.G., 1993. Principles of concurrent engineering. In: Parsaei, H.R., Sullivan, W.G. (Eds.), Concurrent Engineering: Contemporary Issues and Modern Design Tools. Chapman and Hall, London, pp. 1–23.

Kim, W.C., Mauborgne, R., 2005. Blue Ocean Strategy: How to Create Uncontested Market Space and Make the Competition Irrelevant. Harvard Business Press, Boston, Massachusetts.

Kitto, K.L., 1995. Designing Concurrent Engineering and Computer Integrated Manufacturing into a Manufacturing Engineering Technology Program. Available from: http://fie.engrng.pitt.edu/fie95/3c5/3c52/3c52.htm

Koudelam, K., 2006. Advanced Research Laboratory at the Pennsylvania State University. Available from: http://www.arl.psu.edu/capabilities/mm_compmat.html

Kucuksayrac, E., 2015. Design for sustainability in companies: strategies, drivers and needs of Turkey's best performing businesses. J. Cleaner Prod. 106, 455–465.

Kulak, R.F., Plaskacz, E.J., 1996. Advanced Computing for Automotive Applications Engineering Mechanics. Argonne National Laboratory, Transportation Technology R&D Center, University of Chicago, USA.

Mansor, M.R., Sapuan, S.M., Zainudin, E.S., Nuraini, A.A., Hambali, A., 2014. Conceptual design of kenaf fiber polymer composite automotive parking brake lever using integrated TRIZ-morphological chart-analytic hierarchy process method. Mater. Des. 54, 473–482.

Marques, A.T., Esteves, J.L., Viana, J., Loureiro, N., Arteiro, A., 2012. Design for sustainability with composite systems, Paper Presented at the Fifteenth International Conference on Experimental Mechanics (ICEM15), Porto, 22–27 July.

Mastura, M.T., Sapuan, S.M., Mansor, M.R., Nuraini, A.A., 2016. Application of hybrid sugar palm-filled polyurethane composites in conceptual design of an automotive anti-roll bar. J. Eng. Technol. 7, 102–113.

Matthews, F.L., Rawlings, R.D., 1994. Composite Materials: Engineering and Science. Chapman & Hall, London.

Mayer, R.M., 1993. Design With Reinforced Plastics: A Guide for Engineers and Designers. The Design Council, London.

Pahl, G., Beitz, W., Feldhusen, J., Grote, K.H., 2007. Engineering Design: A Systematic Approach, third ed. Springer, London.

Poolton, J., Barclay, I., 1998. New product development from past research to future applications. Ind. Mark. Manag. 27, 197–212.

Prabhakaran, R.T.D., Babu, B.J.C., Agrawal, V.P., 2006. Design for 'X'-abilities of RTM products—a graph theoretic approach. Concurrent Eng. Res. Appl. 14, 151–161.

Prasad, B., 1996. Concurrent Engineering Fundamentals, Integrated Product and Process OrganizationPrentice Hall PTR, Upper Saddle River, New Jersey.

Sahari, J., Sapuan, S.M., Zainudin, E.S., Maleque, M.A., 2013a. Mechanical and thermal properties of environmentally friendly composites derived from sugar palm tree. Mater. Des. 49, 285–289.

Sahari, J., Sapuan, S.M., Zainudin, E.S., Maleque, M.A., 2013b. Thermo-mechanical behaviours of thermoplastic starch derived from sugar palm tree (*Arenga pinnata*). Carbohydr. Polym. 91, 1711–1716.

Sanyang, M.L., Sapuan, S.M., Jawaid, M., Ishak, M.R., Sahari, J., 2016a. Recent developments in sugar palm (*Arenga pinnata*) based biocomposites and their potential industrial applications: a review. Renew. Sust. Energy Rev. 54, 533–549.

Sanyang, M.L., Sapuan, S.M., Jawaid, M., Ishak, M.R., Sahari, J., 2016b. Effect of sugar palm-derived cellulose reinforcement on the mechanical and water barrier properties of sugar palm starch of biocomposite films. BioResources 11, 4134–4145.

Sanyang, M.L., Sapuan, S.M., Jawaid, M., Ishak, M.R., Sahari, J., 2016c. Development and characterization of sugar palm starch and poly(lactic acid) bilayer films. Carbohydr. Polym. 146, 36–45.

Sapuan, S.M., 2001. A knowledge based system for materials selection in mechanical engineering design. Mater. Des. 10, 687–695.

Sapuan, S.M., 2010. Concurrent Engineering for Composites. UPM Press, Serdang, Malaysia.

Sapuan, S.M., 2015. Concurrent engineering in natural fibre composite product development. Appl. Mech. Mater. 761, 59–62.

Sapuan, S.M., Abdalla, H.S., 1998. A prototype knowledge-based system for the material selection of polymeric-based composites for automotive components. Compos. Part A Appl. Sci. Manuf. 29A, 731–742.

Sapuan, S.M., Abdalla, H.S., Nash, R.J., 1995. Proposed design and manufacturing techniques of polymeric based composite pedal box system, Proceedings of the Eleventh National Conference on Manufacturing Research, Leicester, 12–14 September, pp. 77–81.

Sapuan, S.M., Mansor, M.R., 2014. Concurrent engineering approach in the development of composite products: a review. Mater. Des. 58, 161–167.

Sapuan, S.M., Mujtaba, I.M., 2010. Development of a prototype computational framework for selection of natural fibre reinforced polymer composite materials using neural network. In: Sapuan, S.M., Mujtaba, I.M. (Eds.), Composite Materials Technology: Neural Network Applications. CRC Press, Boca Raton, pp. 317–339.

Sapuan, S.M., Osman, M.R., Nukman, Y., 2006. State of the art of the concurrent engineering technique in the automotive industry. J. Eng. Des. 17, 143–157.

Schwartz, M.M., 1997. Composite Materials—Properties, Non Destructive Testing and Repair, Polymers, Ceramics, Metal MatricesPrentice Hall PTR, Upper Saddle River, New Jersey.

Sobey, A.J., 2010. Concurrent Engineering in the Context of the Composite Leisure Boatbuilding Industry. University of Southampton, UK, PhD Thesis.

Szczerbicki, E., Drinkwater, M., 2004. Concurrent engineering design for environment. Cybern. Syst. 35, pp. 7–8, 667–681.

Vallero, D., Brasier, C., 2008. Sustainable Design: The Science of Sustainability and Green Engineering. John Wiley & Sons, Inc, Hoboken, New Jersey.

Wiendahl, H.P., Stritzke, H., 1998. Logistic oriented product design. J. Mater. Process. Technol. 76, 12–15.

Wright, I., 1998. Design Methods in Engineering and Product Design. McGraw-Hill, London.

CHAPTER 2

Concurrent Engineering, Product Design, and Development

INTRODUCTION

Composite materials have made a significant impact on weight reduction in aerospace industry. Similar contribution is found in automobiles, sporting goods, building structures, and other industrial applications. In general, products made from composites are not only light in weight but also demonstrate other desirable properties, such as stiffness and strength. In sports, for instance, composites have found their applications in golf club, bicycle in cycling, and in tennis, badminton, and squash racquets. In the past, these racquets were mainly made from wood. The era of wood in racquets is over and later on aluminum became the alternative. Nowadays, glass, carbon, Kevlar, or ceramic fiber reinforced epoxy composites are used to produce racquets by virtues of their strength, low weight, durability, and aesthetics. In automobile, many components have now been replaced with composite materials, which were previously made from steel, aluminum, plastics, or wood. Examples include air intake manifolds, accelerator and clutch pedals, drive shaft, bumper beam, door panel, suspension, spoiler, and some engine components. This move has led to a lightweight vehicle in the interest of weight and fuel savings.

THE IMPORTANCE OF PRODUCT DEVELOPMENT

The development in materials goes hand in hand with the improvement in manufacturing technology. Development in materials, as well as in product design, has been assisted with respected management technique, notably CE, as expounded in detail in the previous chapter. CE is regarded as an umbrella for the best product development process (Sapuan and Mansor, 2014). In fact, product development is a key process in concurrent engineering (CE). Producing a product within the agreed time frame between all the stakeholders involved, such as manufacturers and suppliers is the important selling point of CE. It can rightly be claimed that CE is focusing on milestone rather than

CONTENTS

Introduction.................29

The Importance of Product Development.29

Definitions of Product Design and Development...............31

Different Design Process Models Developed by Design Experts............31

Description of Each Stage of Total Design Model..............43

Clarifying Need.....................44

Market Investigation.............46

Product Design Specifications........................46

Conceptual Design................47

Embodiment Design.............47

Detail Design.........................49

Manufacture..........................49

The Scope of Design for CE of Composites..49

Classification Characteristics of Design Tasks in Composites.......................49

Conclusions................54

References..................54

Composite Materials. http://dx.doi.org/10.1016/B978-0-12-802507-9.00002-7

conventionally focusing on process. Product development is a practical task that requires the full-hearted involvement of a product development team to develop a product right from a very vague design brief or customer needs through specifications and up to the fabrication of a working prototype and beyond. It is in fact, a process of translating the needs of the users into product through product design and manufacture. In CE, product development means integration of various functions within and from outside the organization, such as engineering designers, manufacturing engineers, suppliers, technical support personnel, and administrative and sale personnel. It ensures good communication among them through advanced technology and ICT tools and through regular meetings. Product development is not only looking after the product until it is ready to be marketed but also other after sale issues, such as warranty, disposal, maintenance, environment, and recycling.

Many models on product development and design have been proposed and reported in established literature. These models will be explained in detail in section to come. The main core activities in a typical design process model include conceptual design, embodiment design, and detail design. Ashby and Johnson (2002) suggested that inputs or major contributing factors to this design process (concept, embodiment, and detail design) include the market, science and technology, sustainability and environment, economics and investment climate, and aesthetics and industrial design. The task of carrying out product design is the work of a professional called a designer. Designer is also referred to as design engineer, engineering designer, or design and development engineer (Haik, 2003). In CE, the engineering designer cannot work in isolation, but he has to work in collaboration with other personnel from within and outside the organization as strongly promoted by CE experts. By employing CE during product development process, many design problems can be solved at the very early stage of design process (Sapuan et al., 2003). The amount of money spent and manpower required in product development is increasing as the development progresses. During the conceptual design the amount of resources is lesser compared to that of manufacture. It implies that it is still very economical to do any changes in design during this stage. Therefore CE promotes cost reduction by engaging other personnel early in the design process, where any changes to be made during this stage are very desirable.

Carter and Baker (1992) identified five factors that change product development environment and they are pivoted around developing products to meeting customer needs at highest quality, at competitive price in the least time. These five forces of changes are technology, tools, tasks, talent, and time (5*T*s). In CE environment, these 5*T*s have to be nurtured and kept controlled. The multidisciplinary team has to ensure that these 5*T*s are used to fulfil the customer satisfaction.

DEFINITIONS OF PRODUCT DESIGN AND DEVELOPMENT

It is appropriate to examine some definitions of several key terms used in this chapter.

A *product* is defined in the *American-Heritage* dictionary as "something that is produced naturally or by human effort." *Design* is stated by Sapuan et al. (2000) as the creation of a component, artefact, mechanical system, device, electronic circuit, software program, or engineering process that meets a set of desired specifications. *Development* is defined by Nuese (1995) as the systematic application of advanced knowledge or technology toward the production of useful materials, devices, systems or methods. *Product design* can be defined as a systematic process that involves idea or concept generation, concept development and evaluation, manufacturing and testing or implementation of an artefact or service. The scope of product design goes beyond industrial design. Industrial design is more concerned with the "art" of a product, such as aesthetics, form, color, texture, and feel. Product design is regarded as a blend of marketing, product management, industrial design, and engineering. Mazumdar (2002) defined *product development* as:

> *A process for translating customer needs into product design and manufacturing which involves managing mutual dependencies between all stages of the product life cycle, including design, manufacturing, distribution, technical support, and disposal or recycle stages.*

DIFFERENT DESIGN PROCESS MODELS DEVELOPED BY DESIGN EXPERTS

Total design as coined by Pugh (1991) is regarded as a core activity in product development. The total design model is depicted in Fig. 2.1. It starts with market investigation, product design specifications (PDSs), conceptual design, detail design, manufacture, and sale. This model can be conveniently used for engineering and consumer products including in composite field.

Sapuan and Maleque (2005) used total design model in the development of a household appliance, that is, a telephone stand from woven banana pseudo-stem fiber reinforced epoxy composites. Similarly, Hambali et al. (2009, 2010), used total design approach in the development of automotive bumper system from composites. They developed a CE framework comprising materials selection, concept design selection, and manufacturing process selection. The major tool used in their study was analytical hierarchy process (AHP).

In other development, Davoodi et al. (2008a,b), carried out design and analysis of automotive bumper system made from glass fiber reinforced polymer composites. They used several engineering design methods to come up with

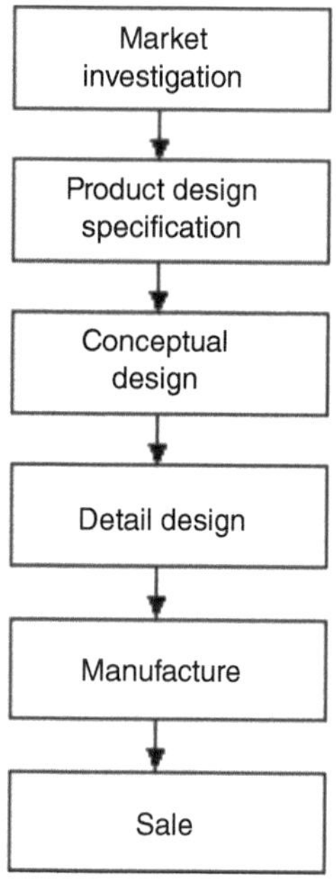

FIGURE 2.1 Total design model by Pugh (1991).

the best design concept after carrying out extended concept generation studies. In the more recent work of Davoodi et al. (2011) design concept development and selection were performed based on different engineering design methods for natural/glass fiber hybrid polymer composites also for automotive bumper system. The work of Sapuan et al. (2005) was also concerned with the development of automotive bumper system from composites but the focus was on the conceptual design of automotive bumper fascia. Total design approach was employed in the study. Various idea generation techniques, such as mind-mapping, morphological chart and analogy were used to generate design concepts.

Sapuan (2005) and Imihezri et al. (2005, 2006) performed design studies based on total design model for the development of automotive pedal system using composites. The activities performed include design concept development and evaluation, materials selection, design analysis, design to cost, and mold flow analysis. The major advantages resulting from these studies include weight and cost reduction. Mansor et al. (2013, 2014), performed materials selection and design concept development and evaluation for automotive hand brake lever from natural fiber/glass fiber hybrid polypropylene composites. AHP was used as the major tool together with morphological chart, and TRIZ method product development from hybrid composites.

Traditionally, design effort was carried out in series; the next stage is not commenced until the earlier stage is not completed. Although, the total design model proposed by Pugh (1991) seemed to imply the serial nature of the model, Pugh (1991) has stressed that; the activities should be performed concurrently. Sapuan and Gayretli (1996) studied the relation between CE and total design. They attempted to look to what extent the principles of CE have been considered in various designs employing total design model as the platform. They concluded that, many design works have overlooked the principles of CE in the development.

In CE, various disciplines within an organization should work in a team, simultaneously and they should be involved in product development from the very early stage of design (Hambali et al., 2011). Through such teamwork approach, the interaction and communication between the team members are ensured. CE concept normally produced higher quality product, at lower cost and shorter time (Sapuan et al., 2006). For instance, a group of engineering designers, while designing a handmade multipurpose table to be produced from woven banana pseudo-stem fiber reinforced epoxy composite, are in need to specify a 0.3 mm tolerance for the surface of the table. Under CE, they are free to invite the manufacturing engineers who will later on be producing this component, during this design stage (Hambali et al., 2011). In this manner, the "throw over the wall syndrome" will not happen.

Fig. 2.2 shows the design process model reported in VDI Design Handbook 2221 (VDI, 1987). In this model, design activities are classified into three categories, namely, stages, results, and phases. This division seems to make the design process more complicated. As far as composite design is concerned, this model may not be convenient to use, because, the complexity of the task may increase. Composite material itself is already complex in its arrangement, mixture, and processing. Liu et al. (2016) pointed out that working with composites is not a straightforward task due to the complexity in the design optimization process, such as nonlinearity in composite properties, different working conditions, and multitude of design variables involved. They carried out design optimization of composite automotive bumper system mainly using Kriging modeling technique and the Particle Swarm Optimization (PSO) algorithm, in addition to standard analysis technique, such as finite element analysis to solve very complex composite design problems.

In the previous model by Pugh (1991), it was presented in a very simple manner, to the extent that embodiment design is not included. In the model in Fig. 2.2, specification is considered as "results" whereas in Pugh (1991) model, specification or more specifically PDS are included as one of the main tasks. The model developed by Pugh (1991) gained more popularity among the product designers, among others, due to the strong emphasis on PDS, merging between conceptual and embodiment designs as conceptual design only, and

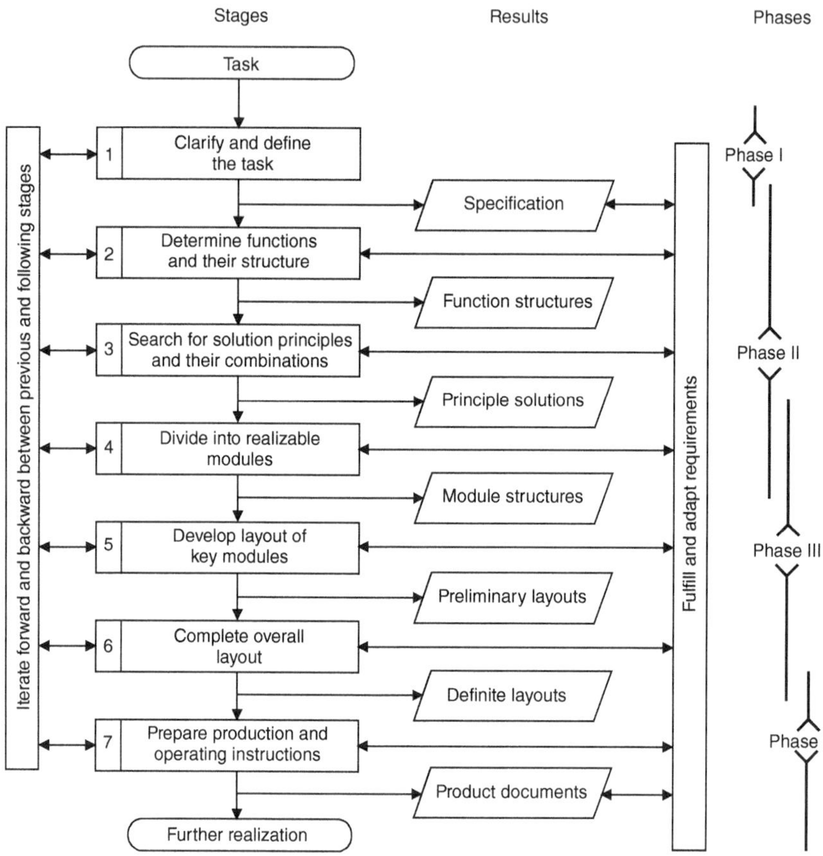

FIGURE 2.2 VDI Design Handbook 2221 (VDI, 1987).

Reproduced with permission of the Verein Deutscher Ingenieure e. V., Standard VDI 2221 (1987): Systematic approach to the design of technical systems and products (tranls. By K. Wallace), VDI e. V., Düsseldorf.

the introduction of well-respected Pugh concept evaluation method. Fig. 2.3 shows the model which is sometimes called German method developed by Pahl and Beitz (Pahl et al., 2007). Their work was originally translated and published in English in 1984. It is well accepted that this work is the pioneering work in engineering design method. It is because along with this model, Pahl and Beitz (Pahl et al., 2007) wrote a very comprehensive and detailed book, and sometimes this work is regarded to be so much over and above what is normally needed. Earlier than this, Johnson (1978) and French (1971) developed similar model but their work was not so popular as compared to that of Pahl and Beitz (Pahl et al., 2007). Their models are shown in Figs. 2.4 and 2.5.

Similar model was also developed by Dym (1994) as depicted in Fig. 2.6. It seems that this model is the enhancement and modification of previous

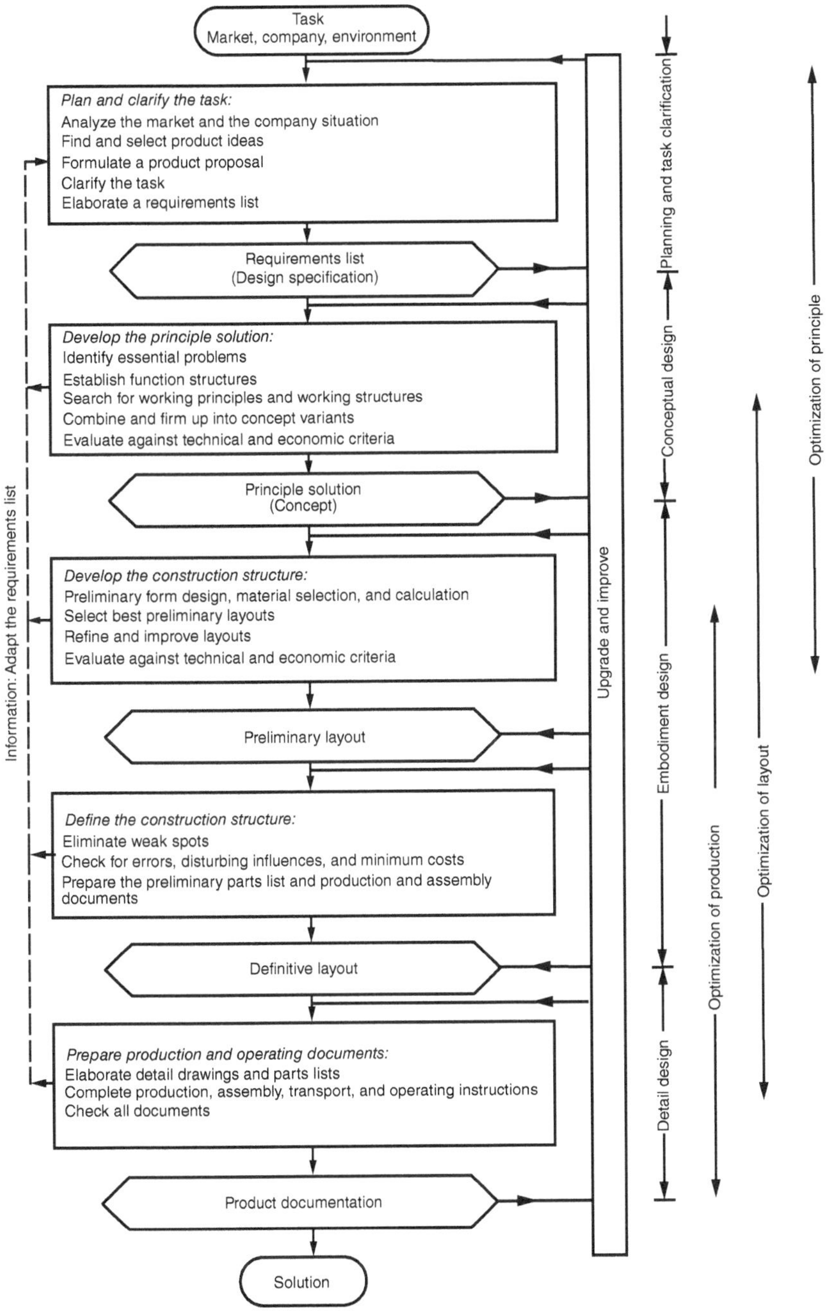

FIGURE 2.3 Pahl and Beitz model (Pahl et al., 2007).

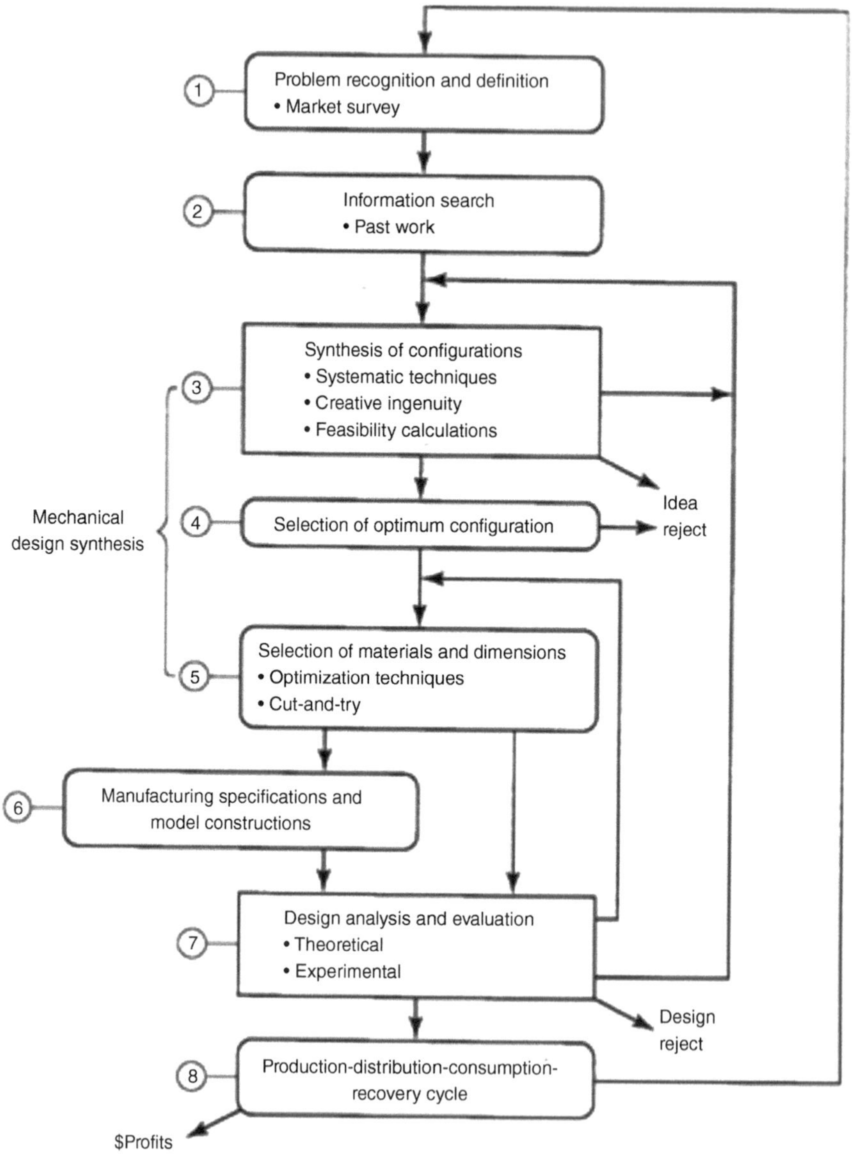

FIGURE 2.4 Design process map by Johnson (1978).

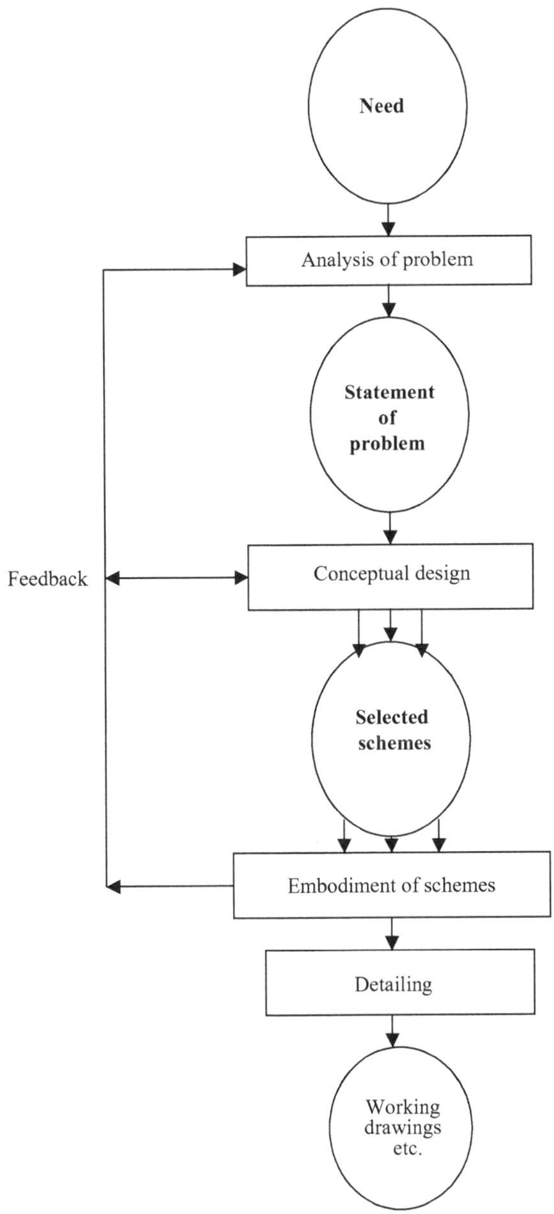

FIGURE 2.5 Engineering design process model proposed by French (1971).

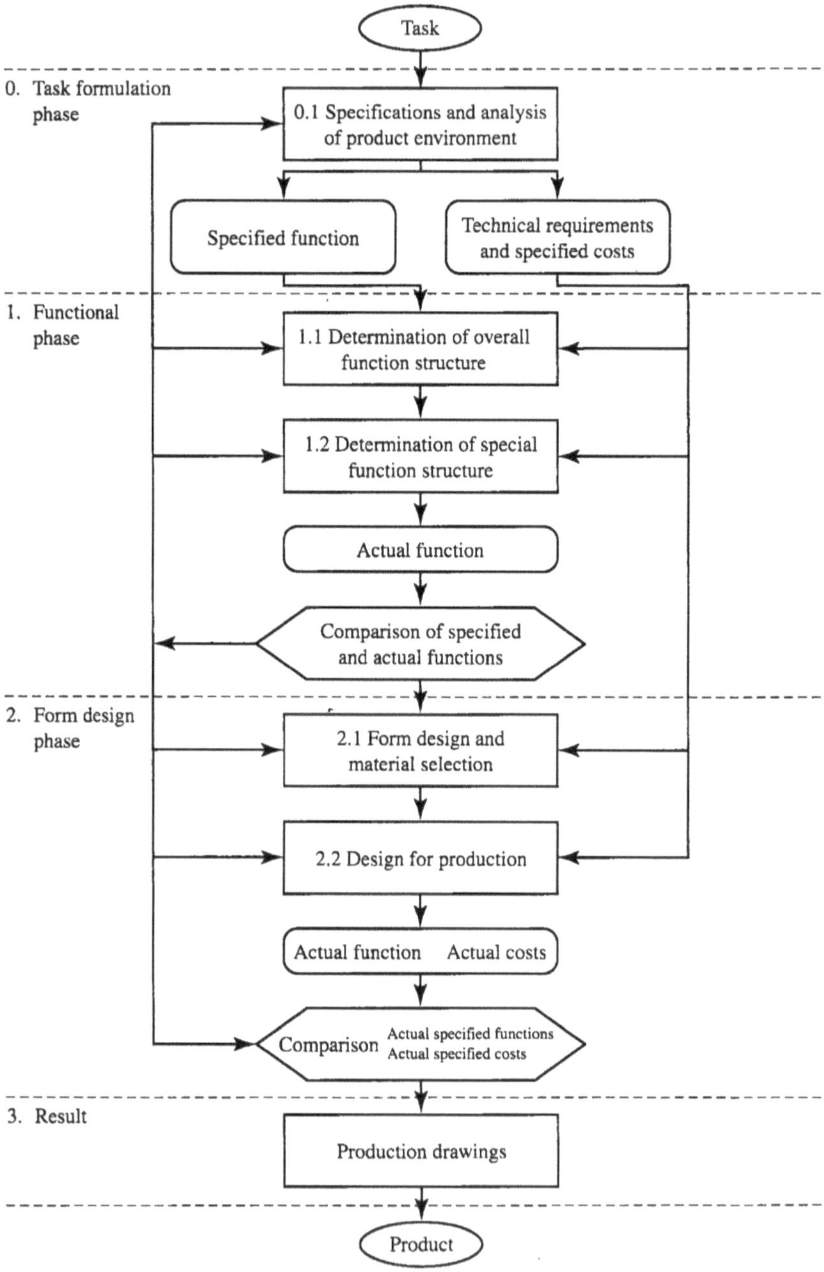

FIGURE 2.6 Design process map by Dym (1994).

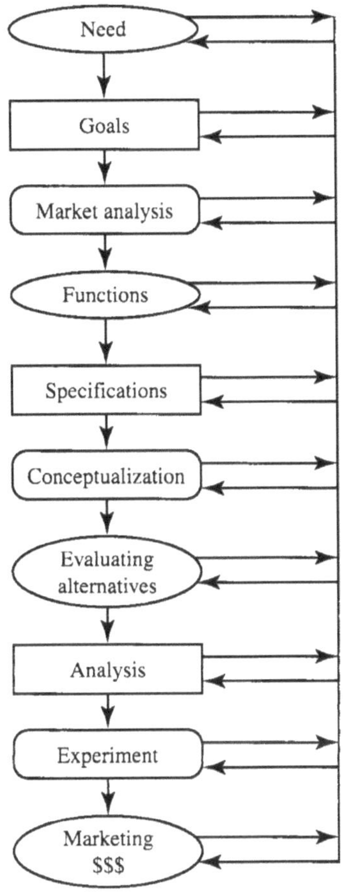

FIGURE 2.7 Design process model by Haik (2003).

model. In 1994, when this book was published, the engineering design community was very familiar with the well-established models developed by Pahl and Beitz in 1984 (Pahl et al. 2007), French (1971), and Pugh (1991) and the model cannot leave a big impact on the design community. A more recent model by Haik (2003) (Fig. 2.7) is just another version of the previous model and many information in this model had been included in the earlier models. A similar model by Ashby (2005) shown in Fig. 2.8 is developed in conjunction with materials selection model. The strength of Ashby (2005) lies in the use of materials selection chart, which is considered a standard technique in materials selection system. Since this model is designed for materials selection purpose, perhaps this model is very suitable to be used for composite product

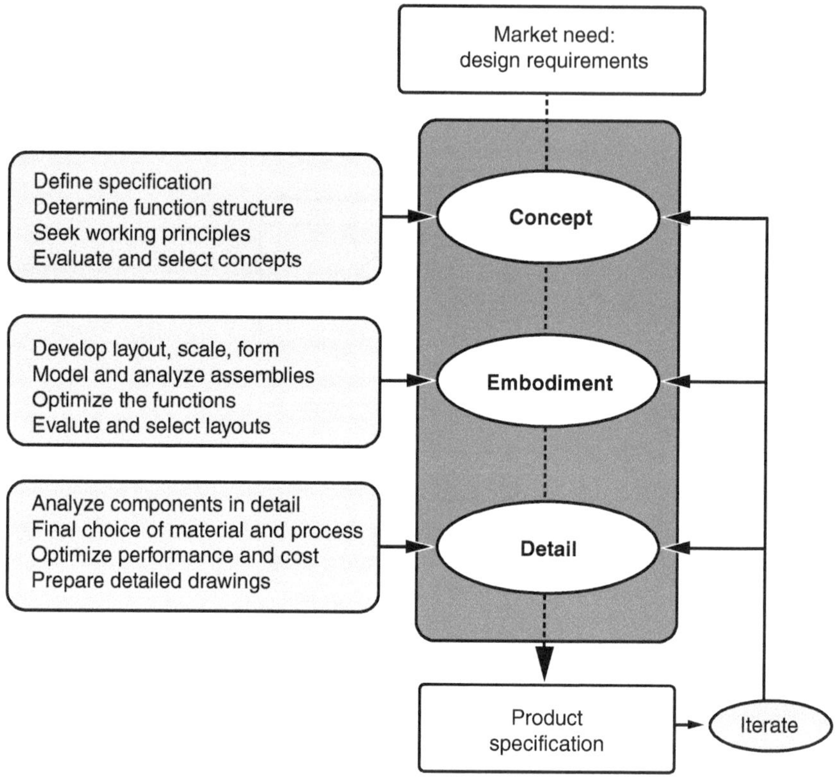

FIGURE 2.8 Design process by Ashby (2005).

development. Fig. 2.9 shows the engineering design process model developed by Ertas and Jones (1993). It is quite similar to that of Dym (1994).

Fig. 2.10 shows the design model adapted from Strong (2006). In realizing a new product, design is considered in total, that is, from customer needs, PDS, conceptual design, detail design, and manufacture. During detail design stage, Strong (2006) had broadly divided the activities into two major categories: component design and process design. The link between these two categories is an important element of CE (Strong, 2006). Examples of component design given by Strong (2006) include layout/drawing, constraints, analysis, and materials selection, while examples of process design include manufacturing method selection, sequencing, machine/tool selection, system layout, integration of system, and manufacturing procedures. However, in Pugh total design model (Pugh, 1991), CE is not only relevant during detail design stage but it is also applicable during other stages in total design process, such as market investigation, PDS, and conceptual design.

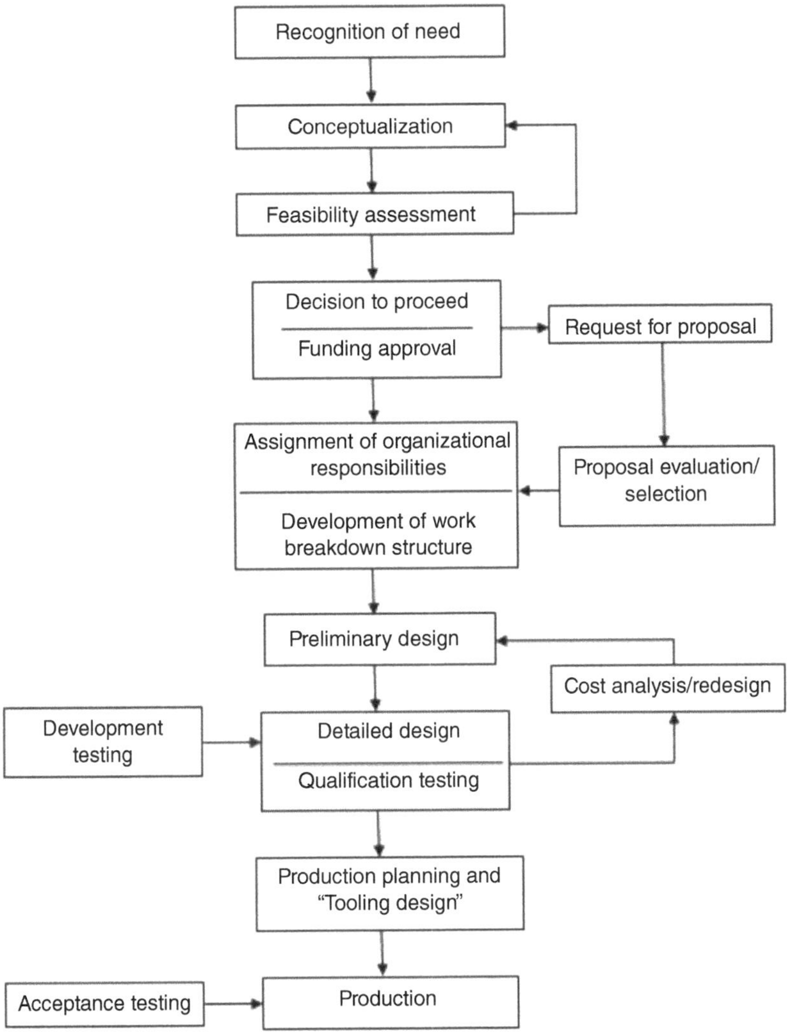

FIGURE 2.9 Engineering design process model by Ertas and Jones (1993).

According to Strong (2006), CE requires that while designing a component, the designer must consider the manufacturing process in order that the component designed is compatible with the manufacturing method. For example, if the component is to be fabricated using hand lay-up process, the designer would want to include design features that facilitate ease of fabrication to reduce fabrication complexity because labor cost is a key cost element in the hand lay-up process.

Considering manufacturing issues during the design stage can benefit design engineers in terms of cost and weight reduction (Mårtensson et al., 2015) and

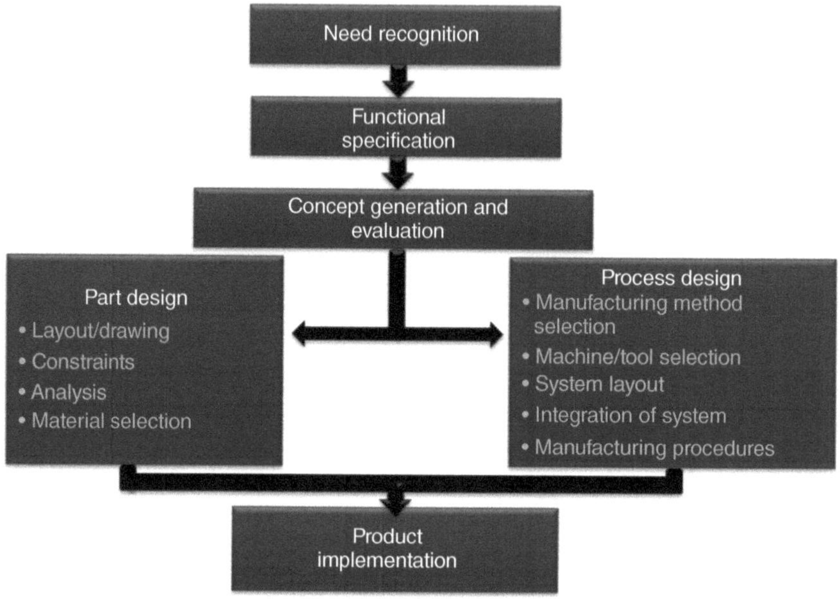

FIGURE 2.10 Design process model adapted from Strong (2006).

reducing the potential failure during service (Zein et al., 2016). By putting constraints in the manufacturability of composites, such as the manner of the sequence of plies is stacked in laminated composites, it restricted design freedom in term of tailor-made design but as far CE is concerned it is a commendable practice (Mårtensson et al., 2015). Manufacturability rules are laid for composite experts on plies stacking sequence. Examples of such rules include:

1. A maximum of four consecutive plies are allowed to have the same orientation *and*
2. only plies with symmetric sequences are permissible (Zein et al., 2016).

According to Mayer (1993) one major problem in design with composites is the limited availability of design data. The major reason for this is the large permutation of fibers, matrices, and manufacturing processes. In design with composites, it is invariably related to materials selection (Ahmed Ali et al., 2015; Hambali et al., 2010). Ashby (2005) provided a good framework of product development activities that provide useful guidelines into design of composites. In Ashby's framework, both geometrical design and materials selection process are concurrently embedded in the design phase of the product (Fig. 2.11).

Another important design consideration in composite field is when dealing with biocomposites. Biocomposites are highly potential substitution materials

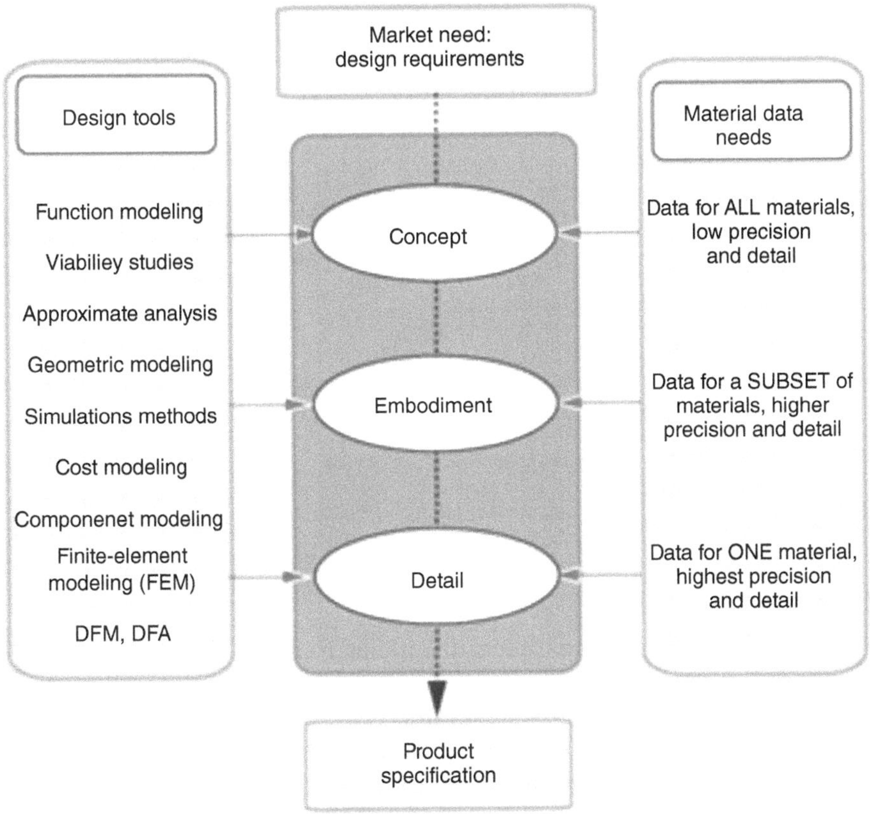

FIGURE 2.11 Design process model with consideration of materials (Ashby, 2005).

for synthetic composites and traditional engineering materials. Their main advantages include low cost, low weight, and sustainable resource (renewable and recyclable). However, biocomposites are mostly applied for nonstructural products due to lower material structural properties (strength and Young's modulus). One potential solution is the use of innovative product design approach in leveraging the properties of biocomposites (Mansor et al., 2014). It helps increase its application to wider range of products especially involving higher load bearing requirements (see Fig. 2.12).

DESCRIPTION OF EACH STAGE OF TOTAL DESIGN MODEL

In this section, a discussion on each stage of task or core activity in a typical design process model is presented. When we look at each model discussed earlier, each model has its own set of tasks and in this section only very common tasks are discussed.

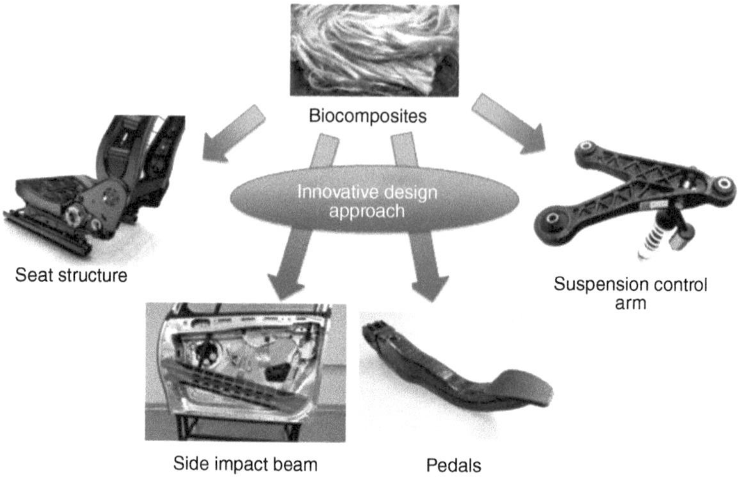

FIGURE 2.12 Innovative design approach in product development.

Clarifying Need

A designer has to clarify the need of the design from the client or should understand what the customers want, so that design effort can be commenced in a clear direction. A design team should meet to discuss the need among themselves and with the relevant parties, such as clients or marketing personnel who carried out the market need survey, with other CE teams from other functions within the organization or from the suppliers. Design can be divided into original design or evolutionary design (Ashby et al., 2007). Original design is a design that is performed from scratch and evolutionary design is an enhancement from earlier design. In both cases, the design need is often ill defined and requires confirmation and clarification from relevant parties (Cross, 1994). Another reason for the need to clarify the need is due to conflicting objectives of the design. Resolution of conflicting objectives, which may arise from budget limitation, lack of clarity, misunderstanding, and different visions among the design team members, needs to be done in order to proceed with the design work.

Capturing design intent (Sapuan, 2001) is the act of capturing intelligence in the design by means of computer system (i.e., about the way to create the model in terms of sketches, features, and dimensions and how the model should behave when it is changed) and it has become a very important topic in product development and CE. It is very important that design intent be documented and developed during the early stage of the design process for the sake of record keeping and to ensure that the goals of the design are achieved.

It is reported in Carter and Baker (1992) that there are five requirements of good design and they are:

1. A good design should clearly represent the customer's needs and wants for the product.
2. A good design should be based on engineering requirements and targets.
3. A good design should follow a functional model of the product.
4. A good design should incur a decreasing number of design changes throughout the design process.
5. A good design should use technology that is mature.

They further reported that in the CE environment, the CE team has a method to come up with good design and it is termed "collaborating on knowledge," in the context of total development process. Pahl et al. (2007) in discussing such "collaborating on knowledge" have developed a "wheel" as shown in Fig. 2.13, and if such collaboration is successfully implemented, it can become a "wheel of fortune."

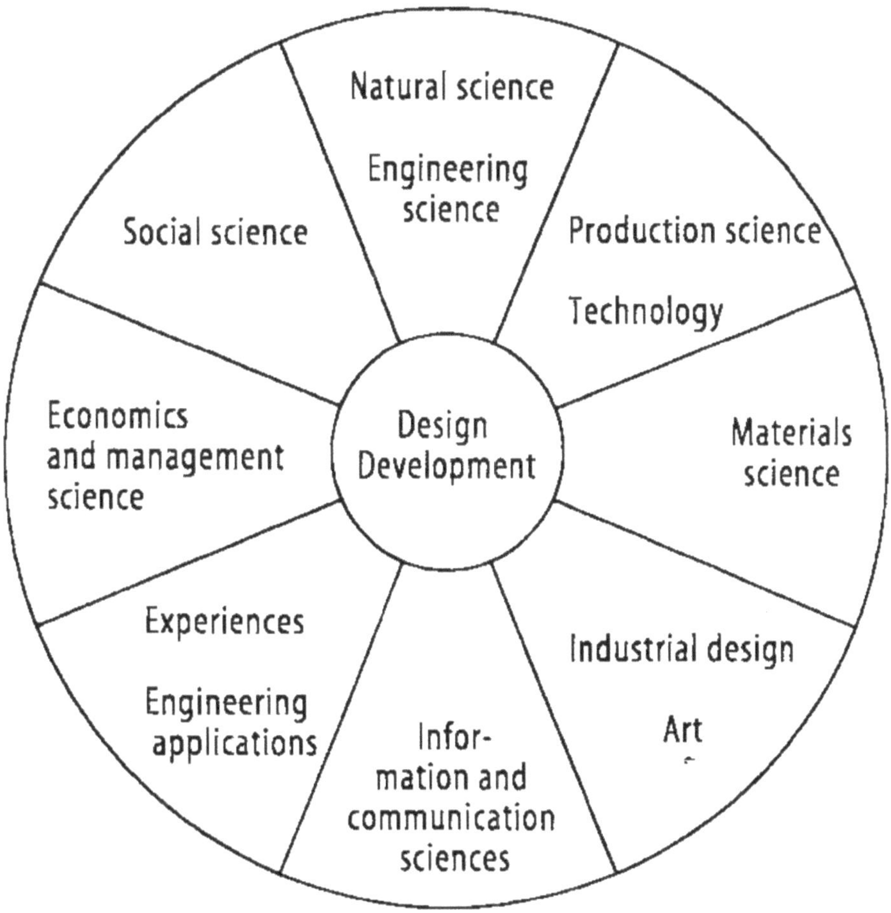

FIGURE 2.13 Design "wheel" (Pahl et al., 2007).

Market Investigation

In composite materials product development, particularly for polymer-based composites, gathering information on similar products is no longer difficult. It may get more difficult when we are dealing with more advanced composites, such as ceramic matrix composites, metal matrix composites, or "smart" composite materials. Market investigation is basically concerned with information gathering prior to performing the design task. Market investigation may refer to carrying out market survey regarding the products in terms of customer preferences and the performance of competitors' products.

Product Design Specifications

PDS is a formal document that is developed by the product design team after detailed market investigation is carried out. It is a dynamic document; it can be modified throughout the design process. PDS should not be confused with product specification given in Fig. 2.8 by Ashby (2005). Once the design process is completed, PDS becomes product specification. Experts like VDI Design Handbook 2221 (VDI, 1987), Pugh (1991), Pahl et al. (2007), Dym (1994), and Haik (2003) paid so much attention to PDS in product design. Pugh (1991) and Haik (2003) regard PDS development to be as important as core activities in design process model. However, Pahl et al. (2007) and VDI Design Handbook 2221 (VDI, 1987) are of the opinion that specification is the outcome of design planning and design clarification.

Hambali (2009) used total design model in the development of a polymer composite bumper beam and a PDS was comprehensively developed based on information obtained from published literature and competitors' products. Such PDS is partially reproduced as follows.

1. Performance
   a. The component must be able to absorb enough energy to meet the original equipment manufacturers (OEMs) internal bumper standard.
   b. The component must be able to stay intact or rigid at high-speed impact and prevent damage to the bodywork in minor impacts.
   c. The component must be able to protect the car body from damage in low-speed collisions.
   d. The component must be capable of collapsing evenly in a high-speed impact.
2. Safety
   a. The component must be free from injurious edge.
   b. The system must comply with all relevant design safety requirements.
3. Maintenance
   a. The component must be easy to be serviced and repaired.
   b. The component must be easy to be disassembled.

   c. Special tools must not be required for maintenance.
   d. The component must be easy to be attached to the vehicle body or chassis.
4. Environment
   a. Component must be environmentally friendly with minimum use of materials, energy, and minimum release of toxic materials.
   b. Component must perform and not be damaged by temperature in the range of -40 to $70\,°C$.
   c. The component must be capable to be used in all weather conditions.
   d. The component must be resistant to corrosion.
5. Materials
   a. The density of materials must be less than the density of steel, which has the density of $7850\ kg/m^3$.
   b. The materials used for component must be polymer composites.
   c. The materials used must have good resistance to chemicals and oils.
   d. The materials used must be water resistant.

Conceptual Design

Conceptual design is defined by Pahl et al. (2007) as:

> *The part of the design process where – by identifying the essential problems through abstraction, establishing function structures, searching for appropriate working principles and combining these into a working structure – the basic solution path is laid down through the elaboration of a solution principle.*

Betteridge and Maropoulos (1995) stated that conceptual design is a key activity in a product development cycle and it greatly influences the quality, cost, manufacturability, and life cycle characteristics of the product. Conceptual design is concerned with the generation of concept and it is sometimes called conceptualization. Conceptual design is normally related to creativity. Various methods have been proposed in generating the ideas, such as gallery method, TRIZ (Mansor et al., 2014), morphological chart (Sapuan, 2006), brainstorming, synectics, and analogy (Sapuan et al., 2005). These methods can be adapted in any types of design including composite product design. Fig. 2.14 shows a flow chart representing the integration of TRIZ and morphological chart in product development.

Embodiment Design

It is very difficult to draw the boundaries between conceptual design and embodiment design because the activities within these two stages are overlapping (see Fig. 2.15). Pugh (1991) merged embodiment design into the conceptual design. To add to this confusion, engineers in different parts of the world

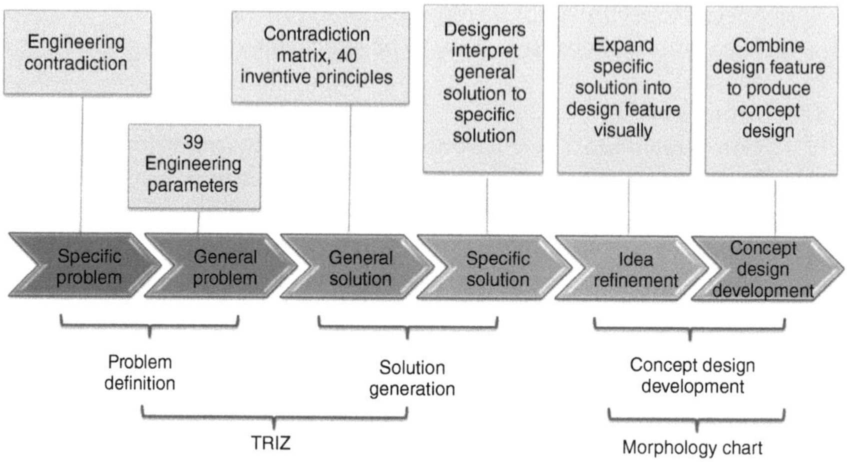

FIGURE 2.14 Integration of TRIZ and morphological chart in product development.

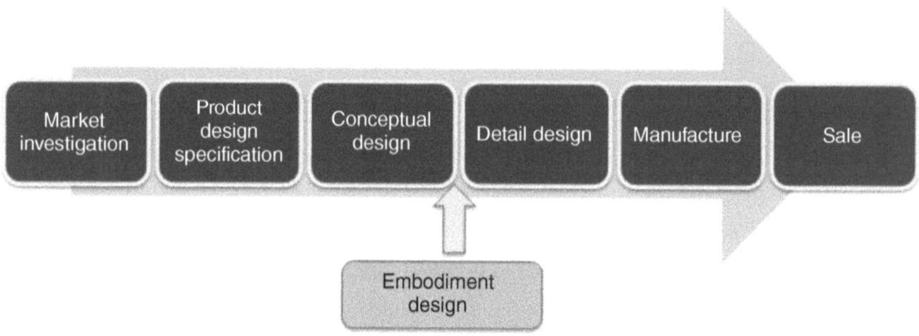

FIGURE 2.15 Modified total design model.

have interpreted embodiment design in their own ways. Wright (1998) defined "embody" as to give a concrete or discernible form to an idea or concept. Embodiment design is concerned with the preliminary layout and configurations of the product. Dieter (2000) compared it as putting meat on the bones, that is, the stage where design concept is invested with physical form. Among the activities that are included within the context of embodiment design are (Dieter, 2000):

1. Product architecture, such as arrangement of physical elements to carry out function.
2. Configuration design, such as preliminary materials and manufacturing selection and modeling of parts.
3. Parametric design, such as robust design tolerances, final dimensions, and design for manufacture.

Detail Design

Ertas and Jones (1993) defined detail design as:

> *the development of a system of drawings and specifications that completely describes a proven and tested design so that it can be manufactured.*

In composites, similar to any other types of design, the use of drawings, are very important. Drawings, such as installation drawing, assembly drawing, and detail drawing should be prepared during this stage of design.

Manufacture

Manufacturing function is involved in the design and operation of the production system in order to produce a product. Manufacturing with composite materials is normally different from metals from the points of view of machinery, materials arrangement, and processing.

THE SCOPE OF DESIGN FOR CE OF COMPOSITES

Classification Characteristics of Design Tasks in Composites

Pahl et al. (2007) gave the classification of design tasks that a designer is normally obliged to and they are described later in the chapter with specific reference to composite industry.

Origin of Task

Origin of task comprises product planning, customer order (product) (Fig. 2.16), design task (assembly), and production and field test. The work of designers is governed by what has been planned by a product-planning department after thorough market analysis is carried out. For instance, after carrying out detailed market analysis, there is a need to embark on a mass-produced automotive composite bumper beam. Then designers have been left with a large solution space, that is, on what manufacturing methods (RTM, compression molding, or injection molding) or material choices (glass, carbon, or Kevlar composites). Based on the manufacturing method and material selected, a suitable design can accordingly be made (Sapuan and Abdalla, 1998). In the case of customer order, for instance, the customer placed an order for a composite septic tank, normally, this type of product is a specific one-off, designers should design such products based on built up from previous developments and orders to limit any risks. Sometimes, a composite designer is posed with the task of designing only part of a product and there is a need to interact with other design teams. If the designer is asked to design a composite accelerator pedal, he needs to interact with design teams who design composite clutch pedal, composite mounting bracket, and composite brake pedal.

FIGURE 2.16 Origin of task in design: customer's request.

Organization

Companies producing composite components may be classified as product- or problem-oriented companies and the task of designers is tailored toward the nature of the companies. Some companies are specialized in producing selected standard composite products, such as pultruded composite pipe, rods (Fig. 2.17), and slab. They can fabricate such products and sell them as standard components. Some companies are awarded a project on the development of large scale project, such as development of electric post to replace the current materials (normally concrete) for electrical power transmission. For such a project, designers' tasks are split into stress analyses, manufacturing process selection, material selection, and design optimization. In this type of project, the company splits the responsibility according to those functions and good coordination is essential. The CE approach is better fitted within these types of organizations.

Novelty

If designer is asked by the planning department to design, for instance, a new metal matrix composite automotive bumper beam, that task is considered a

FIGURE 2.17 Pultruded composite rods: standard components.

novel or original design or sometimes a "crazy" idea and the task is very challenging as not much references are available to support his task. This design approach is revolutionary in nature. Adaptive design follows the known and established solution principles (undertake original designs of assembly or components) but the emphasis is on geometrical, analytical, production, and material issues. Normally, the sizes and arrangements of components can be varied within the limits set by previously designed product structures and this is termed variant design. Generally, design with composites follows this route, which requires evolution of the design rather than revolution.

Production

Designers for one-off or small batch products have to fulfil different requirements from large batch or mass production. In one-off products, normally in composite industry, hand lay-up is generally employed (see Fig. 2.18) and for mass production, injection molding, resin transfer molding, or compression molding methods can be used. In the former, designers are involved in the design of products that are produced manually, and require highly skilled operators. The product requires careful design of all physical processes to minimize risk. For the latter, technical and economic consideration must be checked prior to full-scale production and the use models and prototypes is necessary.

Branch

In composite material, design and development can cater various branches of engineering. Design of pressure vessel from filament wound composites may require knowledge of chemical process. Development of laminated composites

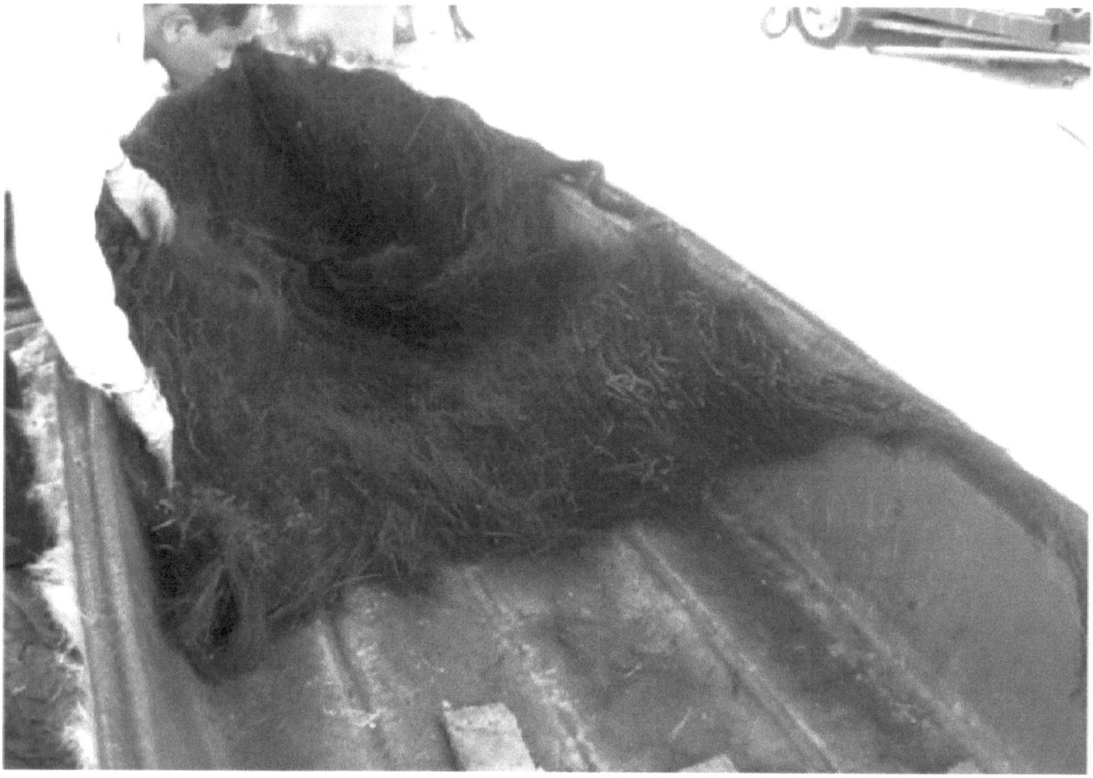

FIGURE 2.18 Fabrication of composite boat using hand lay-up process.

is normally geared toward serving the aerospace industry. In civil engineering designers are normally concerned with the design of reinforced concrete composites and composite bridges. In road safety area, innovative design in new areas, such as that of motorcycle helmet from biocomposites is an important development (see Fig. 2.19). Depending on the branch of engineering, composite designers have to work according to the need of their respective branch.

Complexity

Complexity can be interpreted in different ways by different industries. Developing automotive composite components (Imihezri et al., 2006) from injection molding process has some complexity in that, in designing automotive components from composites, it is quite different from designing with metal counterparts. Many features, not normally needed in metal, can be very important in composites, such as inclusion of ribs (Fig. 2.20), bosses, undercuts, and inserts.

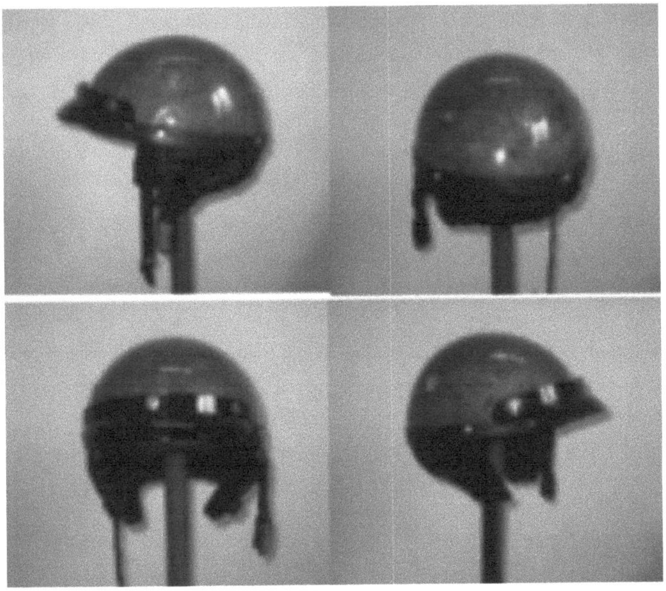

FIGURE 2.19 Biocomposite motorcycle helmet.
Courtesy Dr. Yuhazri Yaakob, UTeM, Melaka, Malaysia.

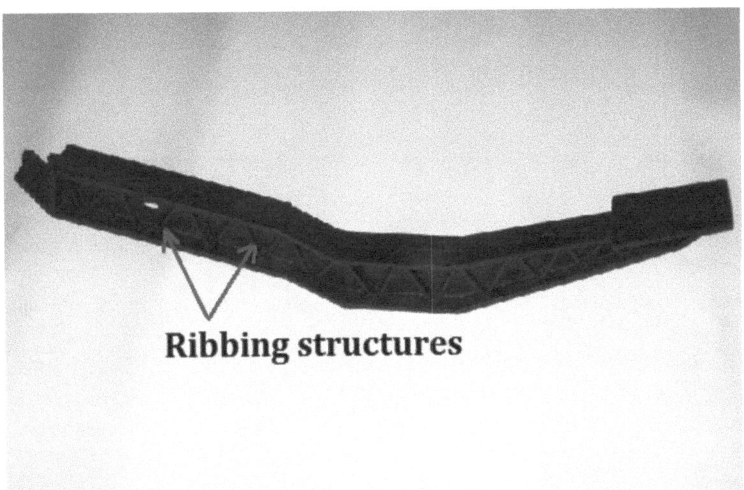

FIGURE 2.20 Ribbing structures in injection molded composite clutch pedal.

For aircraft components, continuous laminated composites are normally used and the manufacturing method suitable for this purpose is hand lay-up. The process is complex because very skilful operators are needed to carry out the design and fabrication.

Goals

The goals of a specific product influence the task to be carried out by the designers. In aerospace industry, the design and development of composite material components must give priority to function and safety. However, in developing children's toys from injection molded glass fiber reinforced thermoplastic composites, the minimization of cost is the main goal. For automotive components, in addition to reducing cost, weight saving should be considered important. For different industries, the goals are also different and the designers for each industry must fulfil their task according to the goals of the industry they are serving.

CONCLUSIONS

In this chapter, engineering design process models developed by various design experts are presented. These models are reviewed in this chapter, because they can be used in the development of products from composites. The chapter paid particular attention to total design model developed by Pugh as this model is easy to use and at the same time is very practical in terms of design practice. The core elements in total design model are individually explained. This chapter provides a basis for further understanding on CE and design in subsequent chapters.

References

Ahmed Ali, B.A., Sapuan, S.M., Zainudin, E.S., Othman, M., 2015. Implementation of expert decision system for environmental assessment in composite material selection. J. Cleaner Prod. 107, 557–567.

Ashby, M.F., 2005. Materials Selection in Mechanical Design. Elsevier Butterworth-Heinemann, Oxford.

Ashby, M.F., Johnson, K., 2002. Materials and Design: The Art and Science of Material Selection in Product Design. Butterworth-Heinemann, Oxford.

Ashby, M., Shercliff, H., Cebon, D., 2007. Materials: Engineering, Science, Processing and Design. Butterworth-Heinemann, Oxford.

Betteridge, M., Maropoulos, P.G., 1995. A review of conceptual design research and industrial practice in concurrent engineering, Proceedings of the Eleventh National Conference on Manufacturing Research, Leicester, 12–14 September, pp. 693–697.

Carter, D.E., Baker, B.S., 1992. Concurrent Engineering: The Product Development Engineering for the 1990s. Addison-Wesley Publishing Company, Reading, Massachusetts.

Cross, N., 1994. Engineering Design Methods: Strategies for Product Design, second ed. John Wiley & Sons, Chichester, England.

Davoodi, M.M., Sapuan, S.M., Ahmad, D., Aidy, A., Khalina, A., Jonoobi, M., 2011. Concept selection of car bumper beam with developed hybrid bio-composite material. Mater. Des. 32 (10), 4857–4865.

Davoodi, M.M., Sapuan, S.M., Yunus, R., 2008a. Conceptual design of a polymer composite automotive bumper absorber. Mater. Des. 29 (7), 1447–1452.

Davoodi, M.M., Sapuan, S.M., Yunus, R., 2008b. Development of fibre reinforced epoxy composite bumper absorber for automotive bumper system. J. Polym. Mater. 25 (1), 15–21.

Dieter, G.E., 2000. Engineering Design: A Materials and Processing Approach, third ed. McGraw-Hill Companies, Inc, Boston.

Dym, C.L., 1994. Engineering Design: A Synthesis Views. Cambridge University Press, Cambridge, UK.

Ertas, A., Jones, J.C., 1993. The Engineering Design Process, second ed. John Wiley and Sons, Inc, New York.

French, M.J., 1971. Engineering Design: The Conceptual Stage. Heinemann Educational Books, London.

Haik, Y., 2003. Engineering Design Process. Brooks/Cole—Thomson Learning, Pacific Grove, USA.

Hambali, A., 2009. Selection of Conceptual Design Using Analytical Hierarchy Process for Automotive Bumper Beam Under Concurrent Engineering Environment, Universiti Putra Malaysia, PhD Thesis.

Hambali, A., Sapuan, S.M., Ismail, N., Nukman, Y., 2009. Application of analytical hierarchy process in the design concept selection of automotive composite bumper beam during the conceptual design stage. Sci. Res. Essays 4 (4), 198–211.

Hambali, A., Sapuan, S.M., Ismail, N., Nukman, Y., 2010. Materials selection of polymer based composite bumper beam using analytical hierarchy process. J. Cent. South Univ. Technol. 17 (2), 244–256.

Hambali, A., Sapuan, S.M., Ismail, N., Nukman, Y., 2011. Concurrent decisions on design concept and material selection using analytical hierarchy process at the conceptual design stage. Concurrent Eng. Res. Appl. 19 (2), 111–121.

Imihezri, S.S.S., Sapuan, S.M., Hamdan, M., Sulaiman, S., 2005. A study on comparison of V and X ribbings for polymeric based composite automotive clutch pedal using mould flow analysis software. Mater. Des. 26 (2), 157–166.

Imihezri, S.S.S., Sapuan, S.M., Sulaiman, S., Hamdan, M.M., Zainuddin, E.S., M.R. Osman, M.R., Rahman, M.Z.A., 2006. Mould flow and component design analysis of polymeric based composite automotive clutch pedals. J. Mater. Process. Technol. 171 (3), 358–365.

Johnson, R.C., 1978. Mechanical Design Synthesis. Krieger, Huntington, NY.

Liu, Z., Lu, J., Zhu, P., 2016. Lightweight design of automotive composite bumper system using modified particle swarm optimizer. Compos. Struct. 140, 630–643.

Mansor, M.R., Sapuan, S.M., Zainudin, E.S., Nuraini, A.A., Hambali, A., 2013. Hybrid natural fiber/glass fiber reinforced polymer composites material selection using Analytical Hierarchy Process for automotive gear lever design. Mater. Des. 51, 484–492.

Mansor, M.R., Sapuan, S.M., Zainudin, E.S., Nuraini, A.A., Hambali, A., 2014. Conceptual design of kenaf fiber polymer composite automotive parking brake lever using integrated TRIZ-morphological chart-AHP method. Mater. Des. 54, 473–482.

Mårtensson, P., Zenkert, D., Åkermo, M., 2015. Effects of manufacturing constraints on the cost and weight efficiency of integral and differential automotive composite structures. Compos. Struct. 134, 572–578.

Mayer, R.M., 1993. Design with Reinforced Plastics: A Guide for Engineers and Designers. The Design Council, London.

Mazumdar, S.K., 2002. Composite Manufacturing, Materials, Product, and Process Engineering. CRC Press, Boca Raton, Florida.

Nuese, C.J., 1995. Building the Right Things Right. AMACOM, New York.

Pahl, G., Beitz, W., Feldhusen, J., Grote, K.H., 2007. Engineering Design: A Systematic Approach, third ed. Springer-Verlag London Limited, London.

Pugh, S., 1991. Total Design—Integrated Methods for Successful Product Engineering. Addison-Wesley Publishing Company, Wokingham, England.

Sapuan, S.M., 2001. A knowledge-based system for material selection in mechanical engineering design. Mater. Des. 10 (8), 687–695.

Sapuan, S.M., 2005. Concurrent design and manufacturing process of automotive composite components. Assembly Autom. 25 (2), 146–152.

Sapuan, S.M., 2006. Using the morphological chart technique for the design of polymeric-based composite automotive pedals. Discov. Innov. 18 (4), 311–317.

Sapuan, S.M., Abdalla, H.S., 1998. A prototype knowledge-based system for the material selection of polymeric-based composites for automotive components. Compos. Part A Appl. Sci. Manuf. 29A (7), 731–742.

Sapuan, S.M., Gayretli, A., 1996. The improvement of the link and integration between the simultaneous engineering philosophy and the total design studies. AEESEAP J. Eng. Educ. 26 (1–2), 52–61.

Sapuan, S.M., Maleque, M.A., 2005. Design and fabrication of natural woven fabric reinforced epoxy composite for household telephone stand. Mater. Des. 26 (1), 65–71.

Sapuan, S.M., Maleque, M.A., Hameedullah, M., Suddin, M.N., Ismail, N., 2005. A note on the conceptual design of polymeric composite automotive bumper system. J. Mater. Process. Technol. 159 (2), 145–151.

Sapuan, S.M., Mansor, M.R., 2014. Concurrent engineering approach in the development of composite products: a review. Mater. Des. 58, 161–167.

Sapuan, S.M., Mollah, M.A.H., Nukman, Y., 2000. Concurrent Engineering and Product Design and Development. Prentice Hall, Petaling Jaya, Selangor, Malaysia.

Sapuan, S.M., Nukman, Y., Hanim, A.R.S., 2003. Concurrent engineering teaching and research in Malaysian universities. Sudan Eng. Soc. J. 49 (41), 19–26.

Sapuan, S.M., Osman, M.R., Nukman, Y., 2006. State of the art of the concurrent engineering technique in automotive industry. J. Eng. Des. 17 (2), 143–157.

Strong, A.B., 2006. Plastics Materials and Processing, third ed. Pearson Prentice Hall, Upper Saddle River, New Jersey.

VDI, 1987. VDI Design Handbook 2221 Systematic Approach to the Design of Technical Systems and Products (K. Wallace, Trans.). VDI-Verlag, Dusseldorf.

Wright, I., 1998. Design Methods in Engineering and Product Design. McGraw-Hill Publishing Company, Maidenhead, Berkshire.

Zein, S., Madhavan, V., Dumas, D., Ravier, L., Yague, I., 2016. From stacking sequences to ply layouts: an algorithm to design manufacturable composite structures. Compos. Struct. 141, 32–38.

CHAPTER 3

Composite Materials

INTRODUCTION

The competition of technology among nations is to create a better standard of living for the benefit of mankind. Therefore, the mankind is seeking for products of new design, material, and technology. Materials play an important role in the advancement of human life. Materials used in the development of products depend on the availability, performance, and cost. Composite materials have replaced many conventional materials, such as metals, ceramics, woods, and polymers due to the advantages that they offer, such as specific strength and stiffness. Nowadays, the use of composite materials in our lives has become very common and these are found in our furniture, transportation, education, sport and leisure, and building and infrastructural system.

The major reasons for switching toward composite materials are due to their properties and performance that are comparable with conventional materials, such as metals, plastics, and woods while demonstrating their own advantages like light weight, corrosion resistance, and part consolidation. Before going into details on concurrent engineering (CE) for composite material development, it is important to devote a chapter on describing the composite material itself in order to familiarize some readers with this material.

In the recent years, there is a rapid growth in the development and use of composite materials due to the advantages that they offer, such as light weight and corrosion resistance. Composites comprise two or more phases where normally the constituent materials are processed separately and then bonded. The properties of the composites are different from the constituent materials acting alone. One of the constituents is reinforcing phase in the forms of fibers, particles, or sheets and are embedded in another constituent called matrix.

Composites are actually solid materials that are composed of a matrix that surrounds and binds together fibrous reinforcements. Composite materials are considered as today's materials because they are now being applied in various industries ranging from children toys to aircraft body. However, they can be

CONTENTS

Introduction.................57
History of Composites.................58
Ancient History of Composites.........................58
History of Matrix Materials..............................58
History of Fibers..................58
Modern History of Composites.........................59
Polymer Matrix Composites.................59
Matrix Materials...................60
Fibers..................................61
Natural Fiber Composites..........................65
Advanced Composites..........65
Advantages and Disadvantages of Polymer Composites.................66
Advantages..........................66
Disadvantages......................67
Applications of Polymer Composites.................67
Automotive Industry.............67
Aerospace Industry...............69
Building Industry..................70
Marine Application................72

(Continued)

Composite Materials. http://dx.doi.org/10.1016/B978-0-12-802507-9.00003-9

Furniture Products *73*
Sporting Goods and Leisure *75*
Manufacturing Methods of Polymer Composite Materials 75
Injection Molding *76*
Resin Transfer Molding *77*
Hand Lay-Up *78*
Spray-Up *79*
Filament Winding *79*
Compression Molding *80*
Pultrusion *81*
Vacuum Bag Molding/ Autoclave Molding *82*
Metal Matrix Composites 82
Composite Constituents *82*
Fabrication Processes of MMC *82*
Applications *83*
Self-Healing MMC *83*
Ceramic Matrix Composites 84
Composite Constituents *84*
Properties of CMC *84*
Fabrication Processes of CMC *84*
Applications *85*
Nanocomposites 85
Ceramic Matrix Nanocomposites *86*
Metal Matrix Nanocomposites *86*
Polymer Matrix Nanocomposites *87*
Potentials and Advantages of Nanocomposites *87*
Conclusions 88
References 89

regarded as tomorrow's materials due to their potentials that they can offer in the future, such as in the areas of nanocomposites, functionally graded and smart materials. The most commonly used matrix is polymer or plastics (polymer matrix composite, PMC), and to a lesser extent ceramics (ceramic matrix composite, CMC) and metals (metal matrix composite, MMC) are also used. Composites are considered to demonstrate anisotropic behavior, that is, they have different properties along the axes in different directions (Hancox et al., 1980). A wide range of matrices and reinforcements can be combined in various forms with virtually unlimited permutations (Murphy, 1994).

HISTORY OF COMPOSITES

This section briefly describes the history of composites ranging from ancient to modern histories and it follows with the history of matrix materials and fibers.

Ancient History of Composites

Composite materials have already been in existence since the *Paleolithic* age (also known as Old Stone age). It is reported that a 300 ft. height *ziggurat* or temple tower had been built in the city center of Babylon, with the only material used being clay mixed with finely chopped straw (http://i-cias.com/e.o/sumer.htm). Gordon (2006) stated that composite materials had been around since the time of Pharaoh in Egypt and during that time, bricks were reinforced with straw to form composite bricks. Herakovich (1998) reported that Egyptians used fiber composites for laminated writing materials from the papyrus tree. In Japan it is reported that Samurai sword was developed from composite material (Sapuan, 2000). In fact most of the natural objects, such as human body, plants, and animals are composites (Herakovich, 1998) but it is not these types of composites that we are interested in here. According to Kelly and Mileiko (1983) among the earliest application of composites is in the pottery industry where prior to firing, glass and animal hair fibers were used to reinforce another material with the aim to reduce shrinkage and improve strength.

History of Matrix Materials

Phenol, formaldehyde, and poly(vinyl chloride) (PVC) were among the first synthetic polymers developed in early 1900s. Bakelite (made from phenol and formaldehyde) was patented in 1907 and PVC was patented in 1914 (Halary et al., 2011).

History of Fibers

In the Holy Quran, a verse was revealed concerning the use of palm fiber as a twisted rope. Fine glass fibers were developed in France in the 18th century

(Hartman et al. (1994). Glass fiber filaments were introduced and made available in the market in 1935 (Schwartz, 1997). Carbon fiber was initially developed in the late 19th century as filament material for electric light bulbs. High strength and modulus carbon fibers were developed by the Royal Aircraft Establishment, Farnborough, United Kingdom in 1953 (McBeath, 2000). Boron fiber was introduced in late 1950s (Herakovich, 1998). Aramid fibers were initially developed and introduced by DuPont in 1972.

Modern History of Composites

Polymer composites have a long and honorable history in the automotive industry, starting with hand lay-up bodies for limited production of sport cars up to the most sophisticated injection molded air intake manifold (Murphy, 1994). In the early days, Ford had developed an experimental car from hemp fiber filled soya composites as early as the 1940s (Mutel, 2012). Murphy (1994) further stated that glass cloth reinforced allyl composites were used in 1941 in the USA. Polymer composites had been developed after polymer materials were available in the 1950s from the petroleum industry (Mayer, 1993). Gomina (2012) reported that during World War II, in Britain, the material for fuselage of Spitfire aircraft was replaced from aluminum to flax fiber reinforced phenolic composites, due to the insufficient supply of aluminum. Composites were initially commercialized in large quantity, few years later, in the defense industry (Mazumdar, 2002). Then the application was extended to other industries, such as chemical plant, boat hull, sport car bodies, and sporting goods (Murphy, 1994).

POLYMER MATRIX COMPOSITES

This section contains information about matrix types and fiber types that are used in the fabrication of polymer composite components. It also gives detail about advanced composites and natural fiber composites. PMCs or simply polymer composites are the composites where the matrix is made from polymer and it is bonded with suitable reinforcements or fillers, such as glass, carbon, Kevlar, boron, or natural fibers or fillers. PMCs are also sometimes called fiber-reinforced plastics. Polymer composites have been used extensively in aerospace, automotive, marine, building and construction, furniture, sports goods, telecommunication, and railway industries because of light weight, high strength, high stiffness, aesthetics, corrosion resistance, and the ability to cope with extreme stresses over long periods (Sapuan, 1998).

Weaker matrices are made stronger by the addition of fibers where these fibers have enabled the cracks to be stopped or deflected at the matrix–fiber interface. The work of fracture is substantially improved and in turn causes the tensile strength to increase (Gordon, 2006). The main motivation in the use of composite materials is largely attributed to the development and study of high-performance reinforcements, such as Kevlar, carbon, and glass fibers in the forms of continuous or chopped fibers in polymer composites, although reinforcement is possible with particles of cheap fillers, such as sand, talc, and wood dust.

Matrix Materials

Polymer used as matrix in composites is also known resin. The functions of polymer matrices are (Peters, 2004):

- To help in distributing or transferring loads.
- To assist in controlling the chemical properties of composites.
- To carry interlaminar shear.
- To hold all reinforcements together as a binder, thus allowing the applied force to be transmitted to the reinforcement.
- To protect and prevent the fibers from mechanical damage due to crack propagation within the material during fabrication and in the finished component.

There are two major types of polymers used as matrices in composites, namely, thermoplastic and thermosetting polymers. When composite was introduced in the 1940s, thermosetting polymers, such as polyester, epoxy, and phenolic were mainly used.

Thermosetting polymers are composed of long chains of molecules that are strongly cross-linked to other long chains of molecules. As a result of the cross-linking, 3D network structures are obtained. Thermoplastics are not cross-linked but composed of long polymer chains that are entangled resulted from joining together of monomer units (Askeland and Fulay, 2009). Thermosetting polymers can either be liquids or solids at room temperature and they soften when being initially heated then cure or set, while thermoplastics soften and flow when they are heated but they harden again after they undergo the cooling process (Clegg and Collyer, 1993).

Examples of thermosetting polymers used as matrices in composites include epoxies, unsaturated polyesters, polyimides, bismaleimide (BMI), cyanate esters, polyurethane, phenolic, silicones, and vinyl esters. Some specialty polymers are also available, such as alkyds, allyls, furane, and melamines (Murphy, 1994).

For thermoplastic polymers, the examples include polystyrene, polycarbonate, polyacetal, polyamides, polyarylates, acrylonitrile-butadienestyrene (ABS), styrene acrylonitrile (SAN), styrene maleic anhydride (SMA), PET, polybutylene terepthalate (PBT), polyethylene, poly(vinyl chloride) (PVC), and polypropylene (Murphy, 1994). There are higher performance thermoplastics available, such as polyether sulfone (PES), polyetherimide (PEI), thermoplastic polyurethane, liquid crystal olymers (LCPs), polyether etherketone (PEEK), polyether ketone (PEK), polysulfones (PSU), polyphenylene ether (PPE), polyphenylene oxide (PPO), and polyphenylene sulfide (PPS) (Murphy, 1994).

Details about these polymers can be found in many books, such as those written by Murphy (1994) and Mazumdar (2002).

Biodegradable Polymers

Biodegradable polymers are polymers that degrade when they are left in the environment. The most popular bioderived polymer is poly(L-lactic acid) (PLA) (Callister and Rethwisch, 2011). PLA can be used as packaging materials, as well as mulch film used in agriculture sector. However, in the recent years, the research and development of biodegradable polymers are being carried out and one such work is done by Sahari et al. (2013). Starch was derived from the inner part of the stem of the sugar palm tree. Starch from the tree was made into biopolymer in the presence of plasticizers, such as water, glycerol, and sorbitol.

Fibers

Fiber is the second most important component of the composite. The functions of fiber are:

- To contribute high strength and stiffness of the composites.
- To carry majority of the load.

Three main types of commonly used fibers as reinforcements in polymer composites are glass, carbon, and aramid (Kevlar) (Mayer, 1993). Other fibers include boron, polyethylene, ceramic, silicon carbide, and natural or cellulose fibers.

Glass Fibers

Glass fibers are the most common and widely used fibers (90%) in polymer composites for structural application. It is cheaper than carbon and aramid fibers. Majority of glasses available in the markets consist mainly of silica, SiO_2 (Hyer, 1998). Glass fibers or filament are produced by drawing glass in molten state through orifice and approximately 200 filaments form a strand. While in use, strands are made into roving (Fig. 3.1), fabrics, or mat for ease of handling.

There are several types of glass (Murphy, 1994):

- E-glass (E for "electrical") is an all-purpose fiber which can be found in the form of chopped staples, continuous filament, and random fiber mats. It is the most highly used fiber.

FIGURE 3.1 Glass fiber rovings.

- C-glass (C for "chemical") is acid and chemical resistant and is not used to make fiber but to make soft tissue.
- A-glass (A for "alkali") is a typical window glass and the original type of glass fiber. It has high alkali content. In the past, it was used to make fibers but it has now been replaced with other fibers.
- S-glass has the highest strength and stiffness, and it was intended for use in aerospace and defense industries and can be found in rovings and yarn form.
- AR glass is alkali resistant.

Carbon Fibers

Carbon fiber (Fig. 3.2) is also called graphite fiber, although in reality they are two different materials. Carbon fiber is stronger than graphite fiber but in the composite community they are referred to as the same materials (Dorworth et al., 2009). Carbon fibers are originally used as reinforcement of composites in aerospace industry. Now the applications are extended to other industries, such as marine and automotive. It has high tensile strength and modulus, high fatigue strength, and high thermal conductivity (Mallick, 2008). The strength and stiffness of carbon fibers are higher than glass and aramid fibers.

Aramid Fibers

Aramid (aromatic polyamide) fibers are based on linear polymer chains consisting of para-linked aromatic aramides (Schwartz, 1997). Their chemical name is polyparaphenylene terephthalamide (McBeath, 2000). Kevlar is the most widely available aramid fiber. The tensile strength and modulus of aramid fibers are higher than glass fibers but aramid fibers have low

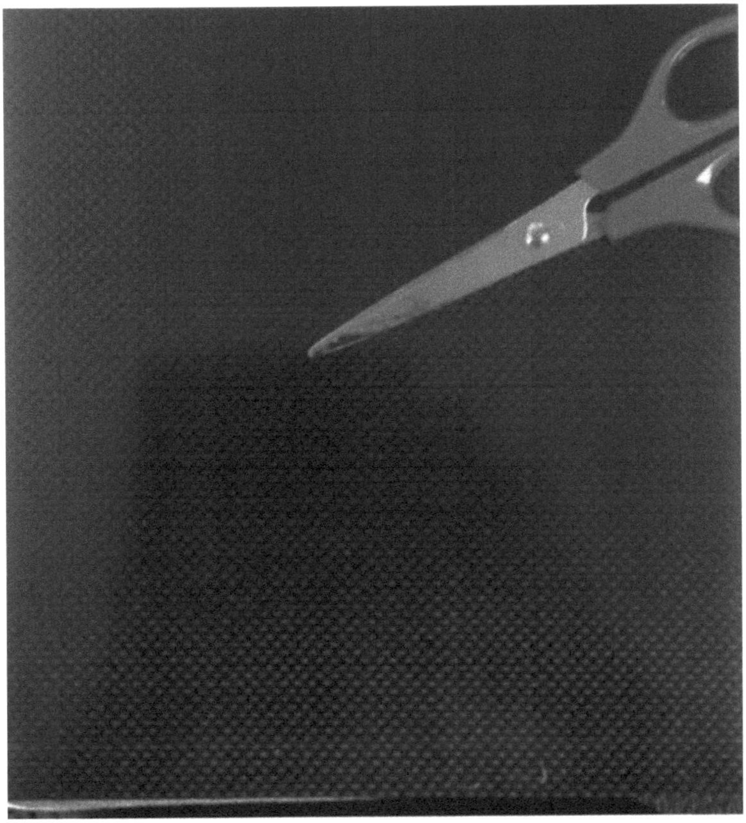

FIGURE 3.2 Woven carbon fiber.

compressive strength. Aramid fibers (Fig. 3.3) are normally used in ballistic body armour due to their very high impact and penetration resistance (McBeath, 2000). These materials are produced by drawing filaments from a chemical brew.

Boron Fibers

In the past, boron fibers were used extensively as reinforcements in structural composites. Boron fibers are mainly used in the aerospace industry and sports goods and these materials are very expansive. The use of boron fibers has now been replaced by carbon fibers.

Natural Fibers

Natural fibers are emerging fibers normally used as reinforcement and filler for polymer composites. Examples of natural fibers used in polymer composites include hemp, sisal, kenaf (Fig. 3.4), oil palm, jute, pineapple leaf,

FIGURE 3.3 Woven aramid fiber.

FIGURE 3.4 Woven kenaf fiber.

banana stem, coir, and bagasse (Sapuan, 2014). The following are the advantages of natural fibers:

- Good insulation against heat and noise
- Renewable, abundant, and continuous supply of raw materials
- Low cost
- Low density
- Biodegradability
- Enhanced energy recovery
- Free from health hazard (cause no skin irritations)
- Acceptable specific strength properties
- High toughness
- Good thermal properties
- Reduced tool wear
- Reduced dermal and respiratory irritation
- Ease of separation

However, natural fiber suffered from the following drawbacks:

- Poor compatibility with hydrophobic polymer matrix
- The lack of an established collection, storage and handling systems that would prevent the degradation of fibers for a long time
- Low processing temperature permissible
- High moisture absorption
- Tendency to form aggregates during processing
- Low resistance to moisture
- Low thermal stability

Natural Fiber Composites

Natural fiber composites find their applications in many nonstructural and semistructural components, such as in the furniture industry, building and construction materials, and automotive components (Fig. 3.5) due to some advantages they offer. Light weight, corrosion resistance, abundantly available, renewable, cheap and comparable specific strength and stiffness properties are among the most desirable attributes that encourage their use (Anuar and Sapuan, 2014). Currently there is a move toward the development of 100% biocomposites, where the fiber is derived from a plant and the biopolymer is also made from the same plant (Sahari et al., 2013).

Advanced Composites

The term advanced composite is related to more expansive and high-performance composites for aerospace applications [Barbero (1998) and Dorworth et al. (2009)]. Schwartz (1997) defined the term advanced

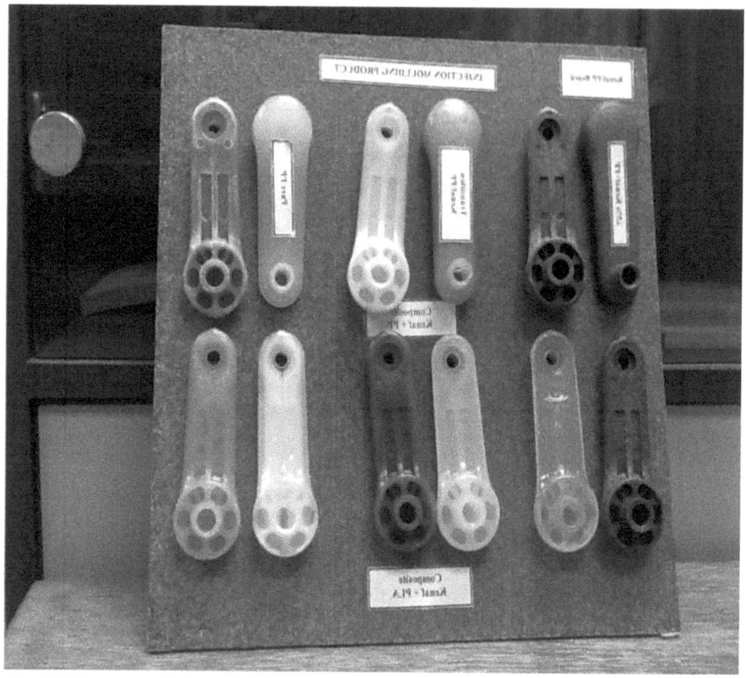

FIGURE 3.5 Car window regulators made from kenaf fiber composites.

composites as composites with high-performance characteristics in terms of strength and stiffness.

ADVANTAGES AND DISADVANTAGES OF POLYMER COMPOSITES

Advantages

Composite materials offer a wide range of advantages to industries, such as:

- Light weight
- High impact resistance
- Chemical corrosion resistance
- Part consolidation
- Repair is possible
- Complex shape can be fabricated easily
- Good moldability
- Fabrication to the net shape possible
- Low manufacturing cost due to low part count
- High specific strength

- Longer fatigue life
- Retain properties at high temperature
- Good electrical and thermal insulation
- High specific stiffness
- Good damping properties
- Structural properties can be tailored to application
- Good wear resistance under heavy load
- Can be pigmented easily
- Electrical insulating
- Nonmagnetic
- Easy to maintain
- Production equipment operates at lower pressures than with metallic materials

Disadvantages

However, polymer composites suffer from some drawbacks, such as:

- Lack of ductility
- Anisotropic (properties are directional dependent): possible weakness of transverse properties
- Variation in properties of parts due to the nature of fabrication
- Susceptible to delamination and splitting
- Residual stresses due to shrinkage during curing process
- Degraded by solvents and moisture
- Design data are not established
- Propagation of crack in fiber direction
- High cost of raw materials
- Manufacturing routes not proven compared to metals
- Difficult to analyze

APPLICATIONS OF POLYMER COMPOSITES

Review of selected publication on the application of composites in various industries is presented in this section.

Automotive Industry

The application of composite materials in automotive components no longer needs any detailed discussion, as the development is so advanced now. Automotive components are not only made from conventional composites, such as glass and carbon composites, but they have also been made from natural fiber composites, nanocomposites, and smart composites. For instance, Daimler Chrysler, Mercedes, Volkswagen, Audi Group, BMW, Ford,

Opel (Njuguna et al., 2011), or even Proton had used natural fiber composites in some selected automotive components. In this section, a review of some research and development work on polymer composites in automotive application is presented.

Sapuan (1998) has carried out extensive research on the development of automotive pedal box system from polymer composite. CE was the major platform of the study. Polyamide 6,6 was selected as the most suitable polymer matrix while short glass fibers were the fibers selected for major components of pedal box system, that is, clutch and accelerator pedals. Design analysis was performed using finite element analysis ALGOR and selection of materials and design to cost were carried out using a knowledge-based system called KEE by Intellicorp. Clutch pedal form 30% glass fiber reinforced polyamide 6,6 (nylon) composites was fabricated using injection molding process.

Hambali (2009) used CE technique in the development of polymer composite bumper beam. In such development, he performed materials selection, design concept selection, and manufacturing process selection for automotive bumper beam. The major tool used in his study was analytical hierarchy process, a multicriteria decision making approach in the selection process. The selection processes in the different knowledge domains were executed concurrently in the spirit of ensuring CE approach as a successful approach.

McBeath (2000) reported on the development of a Williams Formula 1 car using composite materials. In his book, McBeath (2000) describes in detail procedures and guides for developing a competition car from composites. The development was assisted by advanced computer-aided design (CAD) software package. Materials testing and selection were given high priority prior to the fabrication process to ensure the product will perform well during service. Details on pattern making, mold making, component manufacture, composite materials selection are given. Prepreg technology was considered in fabricating such components using the traditional method of hand lay-up and other related processes.

Mutasher (2006) carried out research work on the development and properties of carbon and glass fiber reinforced epoxy composite drive shafts (see Fig. 3.6) intended for automotive application. The drive shafts were fabricated using filament winding process. Mechanical properties, such as tensile, flexural, and impact properties were determined and results of properties of glass, carbon, and hybrid fiber reinforced epoxy composites were presented and compared.

Rwawiire et al. (2015) studied the possibility of utilizing bark cloth/epoxy composites as a material for the manufacture of automotive instrument panels. The hand lay-up method was used in the preparation of the composite.

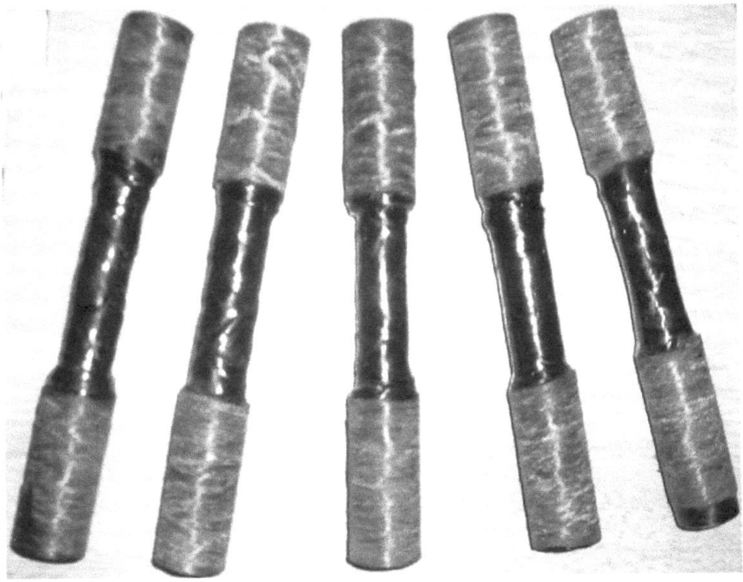

FIGURE 3.6 Samples of automotive drive shafts from filament wound carbon fiber reinforced epoxy composites.

He discovered that the developed biocomposites had an average strength of 33 MPa which is higher than the 25 MPa threshold strength needed for car instrument or dashboard panels, making bark cloth reinforced green epoxy composites a suitable alternative material for interior automotive panels.

Aerospace Industry

Gomina (2012) reported that the aerospace industry accounts for 14% of the composite market and this percentage was expected to increase in the years to come. Examples of the application of composite in the aerospace industry include their use in aircraft, spacecraft, satellites, helicopters, missiles, and booster rockets (Herakovich, 1998). Herakovich (1998) further stated that the properties of composites being sought in the aerospace industry include excellence in strength, stiffness, and fatigue, and tailor-made design. Due to the high requirements needed for a composite to be utilized in the aerospace industry, high-strength fibers are commonly used, such as carbon, aramid, and glass. The polymers that are usually utilized are nonbiodegradable resins, such as epoxies, vinylesters, and phenolic based. These materials are usually disposed off with little effort to recycle and reuse them. The increasing concern on the preservation of the environment and recycling has led to increased research on the use of biobased, ecofriendly materials, such as composites based on natural fibers.

Scarponi (2015) utilized hemp fiber composites as a replacement for traditional woven glass/epoxy laminates used for the manufacture of Naca cowling in an acrobatic ultra-light airplane. He found that the hemp fiber composite showed comparable mechanical performance, about same weight, but easier disposal and better ecofriendly characteristics as compared to their synthetic counterpart. He concluded that the engine cover, produced by the use of hemp/epoxy composites, demonstrates the effective possibility to produce semistructural aeronautical components using natural fiber composites in substitution of glass ones. A slight drawback that was mentioned in this study was the slightly increased weight of the hemp/epoxy composite Naca cowling (3.3%) compared to the traditional glass/epoxy. The author determined that this was due to the conservative approach in the number of plies used as there was lack of a reliable and accurate database for the mechanical properties of natural fiber composites along with the inherent scattering of mechanical properties of natural fibers.

Building Industry

Wood, a natural composite, has long been used by man in the construction of buildings. The limitations of wood in its natural form have restricted the height of buildings that are constructed from it. However, advances in composite technology have overcome these limitations and buildings composed of, almost entirely, wood composites are being built bigger and higher than ever before. On the other hand, concrete structures are increasingly being reinforced with polymer composites to increase their overall strength, durability, and resistance toward wear (Ksiazek, 2015; Neagoe et al., 2015; Yan and Chouw, 2015).

Singh et al. (2011) reported on research work on jute composite door shutters made from sandwich panels. Both corrugated and rigid foam core type jute composite panels were used in the study. Various tests, such as slamming test, edge loading test, shock test, and impact indentation test were conducted before jute composite panels could be used as door shutters.

Senawi (Senawi, R., 2012. Innovative Pultrusion Sdn. Bhd, Senawang, Negeri Sembilan, Malaysia, Personal Communication) is responsible for the development of drain cover from kenaf fiber reinforced polymer composites using pultrusion process. It is a new attempt to develop and commercialize such products. The advantages of this product include light weight, low cost, aesthetically pleasing, utilization of local crop, and the most important is to prevent from theft. In Malaysia, house and building owners have faced problems with metal drain covers because they attracted the attention of thieves for their high cost. It is hoped that the use of composite drain cover (Fig. 3.7) could alleviate such problem.

FIGURE 3.7 **Pultruded composite drain cover.**
Courtesy of Innovative Pultrusion Sdn. Bhd., Seremban, Negeri Sembilan, Malaysia.

Klyosov (2007) has investigated and developed decking from wood–plastic composites. Decking is used as a platform attached to a building, as found in walkways and boardwalks. In his book, details about the development of decking are presented and examples of types of wood used for decking include redwood and cedar. Woods are bonded with some matrices, such as polypropylene, polyethylene, and poly vinyl chloride. Procedures for conducting various testing, such as flexural strength and modulus, flammability, water absorption, shrinkage, slip resistance, and thermal expansion were provided.

Uyup (2008) carried out experimental work on impregnation of bamboo strips with low molecular-weight phenol formaldehyde resin to form composite panel. The bamboo samples were obtained from basal and middle parts of the culm. Weight percentage gain (WPG) and moisture content (MC) of the composite panel were determined. The WPG of the bamboo panel increased as the soaking time was increased. During the drying process, the MC of the bamboo panel decreased drastically when the drying process took place in 3 h but it showed gradual decrease as it was dried for 12 h. Fig. 3.8 shows compression molded phenol formaldehyde/bamboo composite panel (A) longitudinal direction (B) transverse direction.

Irle (2007) conducted a study on development of composite particleboard for panel products where shredded recycled papers were blended with urea formaldehyde particleboard adhesive. Office papers were used as skins and papers from magazines were used as the low density core. The mechanical properties were not very promising and therefore the panels may be suitable to be used as nonload bearing applications, such as insulation and packaging.

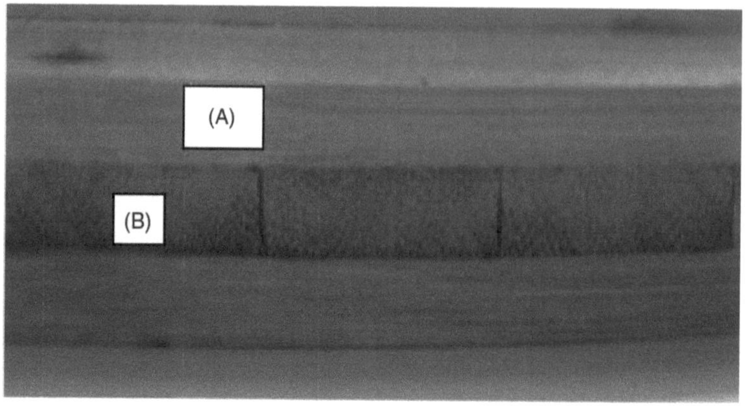

FIGURE 3.8
Compression molded phenol formaldehyde/bamboo composite panel, (A) longitudinal direction and (B) transverse direction.

Hurricane resistant houses were developed from natural fiber reinforced soya bean oil composites (Wool and Sun, 2005). Natural fiber composite beams were fabricated using vacuum assisted resin transfer molding (VARTM) process. Three different fibers were used in this study, namely chicken feather fibers, recycled paper, and E-glass in hybrid form. Four-point bending test was conducted to study the bending behavior of the composites.

Neagoe et al. (2015) researched on the structural response of hybrid beams made of pultruded glass fiber reinforced polymer profiles mechanically connected to reinforced concrete slabs, suitable for building floors, as well as footbridge and marine pier superstructures. Two different models of hybrid glass fiber reinforced polymer-concrete beams were designed, tested, and analyzed by the author. It was discovered that the glass fiber reinforced polymer-concrete hybrid beams prove to be structurally efficient elements with a high flexural capacity to self-weight ratio. The increase in concrete strength improved the ultimate bending capacity and stiffness but the behavior was still limited by the shear deformation that the profile could bear.

Marine Application

Misri (2011) performed design and fabrication of a small boat from hybrid glass-sugar palm fiber reinforced unsaturated polyester composites. He used the total design model comprising market investigation, preparation of product design specification, conceptual design, detail design, and fabrication to develop the boat. The mechanical properties, such as tensile and flexural properties of hybrid glass-sugar palm fiber composites were also determined. The fabrication of composite boat was carried out using hand lay-up process.

Mostafa et al. (2015) developed composite sandwich panels for sail boat hull manufacture using glass-fiber/epoxy composite facing and polyurethane foam core. Mancuso et al. (2015) investigated the flexural behavior of a composite sandwich made of flax fibers reinforced skin facings and an agglomerated cork core that was to be employed as an ecofriendly solution for the making of structural components of small sailing boats. In their study, three different assembly techniques were used; hand lay-up, vacuum bagging, and resin infusion. The researchers discovered that resin infusion resulted in higher shear rigidity and flexural strength at the expense of weight.

Besides being the core material for the manufacture of marine products, polymer composites are also widely used as coatings to enhance the resistance of materials toward corrosion by water and seawater. Steel structures coated with fiber-reinforced polymer (FRP) composites have gained wide acceptance in the marine industry due to their high strength-to-weight ratio, ability to withstand harsh off-shore conditions, resistance against corrosion and erosion, good protection from environmental degradation, and impact loads. Sambyal et al. (2016) developed conducting polymer poly(aniline-*co*-phenetidine)/SiO_2 composites for corrosion protection of mild steel in marine environments. The corrosion resistance of the coatings was observed to be improved remarkably on the incorporation of copolymer composites in the epoxy system. The authors discovered that corrosion test results under accelerated conditions (as per ASTM B117 standards), of the composite coatings revealed almost no extended corrosion and blistering along the scribe mark.

Furniture Products

Sapuan and Maleque (2005) developed a telephone table from woven banana-stem fiber reinforced epoxy composites. Design of the table was performed following the total design model developed by Pugh (1991). Design concepts were generated and the concepts were evaluated using weighted objective method. A mold was made based on the detail drawings drawn and the fabrication was accomplished using hand lay-up process. Several parts of the table were joined by means of mechanical fasteners. An aesthetically pleasing product was achieved.

Another similar work was performed by Sapuan et al. (2007) on the development of a coffee table also made from woven banana pseudo-stem fiber reinforced epoxy composites. The same total design model as performed by Sapuan and Maleque (2005) was used. Several concepts were generated but this time they were evaluated using the Pugh selection method. Fabrication was also done using hand lay-up process. Fig. 3.9 shows the composite table made from banana stem and epoxy.

Gomina (2012) reported some work on composite furniture mainly made from natural fiber composites. Furniture made from natural fiber composites

FIGURE 3.9 Small table made from banana pseudo-stem fiber reinforced epoxy composite.

demonstrated excellent surface finish and this is very important as far as aesthetic effect is concerned. Furniture that were developed using flax fiber reinforced epoxy and polyester composites include coffee table, chair, and armchair fabricated using resin transfer molding and injection molding process.

New materials are always discovered and researched as a reinforcement or matrix in a composite used for furniture applications. Ach Kusairi and Ni'mah (2015) utilized oil palm kernel shells as reinforcement and tapioca adhesive as matrix in the production of particleboards which are widely used in furniture applications. The researchers discovered that some physical properties, such as density and water content complied with the Japanese Industrial Standard for particleboards. However, strength properties need to be improved before they can be utilized as furniture material. Various studies using oil palm trunk fibers to manufacture plywood and particleboards have also been conducted (Baskaran et al., 2015; Hoong et al., 2015; Lee et al., 2015; Nor Hafizah et al., 2014).

Sporting Goods and Leisure

Traditionally, graphite, magnesium, titanium, and advanced aluminum alloys are widely used in golf clubs, tennis racquets, and racing bikes. These materials are prized for their light weight and strength. However, the introduction of carbon fiber in the material world literally caused an explosion in sporting goods manufacture. The material offers compression and lateral stability, as well as light weight, strength, and durability which are invaluable properties in the manufacture of sport equipments. For example, injection molded carbon fiber reinforced nylon composite was reported to be used as material for squash racquet due to its light weight, stiffness, and toughness. These attributes can absorb energy from smashes against the wall and floor (Easterling, 1990). Besides that, it is also widely used in the making of bicycle frames, golf shafts, surfboards, fishing rods, snow skis, snow shoes, ski boots, racing cars, hockey sticks, pole vaults, and jevelins. While composites have traditionally been used to make bicycle frames, the material is slowly finding its way into wheels and other bicycle components. Besides its obvious light weight, the use of composites also allow the fabrication of less parts, easier customization, and in some cases better functionality as well.

Torres et al. (2013) fabricated a longboard using a green-composite sandwich structure with basalt fiber and bioepoxy resin. In addition, this material system was combined with cork as core material for the fabrication of fiber composite sandwich structures. Again, increasing concerns on the preservation of the environment prompted the authors to study more environmentally friendly materials as an alternative to traditional petroleum-based composites. Stress analysis of the board was assessed by FEM simulations and the proposed design proved to satisfy the failure criteria and be acceptable for service conditions.

MANUFACTURING METHODS OF POLYMER COMPOSITE MATERIALS

Mayer (1993) classified polymer composites based fabrication techniques and three classes of materials were identified as:

1. Fibers and polymers are obtained by the user separately and the user carries out the fabrication process, such as in hand lay-up (wet lay-up) process.
2. Molding compounds (semicured composites or intermediate materials) are purchased from the supplier where the mixture of fibers, polymer, and additives is fixed. The user just uses them in injection molding or compression molding process. Examples of these intermediate materials include injection molding compound, sheet molding compound (SMC), glass mat thermoplastics (GMT), dough molding compound (DMC), and bulk molding compound (BMC).

3. After fabrication process, ready-made composites are available that can be assembled to form products. Examples include the items from filament winding and pultrusion process.

However, in this chapter, it is not the intention to elaborate on these three classes of materials but only individual composite manufacturing techniques, such as injection molding, resin transfer molding, hand lay-up, spray-up, filament winding, compression molding, pultrusion, vacuum bagging, and autoclave molding are discussed. Hand lay-up and spray-up processes are known as contact molding. Both vacuum bagging (also known as vacuum bag molding) and autoclave molding are known as prepreg molding. Other techniques that can be used in composite fabrication include extrusion, out-of-autoclave process, calendaring, thermoforming, braiding, structural reaction injection molding, resin film infusion, vacuum infusion, elastic reservoir molding, mandrel wrapping, tube rolling, centrifugal molding, automated tape layering, automated fiber placement, and pulforming but they are not discussed here. Interested readers can read the details of these techniques in other publications, such as in Mayer (1993), Herakovich (1998), Mallick (2008), Hancox et al. (1980), and Dorworth et al. (2009).

Injection Molding

Injection molding (Fig. 3.10) is a composite manufacturing process appropriate for small to medium sized products and gives more design freedom, reduced assembly needs, and good quality consistency as the components and mold

FIGURE 3.10 Injection molding machine.

can be made with no subsequent trimming or machining (Sapuan, 2007). In this process, pallets or granules comprising polymer (with chopped fibers) are melted and injected into a mold after being fed through a hopper. The pallets or granules called injection molding compound (IMC) have been previously formulated and prepared using twin screw extruder and internal mixer. The composite is formed, allowed to cure, and ejected from the machine. Prior to injection into the mold, IMC granules were fed into a heated barrel and passed through a reciprocating screw (shear action) and then the granules were allowed to melt (Mayer, 1993).

Resin Transfer Molding

The resin transfer molding (RTM) process (Fig. 3.11) has been widely accepted for fabricating simple and complex shapes of automobile, aircraft, and spacecraft components with high strength and stiffness to weight ratios of the composite products. This process is widely acceptable in various industries, that is, wherever low manufacturing cost with high mass production is required compared to other processes like compression molding and hand lay-up. Reinforcement called preform is arranged to the intended configuration, and placed dry in mold cavity of the bottom half of a two-part mold. The mold is closed and then a catalyzed resin is transferred into the closed mold via a sprue. The matrices that can be used in the RTM process include polyester, vinyl ester, epoxy, phenolic, acrylic, and methacrylates. Reinforcements can be in the form of woven roving, continuous strand mat, unidirectionals, cloth fabric, and three-dimensional textile performs.

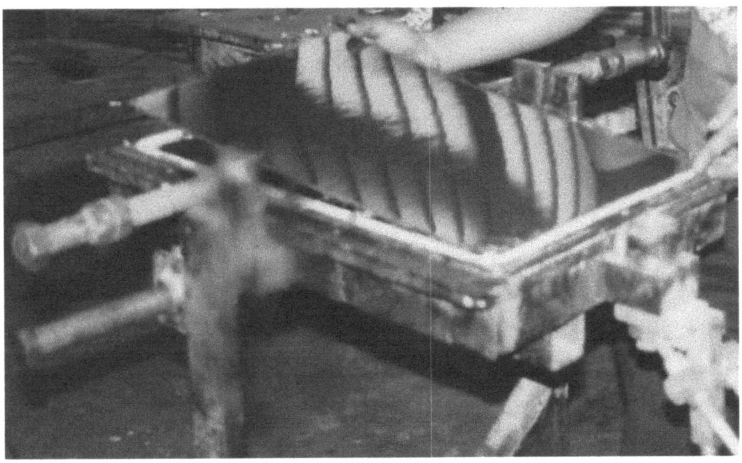

FIGURE 3.11 Resin transfer molding process.

Hand Lay-Up

Hand lay-up (Fig. 3.12) is one of the simplest, least expansive, and most labor-intensive composite manufacturing process. It is a traditional method of composite manufacturing and it was normally used to produce large one-piece components (McBeath, 2000). It had been around prior to the era of robotics and automation. Hand lay-up is sometimes called wet lay-up. It is also called contact molding due to the fact that the operator is in contact with the mold surface by means of a roller. Hand lay-up is an open molding process in which mold is used on one side of the component while the other side of the component is exposed.

Mold release agent is initially placed on the surface of the mold to prevent the component from sticking to the mold. Then a gel coat made of a resin is placed on the mold surface. The purposes of this coat are to provide smooth and hard surface and to prevent the fiber from appearing on the surface of the mold (Mayer, 1993). Then fibers and matrix are laid one layer after the other

FIGURE 3.12 Hand lay-up process.

to the desired thickness. The consolidation of composites to obtain fully wet-out of the fibers and removing of air are done by means of a roller. Finally the composites are cured.

Spray-Up

Spray-up or spray-up molding is one of the contact molding processes. In spray-up molding, a spray-gun is used to apply resin and reinforcement in the form of chopped roving (Sapuan et al., 2006). Chopped roving is continuously supplied into the spray-gun and mixed with resin, and the mixture is in turn sprayed onto the mold. Air stream is used to feed the composites onto the mold. The mold in this process is the same as that used in hand lay-up process. Spray gun can be used manually or using a robot (Murphy, 1994). Consolidation is performed using a hand roller.

Filament Winding

Filament winding is one of the composite fabrication methods utilizing continuous fiber reinforcement (rovings or tape) and resin in liquid form. Fibers are drawn through liquid resin in resin bath to form resin-impregnated fibers and the composites are wound around a rotating mandrel. There are three types of possible winding patterns available, such as hoop, helical (Fig. 3.13), and polar windings and the winding angles are controlled by machine. Each

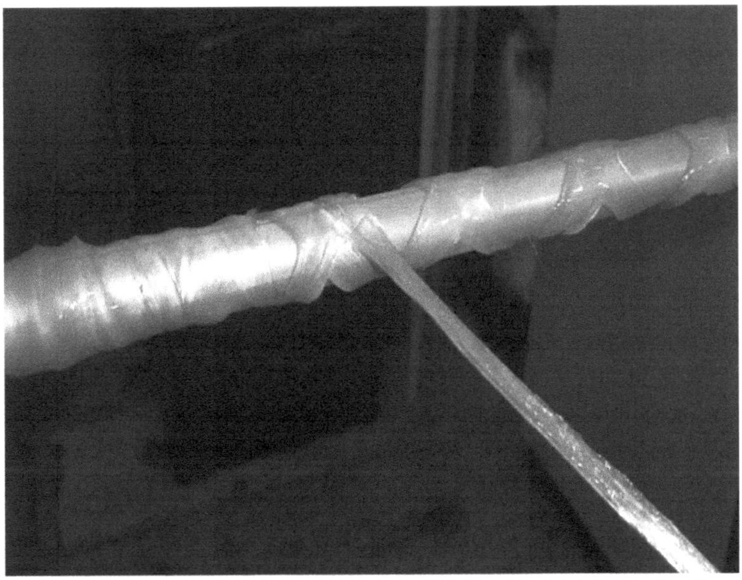

FIGURE 3.13 Helical winding pattern in filament winding process.

winding pattern has a different manner and direction of fiber lay-down. The basic principles of each winding method are the same. Components that can be produced by filament winding method are normally circular and they include tubes, drive shafts, pipes, pressure vessels, chemical storage tanks, fishing rods, compressed natural gas cylinders, and golf club shafts. The continuous fibers can be glass, carbon, Kevlar, and natural fibers while the most common resins include epoxy and polyester.

Compression Molding

Compression molding (Fig. 3.14) employs a mold of matched male and female dies made of metal. It is a manufacturing process with high volume capability. The common intermediate or raw materials for this process are sheet molding compound (SMC) and bulk molding compound (BMC) (Mazumdar, 2002). Compression molding is also known as press molding. Two types

FIGURE 3.14 Compression molding.

of compression moldings are available namely hot press and cold press but hot press is more commonly used. Hot press operates involving heat and pressure and the equipment comprises a press, heated platens, and matched tooling. Molding compound is charged into the mold. Then the male and female halves of the mold are closed. Upon closing, curing process takes place by means of heat and high pressure. The composite component is then ready after being ejected from the mold.

Pultrusion

The term pultrusion is derived from the combination of "pull" and "extrusion." Pultrusion (Fig. 3.15) is a composite manufacturing process that involves pulling of composites in the form of continuous fiber roving or fabrics using a belt or a caterpillar puller system. The composites are passed through and immersed in resin bath to wet out and to form resin impregnated fibers and they are then passed through a series of forming dies. The product is then cut into the required length. The pultrusion process is done to create continuous composite profiles with constant cross section shapes at any length, either solid or hollow, such as bars, rods, I beam, T-beams, channels, and thin-walled tubes (Herakovich, 1998). The matrices used are normally thermosetting polymers, such as epoxy and polyester. The fibers used include glass, carbon, aramid, and natural fibers. Fibers in continuous roving, woven cloth, or mat form are placed in creel and they are pulled. There are two ways of impregnating fibers with resin in resin bath and they are dip impregnation and roller impregnation (Middleton, 2000).

FIGURE 3.15 **Continuous fibers impregnated with polymer in resin bath in pultrusion process.**
Courtesy of Innovative Pultrusion Sdn. Bhd, Seremban, Malaysia.

Vacuum Bag Molding/Autoclave Molding

Vacuum bag molding, also known as vacuum bagging, is one of the composite manufacturing techniques and it is a modification of the wet lay-up process. In this process the laid-up mold is inserted into a bag made of flexible film and then it is evacuated. In this technique, the equipment comprises lay-up mold, peel ply, release film, bleeder, separator, breather, and vacuum bag. At the bottom, a lay-up mold is initially fixed, composite laminate is placed onto it, and then peel ply, release film, bleeder, separator, breather, and vacuum bag are stacked on top of each other and all edges are sealed. After the laminate is laid-up, the bag is evacuated and pressure is applied to eliminate or to reduce voids in laminate, consolidate the laminate, and to ensure good adhesion between fiber and matrix. Vacuum draws out resin, volatiles, and trapped excess air (Jaafar and Shawkataly, 2009). If the laid-up mold inserted into a special bag is placed in an autoclave for cure and consolidation under controlled heat and pressure, it is called autoclave molding. In short, the difference between vacuum bag and autoclave molding is in the manner the consolidation is achieved. The former by vacuum using vacuum bag while the latter by pressure in an autoclave.

METAL MATRIX COMPOSITES

MMCs are defined as composites engineered by a combination of two or more materials, where at least one of it is metal. Engineered MMCs consisting of continuous or discontinuous fibers, whiskers, or particles in a metal achieve combinations of very high specific strength and specific modulus. The many advantages of MMCs are the increased temperature strength, fatigue strength, wear resistance, corrosion resistance, and electrical and thermal conductivities.

Composite Constituents

Structurally, MMCs consist of continuous or discontinuous fibers, whiskers, or particles in an alloy matrix, which reinforce the matrix or provide it with requisite properties not achievable in monolithic alloys (Rohatgi, 1993). Various metals have been utilized as major constituents in MMCs and among them are aluminum, copper, iron, titanium, magnesium, etc. (Abyzov et al., 2015; Munir et al., 2015; Rhee et al., 2015; Verelli and Hassoun, 2015).

Lighter metals, such as aluminum, magnesium, or titanium are usually used in structural applications while for high temperature purposes, cobalt and cobalt–nickel alloy matrices are usually utilized as the metal.

Fabrication Processes of MMC

There are various methods of producing MMCs. By altering the manufacturing method, the processing, and the finishing, as well as by the form of the reinforcement components it is possible to obtain different characteristic profiles, although the same composition and amounts of the components are involved (Kainer, 2006). Among the processes that are used are the melting metallurgical process, powder metallurgical process, hot isostatic pressing of powder mixtures and fiber clutches, joining and welding of semimanufactured products, etc. Different properties and profiles of the final MMC can be obtained by changing the process methodology and parameters even with the same composition and amount of matrix and binder.

Applications

MMCs can be used in various areas, such as automative engineering, aviation industry, and other light and heavy industries. Light alloy composites are already used in the fabrication of valve trains, piston rods, piston and piston pins, cylinder head, crankshaft main bearing, and part-strengthened cylinder blocks in the automotive industry (Kainer, 2006). Titanium matrix composites have been proven to be viable as a material for propulsion components in the aviation industry. Piston rods in the F-119 engine of F-22 fighter aircraft are made from titanium matrix composites (Singerman and Jackson, 1996). MMCs possess high-temperature capability, high thermal conductivity, low CTE, and high specific stiffness and strength and these potential benefits generated optimism for MMCs for critical space system applications in the late 1980s (Rawal, 2001).

Aluminum–graphite composites are used in power electronic modules because of their high thermal conductivity, the adjustable coefficient of thermal expansion, and the low density. Dymalloy, an MMC consisting of 20% copper and 80% silver alloy matrix with type 1 diamond is used in microelectronics as a substrate high power and high density multichip modules, where it aids with removal of waste heat (Kerns et al., 1995). MMCs are also commonly used in the manufacture of high- and low-performance cutting tools. Cobalt matrix is usually used in the high-performance tools while other metals, such as bronze are used in low-performance tools.

MMCs have also been used in the manufacture of bicycle frames. However, due to its restrictive cost, they are usually used for specialized applications or top of the line products.

Self-Healing MMC

In recent years, interest has increased in the area of self-healing inorganic materials. The inspiration for this property comes from biological systems that can self-repair damage (Ferguson et al., 2014; Nosonovsky and Rohatgi, 2011). Even though the applied research in this area is in its relative infancy, the potential of these materials is enormous as various fatigue and strength failures can be alleviated if the material is self-healing.

As mentioned by Ferguson et al. (2014), the challenge of creating a self-healing MMC is a daunting task as metals have high melting temperatures and the self-repair process must take place at high temperatures or specialized environments. A technique that has been employed for self-healing in MMCs is by delivering low melting temperature alloys to damaged regions as a healing agent. Another method is by incorporating memory alloys into MMCs to enable the MMC to return to its original shape after heating (Ferguson et al., 2015; Misra, 2013).

CERAMIC MATRIX COMPOSITES

CMC is a new class of composite materials developed by material experts.

Composite Constituents

Ceramic matrix should, among others, be able to infiltrate a bundle of fibers, whiskers, or particulate perform, form a mechanical or frictional bond with the reinforcement, not damage the fiber physically, have good toughness and have good resistance to creep, fatigue, and impact (Chawla, 1993). Among the ceramic matrix materials reported to be used in CMC include silicone carbide, silicone nitride, alumina, mullite, boron nitride, boron carbide, carbon, and sialons.

Reinforcements in CMCs normally are in the forms of continuous fiber, short fiber, whisker, or particle. The fiber types used as reinforcements include glass fibers, boron fibers, carbon fibers, oxide fibers, nonoxide fibers, and whiskers (Chawla, 1993).

Properties of CMC

In CMC, in the study of mechanical behavior of CMCs, transfer of the applied load to the fiber that occurs by means of shear strains in the matrix is important and a model that describes this is called shear-lag model. The model can be applied for PMC and MMC as well. As far as mechanical testing of CMCs is concerned, testing like flexural and tensile are important. Other important properties are fatigue and creep (Chawla, 1993).

Fabrication Processes of CMC

CMCs can be made either by conventional powder processing techniques or by unconventional techniques (Matthews and Rawlings, 1994). Conventional techniques include cold pressing and sintering, hot pressing, reaction bonding processes, combined hot-pressing and reaction bonding method, and whisker reinforced composites by hot pressing. Cold pressing of the matrix powder and fiber followed by sintering is a method to produce CMCs and complete removal of the organic binder must occur before a fully sintered body is obtained. Hot pressing of CMCs is used in the consolidation stage of CMCs. Reaction bonding process is used with silicon carbide or silicon nitride matrices. In this process little or no matrix shrinkage occurs during densification. Combined hot-pressing and reaction bonding method is a hybrid method to produce CMCs. In whisker reinforced composites by hot pressing, whiskers or short fibers are used as reinforcement in place of continuous fibers (Chawla, 1993).

Unconventional techniques include infiltration, directed oxidation, or the Lanxide process, in situ chemical reaction techniques, sol–gel and polymer pyrolysis, and self-propagating high temperature synthesis (SHS). Infiltration of a preform made of a reinforcement can be done with a matrix material in solid, liquid, or gaseous form. Another version of liquid infiltration is the directed oxidation process, or the Lanxide process, developed by Lanxide Corporation. In situ chemical reaction techniques include chemical vapor deposition (CVD), chemical vapor impregnation (CVI), and different types of reaction bonding techniques. Sol–gel and polymer pyrolysis techniques can be used to make ceramic matrix materials in the interstices of a fibrous perform. Self-propagating high temperature synthesis can be used to produce a variety of refractory materials (Chawla, 1993).

Applications

Majority of the applications of CMC are in the aerospace industry. In the aerospace industry, CMC are used in turbine engine exhaust, reentry thermal protection material for spacecraft, aircraft brake assembly, and rocket propulsion components. Nonaerospace applications of CMCs include medical implants, automotive engines, automotive brake assembly, and dies and tool bits (Chawla, 1993).

NANOCOMPOSITES

Nanocomposites are composites in which at least one of the phases shows dimensions in the nanometer range (1 nm = 10^{-9} m) (Roy et al., 1986). In general terms, nanocomposites can encompass porous media, colloids, gels, and copolymers but they usually represent the solid combination of a bulk matrix and nanodimensional phase(s) differing in properties due to dissimilarities in structure and chemistry. Different investigators have employed various equipments and techniques for the characterization of nanocomposites, including atomic force microscopy (AFM), scanning tunnelling microscopy (STM), Fourier transformed infrared spectroscopy (FTIR), X ray photoelectron spectroscopy (XPS), nuclear magnetic resonance (NMR), differential scanning calorimetry (DSC), scanning and transmission electron microscopy (SEM/TEM), etc. (Camargo et al., 2009). Nanocomposites are the object of much research due to their huge potential of overcoming the limitations of conventional composites and also monolithic materials. Nanocomposites are often heralded as the material of the future as they possess unique properties besides having the potential to far outstrip the performances of microcomposites and other conventional composites.

A unique property of nanocomposites is their exceptionally high aspect ratio. They have a much higher surface to volume ratio compared to conventional composites which accounts for its higher mechanical strengths. Particles, sheets, or fibers can be used as the reinforcement material. The discovery of carbon nanotubes in 1991 (Iijima, 1991) and its ensuing industrial and practical success as a reinforcement material in nanocomposites showcases the abilities of nanocomposites. The use of carbon nanotubes in a nanocomposite greatly increases the material's electrical and thermal conductivity. The applications of carbon nanotube reinforced nanocomposites are wide and include medical (Yu et al. 2015b), automotive, energy (Li et al., 2015; Yang et al., 2015), and various other sectors.

As in the case of microcomposites, nanocomposite materials can be classified, according to their matrix materials, into three different categories (Camargo et al., 2009):

- Ceramic Matrix Nanocomposites (CMNC),
- Metal Matrix Nanocomposites (MMNC), and
- Polymer Matrix Nanocomposites (PMNC).

Ceramic Matrix Nanocomposites

As with ceramic composites, the reinforcement material composes of ceramics, such as oxides, nitrides, borides, silicides, etc. Ceramic nanocomposites are made utilizing second phase dispersion into ceramic matrices with nanosized particulates, platelets, whiskers, or fibers. This was an effort to mitigate the limitations of conventional ceramics where efforts to introduce micron-sized ceramics did not succeed. The dispersion can be characterized as either intragranular or intergranular. The key issue in developing inorganic composite materials with the desired properties is obtaining a uniform distribution of the nanoreinforcement in the matrix (Saheb et al., 2014). The most common methods used for the preparation of ceramic matrix nanocomposites are the conventional powder method, polymer precursor route, spray pyrolysis, vapor techniques (CVD and PVD), and chemical methods.

Metal Matrix Nanocomposites

Metal matrix nanocomposites refer to materials consisting of a ductile metal or alloy matrix in which some nanosized reinforcement materials are implanted.

These materials combine metal and ceramic features, that is, ductility and toughness with high strength and modulus. Thus, metal matrix nanocomposites are suitable for the production of materials with high strength in shear/compression processes and high service temperature capabilities (Camargo et al., 2009).

The greatest challenges facing the development of MMNCs for wide application are the cost of nanoscale reinforcements and the cost and complexity of synthesis and processing of nanocomposites using current methods (Rohatgi and Schultz, 2007). The methods that have been used to synthesize metal matrix nanocomposites include powder metallurgy, deformation processing, vapor phase processing, and in some cases solidification processing.

Polymer Matrix Nanocomposites

Almost all types of polymers, such as thermoplastics, thermosets, and elastomers have been used to make polymer nanocomposites (Anandhan and Bandyopadhyay, 2011). However, they have some disadvantages, such as low modulus and strength compared to metals and ceramics. In this context, a very effective approach to improve mechanical properties is to add fibers, whiskers, platelets, or particles as reinforcements to the polymer matrix (Camargo et al., 2009). For example, polymers have been filled with several inorganic compounds, either synthetic or natural, in order to increase heat and impact resistance, flame retardancy, and mechanical strength, and to decrease electrical conductivity and gas permeability with respect to oxygen and water vapor (Fisher, 2003).

Many polymer nanocomposites can be fabricated and processed in ways similar to that of conventional polymer composites, making them particularly attractive from a manufacturing point of view (Hussain et al., 2006). Besides that, the biodegradability of polymer resins can be improved significantly with the addition of different reinforcements, particularly reinforcements in the form of natural fibers. According to Camargo et al. (2009), the most important methods to prepare polymer matrix nanocomposites are: (1) intercalation of the polymer or prepolymer from solution; (2) in situ intercalative polymerization; (3) melt intercalation; (4) direct mixture of polymer and particulates; (5) template synthesis; (6) in situ polymerization; and (7) sol–gel process.

Potentials and Advantages of Nanocomposites

Research has not even begun to uncover the vast potential of nanocomposite materials. It is already being used in almost all sectors of industry and breakthroughs are still being made as newer materials are discovered as both reinforcement and matrix in nanocomposites. Advances in processing and manufacturing are also rapid. Ceramics as a whole have high thermal and chemical resistance but are naturally brittle and suffer from poor resistance to fatigue, creep, and thermal shock. Palmero (2015) mentioned that there are several reasons for which nanocomposite ceramics can present enhanced mechanical properties as respect to monolithic materials. The incorporation of high-strength nanofibers into ceramic matrices has allowed the preparation of advanced nanocomposites with high toughness and superior failure characteristics compared to the sudden failures of ceramic materials (She et al., 2000). In the case of metal matrix nanocomposites, incorporation of as little as 1 vol.% of nanosize ceramics has led to a much greater increase in the strength of aluminum and magnesium base composites than was achieved at much higher loading levels of micron-sized additions (Rohatgi and Schultz, 2007).

An advantage of nanocomposites is that they are environmentally friendly. Addition of reinforcements to a wide variety of polymer resins produces a dramatic improvement in their biodegradability (Camargo et al., 2009). The nanocomposites can be used in environmentally friendly systems (Kickelbick, 2003) and also in various detecting systems (Abdulla et al., 2015; Zhang et al., 2016). Another advantage is the higher strength to weight ratio of nanocomposites compared to conventional composites. Yu et al. (2015a) discovered that carbon-nanotube modified silane coatings significantly increased the interfacial shear strength of carbon fiber reinforced polymer composites.

Metal and ceramic nanocomposites are expected to generate a great impact over a wide variety of industries, including the aerospace, electronic, and military, while polymer nanocomposites' major impacts will probably appear in battery cathodes, microelectronics, nonlinear optics, sensors, etc. Improved properties include significant enhancements in fracture strength (about 2 times) and toughness (about one-half time); no time-dependent wear transitions even at very low loads; higher high temperature strength and creep resistance; increased hardness with increasing heat treatment temperature; hardness values higher than those of existing commercial steel and alloys; possibility of synthesis of inexpensive materials; and significant increase in Young's modulus (about 105%), shear modulus, and fracture strength (almost 3 times compared to microcomposites) (Camargo et al., 2009).

CONCLUSIONS

This chapter provides information on composite materials in terms of advantages and disadvantages, history, and composite types. Three major types of composite materials including polymer composites, MMC, and CMC were discussed. Recent advances in composite materials technology like nanocomposites were discussed in detail. This chapter is useful in understanding the materials particularly when they are discussed in the context of design and CE in the following chapters.

References

Abdulla, S., Matthew, T.L., Pullithadatil, B., 2015. Highly sensitive, room temperature gas sensor based on polyaniline-multiwalled carbon nanotubes (PANI/MWCNTs) nanocomposite for trace-level ammonia detection. Sens. Actuat. B Chem. 221, 1523–1534.

Abyzov, A.M., Shakhov, F.M., Averkin, A.I., Nikolaev, V.I., 2015. Mechanical properties of diamond copper composite with high thermal conductivity. Mater. Des. 87, 527–539.

Ach Kusairi, S., Ni'mah, L., 2015. Utilization fibers and palm kernel shells and tapioca adhesive as matrix in the manufacture of composite boards as an alternative raw material in furniture industry. Int. J. ChemTech Res. 8 (4), 1645–1655.

Anandhan, S., Bandyopadhyay, S., 2011. In: Cuppoletti, J. (Ed.), Polymer Nanocomposites: From Synthesis to Applications, Nanocomposites and Polymers With Analytical Methods. InTech., Rijeka, Croatia.

Anuar, H., Sapuan, S.M., 2014. Prospect of durian skin fibre and its composites (Keynote lecture), Proceedings of the Postgraduate Symposium on Composite Science and Technology 2014, 28 January, Putrajaya, pp. 1–5.

Askeland, D.R., Fulay, P.P., 2009. Essentials of Materials Science and Engineering, second ed. Cengage Learning, Toronto.

Barbero, E.J., 1998. Introduction to Composite Materials Design. Taylor & Francis, Philadelphia.

Baskaran, M., Hashim, R., Sulaiman, O., Hiziroglu, S., Sato, M., Sugimoto, T., 2015. Optimization of press temperature and time for binderless particleboard manufactured from oil palm trunk biomass at different thickness levels. Mater. Today Commun. 3, 87–95.

Callister, W.D., Rethwisch, D.G., 2011. Materials Science and Engineering, eighth ed. John Wiley & Sons, Inc, Hoboken.

Camargo, P.H.C., Satranarayana, K.G., Wypych, P., 2009. Nanocomposites: synthesis, structure, properties and new application opportunities. Mater. Res. 12 (1), 1–39.

Chawla, K.K., 1993. Ceramic Matrix Composites. Chapman & Hall, London.

Clegg, D.W., Collyer, A.A., 1993. The Structure and Properties of Polymeric Materials. Institute of Materials, London.

Dorworth, L.C., Gardiner, G.L., Mellema, G.M., 2009. Essentials of Advanced Composite Fabrication and Repair. Aviation Supplies & Academics, Inc, Newcastle, Washington.

Easterling, K., 1990. Tomorrow's Materials, second ed. The Institute of Materials, London.

Ferguson, J.B., Schultz, B.F., Rohatgi, P.K., 2014. Self-healing metals and metal matrix composites. J. Mater. 66 (6), 866–871.

Ferguson, J.B., Schultz, B.F., Rohatgi, P.K., 2015. Zinc alloy ZA-8/shape memory alloy self-healing metal matrix composite. Mater. Sci. Eng. 620, 85–88.

Fisher, H., 2003. Polymer nanocomposites: from fundamental research to specific applications. Mater. Sci. Eng. C 23 (6–8), 763–772.

Gomina, M., 2012. Flax and hemp composite applications. In: Reux, F., Morel, A. (Eds.), Flax and Hemp Fibres: A Natural Solution for the Composite Industry. JEC Composites, Paris, pp. 141–162.

Gordon, J.E., 2006. The New Science of Strong Materials or Why You Don't Fall Through the Floor. Princeton University Press, Princeton, Expanded ed..

Halary, J., Laupretre, F., Monnerie, L., 2011. Polymer Materials: Macroscopic Properties and Molecular Interpretations. John Wiley & Sons, Inc, Hoboken.

Hambali, A., 2009. Selection of Conceptual Design using Analytical Hierarchy Process for Automotive Bumper Beam under Concurrent Engineering Environment, Universiti Putra Malaysia, PhD Thesis.

Hancox, N.L., Sajic, P., Tattersall, P., Tetlow, R., Lovell, D.R., 1980. A Guide to High Performance Plastics Composites. The British Plastics Federation, London.

Hartman, D.R., Greenwood, M.E., Miller, D.M., 1994. High Strength Glass Fibers, Owen-Corning Fiberglas Corporation Tehnical Paper, Toledo, OH.

Herakovich, C.T., 1998. Mechanics of Fibrous Composites. John Wiley & Sons, Inc, New York.

Hoong, Y.B., Pizzi, A., Chuah, L.A., Harun, J., 2015. Phenol-urea-formaldehyde resin co-polymer synthesis and its influence on Elaeis palm trunk plywood mechanical performance evaluated by ^{13}C NMR and MALDI-TOF mass spectrometry. Int. J. Adhes. Adhes. 63, 117–123.

Hussain, F., Hojjati, M., Okamoto, M., Gorga, R.E., 2006. Review article: polymer-matrix nanocomposites, processing, manufacturing, and application: an overview. J. Compos. Mater. 40, 1511–1575.

Hyer, M.W., 1998. Stress Analysis of Fiber-Reinforced Composite Materials. McGraw-Hill, Boston.

Iijima, S., 1991. Helical microtubes of graphitic carbon. Nature 354, 56–58.

Irle, M., 2007. Composites based on recycled paper, Proceedings of the International Panel Products Symposium 2007, 17–19 October, Cardiff, pp. 219–223.

Jaafar, M., Shawkataly, A.K., 2009. Properties of bagasse fibre-reinforced unsaturated polyester (USP) composites. In: Sapuan, S.M. (Ed.), Research on Natural Fibre Reinforced Polymer Composites. UPM Press, Serdang, pp. 63–83.

Kainer, K.U., 2006. Metal Matrix Composites: Custom-Made Materials for Automotive and Aerospace Engineering. Wiley-VCH Verlag GmbH & Co. KGaA, Weinhei.

Kelly, A., Mileiko, S.T. (Eds.), 1983. Fabrication of Composites. Elsevier, Amsterdam.

Kerns, J.A., Colella, N.J., Makowiecki, D., Davidson, H.L., 1995. Dymalloy: a composite substrate for high power density electronic components, International Symposium on Microelectronics: Showcasing the Stars of Microelectronics and International Symposium on Ball Grid Array, Los Angeles, 22–26 Oct.

Kickelbick, G., 2003. Concepts for the incorporation of inorganic building 24 blocks into organic polymers on a nanoscale. Prog. Polym. Sci. 28 (1), 83–114.

Klyosov, A.A., 2007. Wood-Plastic Composites. John Wiley & Sons, Inc, Hoboken.

Ksiazek, M., 2015. Use in the building cement composites impregnated with special polymerized sulphur. J. Build. Eng. 4, 255–267.

Lee, S.H., Zaidon, A., Lum, W.C., San, H.P., Peng, T.L., Jinn, C.M., Chai, E.W., Ling, C.K., 2015. Properties of particleboard with oil palm trunk as core layer in comparison to three-layer rubberwood particleboard. J. Oil Palm Res. 27, 67–74.

Li, Y., Wu, J., Chopra, N., 2015. Nano-carbon-based hybrids and heterostructures: progress in growth and application for lithium-ion batteries. J. Mater. Sci. 50 (24), 7843–7865.

Mallick, P.K., 2008. Fiber-Reinforced Composites: Materials, Manufacturing, and Design, third ed. CRC Press, Boca Raton.

Mancuso, A., Pitarresi, G., Tumino, D., 2015. Mechanical behaviour of a green sandwich made of flax reinforced polymer facings and cork core. Proc. Eng. 109, 144–153.

Matthews, F.L., Rawlings, R.D., 1994. Composite Materials: Engineering and Science. Chapman & Hall, London.

Mayer, R.M., 1993. Design with Reinforced Plastics: A Guide for Engineers and Designers. The Design Council, London.

Mazumdar, S.K., 2002. Composites Manufacturing, Materials, Product, and Process Engineering. CRC Press, Boca Raton.

McBeath, S., 2000. Competition Car Composites, A Practical Handbook. Haynes Publishing, Somerset.

Middleton, V., 2000. Filament winding. In: Owen, M.J., Middleton, V., Jones, I.A. (Eds.), Integrated Design and Manufacture Using Fibre-Reinforced Polymeric Composites. CRC Press, Boca Raton, pp. 187–206.

Misra, S.K., 2013. Shape Memory Alloy Reinforced Self-healing Metal Matrix Composites, The University of Wisconsin-Milwaukee, Master of Science in Engineering Thesis.

Misri, S., 2011. Design and Fabrication of Small Boat by Using Sugar Palm Fibre Reinforced Epoxy Composites, Universiti Putra Malaysia, Master of Science Thesis.

Mostafa, A., Shankar, K., Morozov, E.V., 2015. Behaviour of PU-foam/glass-fibre composite sandwich panels under flexural static load. Mater. Struct. 48, 1545–1559.

Munir, K.S., Li, Y., Liang, D., Ma, Q., Wei, X., Cuie, W., 2015. Effect of dispersion method on the deterioration, interfacial interactions and re-agglomeration of carbon nanotubes in titanium metal-matrix composites. Mater. Des. 88, 138–148.

Murphy, J., 1994. Reinforced Plastics Handbook. Elsevier Advanced Technology, Oxford.

Mutasher, S.A., 2006. Mechanical Performance of a Hybrid Aluminium/Composite Drive Shaft, Universiti Putra Malaysia, PhD Thesis.

Mutel, F., 2012. Preface. In: Reux, F., Morel, A. (Eds.), Flax and Hemp Fibres: A Natural Solution for the Composite Industry. JEC Composites, Paris, p. 3.

Neagoe, C.A., Gil, L., Perez, M.A., 2015. Experimental study of GFRP-concrete hybrid beams with low degree of shear connection. Constr. Build. Mater. 101, 141–151.

Njuguna, J., Wambua, P., Pielichowski, K., Kayvantash, K., 2011. Natural fibre-reinforced polymer composites and nanocomposites for automotive applications. In: Kalia, S., Kaith, B.S., Kaur, I. (Eds.), Cellulose Fibers: Bio- and Nano-Polymer Composites, Green Chemistry and Technology. Springer-Verlag, Heidelberg, pp. 661–700.

Nor Hafizah, A.W., Paridah, M.T., Nor Yuziah, M.Y., Zaidon, A., Choo, A.C.Y., Nor Azowa, I., 2014. Influence of resin molecular weight on curing and thermal degradation of plywood made from phenolic prepreg palm veneers. J. Adhes. 90 (3), 210–229.

Nosonovsky, M., Rohatgi, P.K., 2011. Biomimetics in material science. Springer Ser. Mater. Sci. 152, 1–22.

Palmero, P., 2015. Structural ceramic nanocomposites: a review of properties and powders' synthesis methods. Nanomaterials 5, 656–696.

Peters, S.T., 2004. Composite materials and processes. In: Harper, C.A. (Ed.), Handbook of Plastics, Elastomers, and Composites. fourth ed. The McGraw-Hill, New York, pp. 229–320.

Pugh, S., 1991. Total Design: Integrated Methods for Successful Product Engineering. Addison Wesley Publishers Ltd, Wokingham, England.

Rawal, S., 2001. Metal-matrix composites for space applications. J. Mater. 53 (4), 14–17.

Rhee, H., Whittingham, W.R., Oppedal, A.L., Sherif, A.R., King, R.L., Kim, H.J., Lee, C., 2015. Mechanical properties of novel aluminium metal-matrix composites: application to overhead conductors. Mater. Des. 88, 16–21.

Rohatgi, P.K., 1993. Metal-matrix composites. Defence Sci. J. 43 (4), 323–349.

Rohatgi, P.K., Schultz, B., 2007. Lightweight metal matrix nanocomposites—stretching the boundaries of metals. Mater. Matters 2, 16–20.

Roy, R., Roy, R.A., Roy, D.M., 1986. Alternative perspectives on "quasi-crystallinity": non-uniformity and nanocomposites. Mater. Lett. 4 (8–9), 323–328.

Rwawiire, S., Tomkova, B., Militky, J., Jabbar, A., Kale, B.M., 2015. Development of a biocomposite based on green epoxy polymer and natural cellulose fabric (bark cloth) for automotive instrument panel applications. Compos. Part B Eng. 81, 149–157.

Sahari, J., Sapuan, S.M., Zainudin, E.S., Maleque, M.A., 2013. Thermo-mechanical behaviours of thermoplastic starch derived from sugar palm tree (*Arenga pinnata*). Carbohydr. Polym. 91 (2), 1711–1716.

Saheb, N., Qadir, N.U., Siddiqui, M.U., Arif, A.F.M., Akhtar, S.S., Al-Aqeeli, N., 2014. Characterization of reinforcement dispersion in inorganic nanocomposites: a review. Materials 7, 4148–4181.

Sambyal, P., Ruhi, G., Dhawan, R. and Dhawan, S.K., 2016. Designing of smart coatings of conducting polymer poly(aniline-*co*-phenetidine)/SiO2 composites for corrosion protection in marine environment. Surf. Coat. Technol. 303, 362–371.

Sapuan, S.M., 1998. Concurrent Engineering Design System for Polymeric-Based Composite Automotive Components, De Montfort University, UK, PhD Thesis.

Sapuan, S.M., 2000. Bahan Komposit Berasaskan Polimer, Sifat-Sifat, Reka Bentuk, Pembuatan dan Penggunaan. Universiti Putra Malaysia Press, Serdang, Selangor, Malaysia.

Sapuan, S.M., 2007. Light weight polymer composite materials for automotive industry. In: Faiz, M. (Ed.), Specialty Polymers: Materials and Applications. IK International Publishing House Pvt. Ltd, New Delhi, pp. 239–277.

Sapuan, S.M., 2014. Conceptual design of products from natural fibre composites, Proceedings of 2014 International Conference on Mechanical Design, Manufacture and Automation Engineering (MDMAE 2014), 11–12 January, Phuket, pp. 381–384.

Sapuan, S.M., Harun, N., Abbas, K.A., 2007. Design and fabrication of a multi-purpose table using composite of epoxy and pseudo-stem banana fibres. J. Trop. Agric. 45 (1–2), 66–68.

Sapuan, S.M., Maleque, M.A., 2005. Design and fabrication of natural woven fabric reinforced epoxy composite for household telephone stand. Mater. Des. 26 (1), 65–71.

Sapuan, S.M., Rahmat, A.R., Ismail, Y.S., Hassan, M.Y., 2006. Glosari Bahan Komposit. Pusat Penerbitan Universiti (UPENA) UiTM, Shah Alam, Malaysia.

Scarponi, C., 2015. Hemp fiber composites for the design of a Naca cowling for ultra-light aviation. Compos. Part B 81, 53–63.

Schwartz, M.M., 1997. Composite Materials, Volume I: Properties, Nondestructive Testing, and Repair. Prentice Hall PTR, Upper Saddle River.

She, J., Inoue, T., Suzuki, M., Sodeoka, S., Ueno, K., 2000. Mechanical properties and fracture behaviour of fibrous Al_2O_3/SiC ceramics. J. Eur. Ceramic Soc. 20 (12), 1877–1881.

Singerman, S.A., Jackson, J.J., 1996. Titanium metal matrix composites for aerospace applications. In: Kissinger, R.D., Deye, D.J., Woodford, D.A. (Eds.), Superalloys. The Minerals, Metals and Materials Society, Warrendale, PA.

Singh, B., Gupta, M., Tarannum, H., Randhawa, A., 2011. Natural fibre-based composite building materials. In: Kalia, S., Kaith, B.S., Kaur, I. (Eds.), Cellulose Fibers: Bio- and Nano-Polymer Composites, Green Chemistry and Technology. Springer-Verlag, Heidelberg, pp. 701–720.

Torres, J.P., Hoto, R., Andres, J., Garcia-Manrique, J.A., 2013. Manufacture of green-composite sandwich structures with basalt fiber and bioepoxy resin. Adv. Mater. Sci. Eng. 2013, 9, Article ID 214506.

Uyup, M.K.A., 2008. Modification of Bamboo With Phenolic-Resin for the Production of Dimensionally Stable Plybamboo, Universiti Putra Malaysia, PhD Thesis.

Verelli, R., Hassoun, J., 2015. High capacity tin-iron oxide-carbon nanostructured anode for advanced lithium ion battery. J. Power Sources 299, 611–616.

Wool, R.P., Sun, X.S., 2005. Bio-Based Polymers and Composites. Elsevier Academic Press, Waltham.

Yan, L., Chouw, N., 2015. Effect of water, seawater and alkaline solution ageing on mechanical properties of flax fabric/epoxy composites used for civil engineering applications. Constr. Build. Mater. 99, 118–127.

Yang, Z., Fujigaya, T., Nakashima, N., 2015. Homogeneous coating of ionomer on electrocatalyst assisted by polybenzimidazole as an adhesive layer and its effect on fuel cell performance. J. Power Sources 300, 175–181.

Yu, B., Ziang, Z., Yang, J., 2015a. Long-term moisture effects on the interfacial shear strength between surface treated carbon fiber and epoxy matrix. Compos. Part A Appl. Sci. Eng. 78, 311–317.

Yu, Y., Kong, L., Li, L., Li, N., Yap, P., 2015b. Antitumor activity of doxorubicin-loaded carbon nanotubes incorporated poly(lactic-*co*-glycolic acid) electrospun composite nanofibers. Nanoscale Res. Lett. 10, 343–351.

Zhang, D., Chang, H., Li, P., Liu, R., 2016. Characterization of nickel oxide decorated-reduced graphene oxide nanocomposite and its sensing properties toward methane gas detection. J. Mater. Sci. Mater. Electronics 27, 3723–3730.

CHAPTER 4

Concurrent Engineering in Design and Development of Composite Products

INTRODUCTION

Concurrent engineering (CE) represents a key trend in product development over the last few decades, which has changed academic and industrial approaches of looking at the product development process. It is an integration of the new product development process to allow participants making upstream decisions to consider downstream and external requirements. Central characteristics of a concurrent development process are activity overlapping, information transfer in small batches, and the use of cross-functional teams.

DFX was called design for excellence by Barbosa and Carvalho (2013). Design for excellence was previously defined by Bralla (1996) as:

> *Knowledge base to approximate the product design to the maximum of its desirable characteristics such as high quality, reliability, maintainability, safety, easiness of use, environment requirements, reducing the lead time for sale, reducing manufacturing costs and maintenance products.*

CE is one of the important approaches in composite product development. Tailor-made design in composite product development supports CE principles, that is, while designing a product, manufacturing issues are considered. In doing so, quality, time, and cost are improved. Designing the composite products requires a lot of skill with multidisciplinary knowledge. Every aspect of the design, such as design concept, materials selection, and detail design should be considered in a single approach without losing any information and therefore the information should be properly captured. CE should be given paramount consideration in the design and manufacture of composite products due to the fact that most of the composite products are developed based on tailor-made principles; that is, while the design of a product is performed and materials are selected for the product, manufacturing process issues are considered at the early stage of product design process. CE requires simultaneous consideration of the manufacturing process and

CONTENTS

Introduction 95

Previous Work on CE in Composite Product Development 98

Early Work (1989–99) 98

New Millennium Era (2000–09) 101

Recent Work 104

Some Commercially Relevant Examples of Recent Development, Trend, and Advances in the Area of CE for Composites 106

CE Seems to be an Obvious Route to any Sort of New Composite Product Development: How it is Differentiated? 111

CE Have Changed the Ways Industries Work: Industrial Experience of CE for Composites at Steelcase 111

Advantages of CE at Steelcase 114

(Continued)

Composite Materials. http://dx.doi.org/10.1016/B978-0-12-802507-9.00004-0

Previous Works by the Author and his Coworkers in CE for Composites 116
Background *116*
Bumper System *116*
Pedal Box System *121*
Other Automotive Components *126*
Marine Application *131*
Other Applications *131*
Summary 133
References 134

other downstream issues during the product design stage. In this section, a review of the previous work on the CE activities in composite product development is presented.

CE is concerned with the production of higher quality product, being developed at lower cost and shorter time compared to the traditional method. Shorter development time can be achieved through proper life-cycle management. In order to achieve shorter lead time, various life-cycle tools and techniques were developed. According to Prasad (1996) computer tools and technology tools can be employed to reduce cycle time in a CE process provided the right tools and techniques are used.

CE requires that while designing a component the designer must consider the manufacturing process in order that the component is designed to be compatible with the manufacturing method (Strong, 2006). For instance, if the component is to be manufactured using the pultrusion process, the designer would like to include design features that facilitate ease of fabrication to reduce fabrication complexity because manufacturing cost is a key cost element in the pultrusion process.

Prasad (1996) proposed influencing agents of CE and they were categorized into 7 agents (also called 7Ts): tasks, teams, tools, talents, techniques, time, and technology (Fig. 4.1). In composite product development, 7Ts are used in an appropriate manner. For instance, the team comprising tool makers,

FIGURE 4.1 7Ts for concurrent engineering (CE).
Adapted from Prasad, B., 1996. Concurrent Engineering Fundamentals, Integrated Product and Process Organization, vol. 1. Prentice-Hall, Inc., Upper Saddle River, New Jersey.

designers, composite experts, are brought together to develop a composite product. A task to be performed should be known and clear to the team. The team comes from multidiscipline backgrounds with different talents. They are provided with several tools, and techniques in product development and technology, such as advanced computer technology should be at their disposal.

CE is possible through the use of various information technology (IT) tools in the design process of composites, carried out by the design team, mold maker, and processor. They are all linked together electronically or physically, so that each can make a contribution throughout the progress of the project. This paperless product design approach has replaced much of the prototyping of the past. This is preferable since it reduces the rework involved in product design, time and cost of production. The use of such analysis program by product designers is one of the latest developments in the time compression trend within manufacturing.

Fig. 4.2 shows composite fabrication trend, such as tailor-made design, the difference between design with composites and metals, knowledge of materials properties and interdependence between materials properties and fabrication (interdependence).

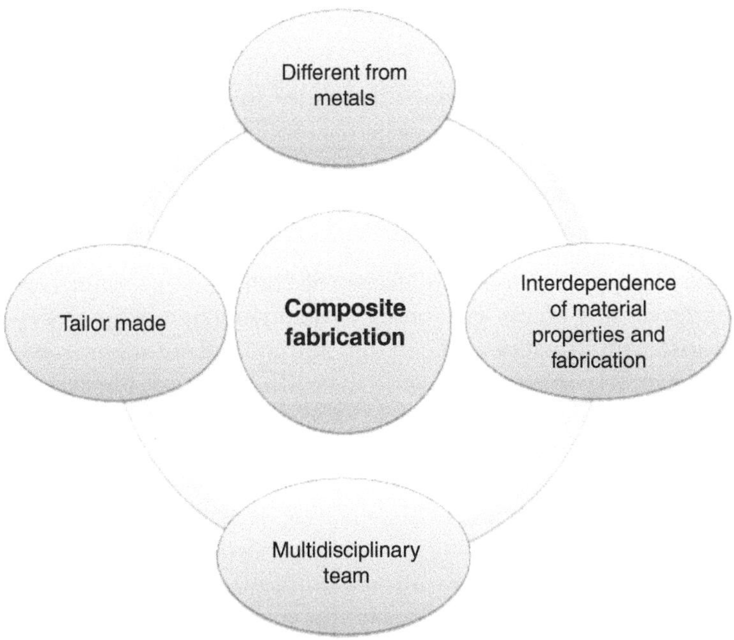

FIGURE 4.2 Composite fabrication characteristics.

PREVIOUS WORK ON CE IN COMPOSITE PRODUCT DEVELOPMENT

Early Work (1989–99)

Marchant (1994) and Johnson and Marchant (1990) reported the need of CE in composite design. Johnson and Marchant (1990) stated that design process for composite structures is broken down into four phases. The first phase is a design brief. During this phase, the project manager is responsible for the project budget and he/she is working closely with the client, purchaser, and user. The second phase involves the preliminary design. It is in this phase that the importance of design management must be appreciated. There are many disciplines involved, several activities and, with composites, many different candidate materials and processes. For materials selection, the requirements that have to be fulfilled include environmental, mechanical, cost, physical, structural, and manufacturing process. During this phase, designers are working closely with the project manager. All designers should ask themselves at the early stages of a design programme: can it be made and can it be assembled? The next stage is detail design and drawings and the major inputs to this stage include costing, product specifications, results of computer analysis and detailed calculations, and results of materials testing. The final stage is manufacturing and the information of materials testing, quality control (QC), and structural testing are important.

Wall and Crawford (1995) adopted design for manufacture (DFM) or CE in the design of thermoplastic composite components in aircraft primary structure. Stress modeling, materials selection, cost analysis, component selection, design optimization, forming, consolidation, and testing are among the development activities and CE was adopted in carrying out such studies.

Srinivasan et al. (1991) used the design of composite pressure vessels as a basis for developing an understanding of the interaction between design and manufacture. The structure of the design and manufacturing process is illustrated in Fig. 4.3. In this development, the earlier events of the design process are grouped under "design for concept" and the later events are grouped under "design for manufacture" and they claimed that this distinction is made to determine the nature of decision support needed at various events of the design process. The design activities include selection of fiber and matrix material and their volume fraction, selection of manufacturing process and detail design. However, in the author's opinion, in the "real" DFM or CE framework, distinction between "design for concept" and "design for manufacture" is probably not needed because "design for concept" is a part of CE activity.

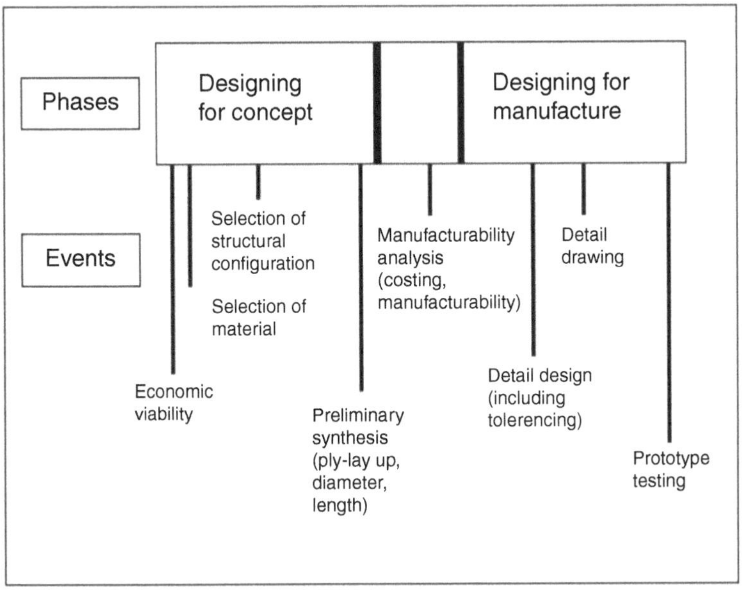

FIGURE 4.3 Design of composite pressure vessel as basis to develop understanding of interaction between design and manufacture by Srinivasan et al. (1991).

Gandi and Thomson (1990) proposed DFM system for developing fiber composite materials. They prepared a flowchart of the CE process and called it DFM algorithm and the flowchart comprises information on:

- Specification
- Materials selection
- Part geometry
- Hybrid composites
- Characteristic and constraints imposed by manufacturing process
- Macromechanical study of composite-based part
- Elastodynamic characteristics of composite-based part
- Finite element analysis
- Part manufacturing specifications

Mayer (1991) reported the use of CE in composites mainly to support the defense industry. The major problem identified in the study was lack of reproducibility in properties of materials, as well as other associated problems. Therefore, CE was proposed to overcome the problems by, among others, net-shape processing concepts, development of database of materials properties, and processing the materials intelligently. In addition, CE had been implemented on a product and process requirements-based approach. As far as process simplification by net-shape concept is concerned, by means of an example of Lanxide process to produce ceramic matrix composite component, CE had successfully reduced tolling, labor, energy, machining, and capital costs. In addition, scrap and intermediate equipment were also removed. The improvement in control system in composite manufacturing was also integral to this improvement process.

Karandikar and Mistree (1992) carried out a study on DFM for filament wound composite pressure vessel. Filament winding is a composite manufacturing process used to produce a hollow circular shaped product by pulling preimpregnated fiber composites, after passing through a resin bath, and winding them over a rotating mandrel. They developed a Decision Support Problems (DSP) which can be used for modeling decisions in design, manufacture, and maintenance of filament wound composite product development. The mathematical and goal programming models developed in this study were aimed at supporting the designers in selecting the best materials, improving the structure, and manufacturing process, and reducing development costs of the components. In can be concluded that DSP had successfully integrated design and manufacturing information in the development of filament wound composite pressure vessels.

Jones (1995) carried out a development study of resin transfer molding (RTM) for composite aircraft design and manufacture based on CE approach. He performed design of component, preform, and tooling and treated them to be equally important parts of a total product development solution. In this context, CE was used in terms of integrating preform, component, and tooling design by modeling, using as common geometry and data sets as possible. Integration of IT tools, such as finite element analysis, computer-aided design (CAD), and customized software was carried out, which was an important element of CE.

Quinn (1999) reported his work concerning the design of polymer composite products, and a design approach was presented by performing a total design process. After the designer clarifies and fully understands the design brief, he/she should have a list of candidate materials, geometries, and manufacturing processes for the intended component and this stage is called a "melting pot." The next step is for him/her to select the most suitable candidates for each category. If there are multiple candidate materials, geometries, and processes, then, these candidates should be compared and analyzed. Once the candidates are finalized, the products can be further developed. However, if the candidates analyzed do not meet the criteria set for the design, the process should be repeated in the "melting pot" until the best candidate material, geometry, and manufacturing process is obtained. Quinn (1999) further stated that "melting pot" comprises ingenuity, guesswork, experience, prejudices, and intuition.

Although in most of the papers that will be reviewed shortly, the use of injection molding for composite product is not mentioned specifically, it is understood that injection molding is a manufacturing processing method for filled polymer composites and it is justified to review such papers particularly to study the relevance of CE in composite development. In injection molding, development of a concurrent mold design system using a knowledge-based system was performed in CE environment (Lee et al., 1997). The aims of the work are to develop a mold development process that facilitates CE-based practice and to develop a knowledge-based design aid for mold design that accommodates manufacturability concerns, as well as requirements of products.

New Millennium Era (2000–09)

Rolfes et al. (2000) introduced an information system for concurrent/integrated engineering (ISCIE) as a solution for the problems associated with product development using sequential engineering. The program was planned for the design and fabrication of aircraft components, such as fuselage from composites to replace the metal counterparts. ISCIE relied on the use of advanced analysis tools for analyzing thermal, stress, failure, and damage tolerance during conceptual and predesign phase. Predesign phase is the phase after conceptual design and before detail design. It is perhaps the same with the widely known phase called embodiment design (Pahl et al., 2007). Advanced structural mechanics tools, that is, finite element analysis software packages were used to analyze thermal, stress, failure, and damage tolerance and knowledge from these analyses were passed to a common numerical procedure. The analysis performed includes three dimensional (3D) thermal analysis of composite and sandwich structures, 3D stress analysis and failure criteria for composite structure, and impact and residual strength of composite structures. By adopting this approach, improvement in terms of cost, time, quality, development risk, and flexibility in dealing with change of requirements were achieved.

In the system developed by Kim et al. (2000) the CE is represented by a system modeling technique as shown in Fig. 4.4. The design process of composite structures consists of various modules, such as material selection, product design and modeling and analysis and simulation. They performed design task based on CE principle in polymer composite product development. The software system called Concurrent Engineering Systems for Composite Design comprised several modules, such as design and analysis using CAD and finite element analysis (FEA), buckling and postbuckling analysis of composite structure, composite thermoelastic analysis, materials selection using materials database and design optimization using knowledge-based system (Fig. 4.2). The integration of various IT tools was performed and a graphical user interface was made available for the users.

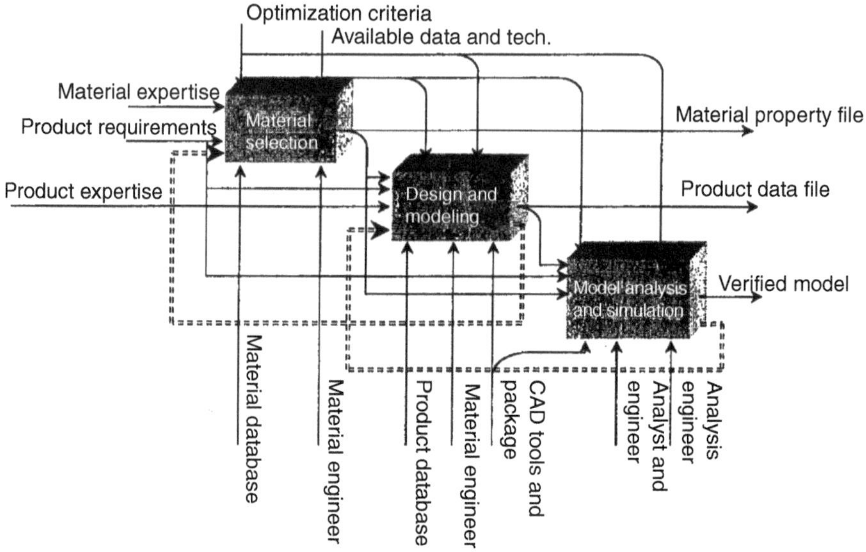

FIGURE 4.4 CE system for composite design by Kim et al. (2000).

In this model, and in fact in many CE models, the modules were developed mainly using IT tools and attempts were made to integrate all the modules into a single system. Design engineers working at different locations could conveniently assess the system and were able to communicate with the rest of the CE team members electronically. However, many systems were developed only at the prototyping level and commercial software dedicated to CE for composites are still not available. Another issue with these kinds of system is that it only caters for a specific type of composites and some are only developed for specific components. It is difficult to find a software system with more general application. The reason is quite obvious because composite material itself comprises different varieties of fibers, matrix types, fiber arrangements, and fabrication processes.

Hale and Vasey-Glandon (2001) reported on the software system called the Parametric Composite Knowledge System (PACKS), which is a CE design system for aerospace composites developed by The Boeing Corporation for commercial and military aircraft composite structures. The system supported design and analysis, as well as composite manufacturing at the Boeing Corporation. It had the capability to reduce the composite manufacturing development cost by more than 50% and it had "communication" with manufacturing function early in the design process to reduce time to market.

A methodology for combining all the design aspects together in concurrent design methodology, which finally led to the achievement of the six-sigma limits, that is, almost defect-free products from the RTM technology was proposed by Prabhakaran et al. (2006). It utilized the advantages of the graph theoretic approach by considering all the design aspects together in a single methodology with the help of matrix algebra. It was basically a virtual design methodology, which decided the process, and the product strength and weakness with the help of a multinomial defined by using the matrix algebra. The design index, developed using the proposed methodology, decided if the overall design was acceptable or not by considering all the aspects of the design related to the product, process, and environment. Finally, a step-by-step procedure was proposed to help generate a new algorithm for software coding (Prabhakaran et al., 2006). In this paper, the acronym DFX was redundantly presented because in DFX, "X" stands for all possible "abilities" but in this paper "DFX-abilities" was written instead.

Babu et al. (2008) used DFX in the design and fabrication of filament wound composite automotive drive shafts. They studied DFX for composites by investigating the elements of DFX, such as design for cost (DFC), design for minimum weight (DFMW), and DFM. This work was the extension of their previous work (Prabhakaran et al., 2006) with a different manufacturing process (RTM). The same graph theoretic methodology was used in this study. It worked based on seven steps, that is, identification of design aspects and their interdependency, identification of design parameters and their interdependency within design aspects, development of graph theoretic representations, representation of the design graph by a variable permanent matrix, development of a subpermanent function (a multinomial) of each design, obtaining overall permanent function (or overall DFX index) per [DFX] and obtaining overall DFX index after some manipulation in formulae. Overall DFX index is used to determine the indices for DFC, DFM, and DFMW of filament wound composite automotive drive shafts.

Edwards (2004) listed 16 significant implications on DFM from the use of RTM process to produce composite products and among them are:

- The more efficient use of materials and more effective way of transferring the resin into the mold.
- The materials are contained in closed mold and they pose minimum health hazards.
- Excellent product quality where products are reproduced consistently. This makes secondary processing like joining and assembly easy to perform.

Belloy and Foucard (2000) studied the surface roughness, the material, the dimensional tolerance, and the material chemical composition, which are attributes required during manufacturing but they are stored in a database during the design stage. The algorithm developed enabled integration of surface quality into a CAD/CAM system to predict surface roughness, flatness, and coaxiality. In this paper, the manufacturing process is considered in the design stage and this is the essence of CE. As far as material is concerned, there is no mention about what type of material was studied. By looking at the process involved, it is more appropriate for metals but in the author's opinion, a similar model may be adopted for composite materials.

Another important work on CE of aircraft composite structure was carried out Boeing, USA. Cost reduction and weight reduction efforts on the Boeing 767-400ERE raked wing-tip structure was performed (Swanstrom and Hawke, 2000). Similar work on cost and weight reduction studies of aircraft components were performed by Choi et al. (2007), where their concern was on the development of a knowledge-based system for weight and cost reduction of composite aircraft wing spoiler during conceptual design stage. The platform for KBS was CATIA V5 where weight and cost reduction studies were performed. The KBS was interfaced with finite element analysis software PASTRAN/NASTRAN.

Recent Work

Perry et al. (2012) studied design for recycling for composite materials and they stated that design for recycling is an important element in CE because in the term design for "X" (which is just another name for CE), "X" can stand for recycling. The research is focusing on recycling of carbon fiber composites. In this study, recycler is considered as one of the members of the CE team who is involved in deciding the "fate" of the materials because the decision whether to reuse or to dispose is in his hand. Perry et al. (2012) pointed out the current and future limitations and legislation related to recycling of carbon fiber reinforced composites. Perhaps due to the high cost of carbon fibers, design for recycling of carbon fibers is given preference to recycling of the matrices. Analysis for a multilifecycle use of the composites was proposed. Findings from this study can be a good means of communication between designers and recyclers in the CE team in developing carbon fiber reinforced composite products.

Reichert et al. (2014) reported their findings of the integration of design and manufacture for glass and carbon fiber reinforced polymer composite full-scale architectural pavilion. Concurrent integration of biomimetic principle, material design, structural analysis, simulation, and computation work, and design optimization was performed to come up with the ideal design. Biomimetic approach was adopted to mimic the structure of the shell of a lobster in developing a composite pavilion.

In the research work of Barbosa and Carvalho (2013), a design analysis model was developed to be used by design engineers in carrying out aircraft component design using automated composites. The stages in the design analysis model include material definition, product design, weight of product, fasteners definition, and tooling conception. The design analysis model was tested in aircraft components, such as fuselage assembly line. Two concepts of fuselage assembly line were compared; the first was a traditional concept mainly developed from aluminum and the second was developed using design analysis model and was mainly made from composites. The latter is more desirable in terms of weight, number of parts, tooling cost, and automation.

Albiñana and Vila (2012) developed a framework for concurrent materials and manufacturing process selection in product design. This concurrent analysis of materials and manufacturing engineering approach was tested with outboard motor brackets for recreational boats. In this study, product attributes and functionalities are listed. Then the analysis of function and design constraints of the products are carried out. The materials selection requirements are also set. At the end of the process, requirement and functionality of the product are identified. Material and manufacturing technique are also determined. As far as material is concerned, 60% long glass fiber reinforced polybutylene terephthalate (PBT) composite was selected as the material for the motor brackets for recreational boats.

A PhD work by Sobey (2010) was concerned with the use of CE in the design of composite leisure boat in the boatbuilding industry. The main model followed in this study was based on the total design developed by experts like Pahl et al. (2007) and Pugh (1991), which involved conceptual design, embodiment design, and detail design. Various design and optimization tools and techniques, such as genetic algorithm, quality function deployment, neural network, Monte Carlo simulation, and grillage methods were used to come up with design and manufacturing solutions. Results from the design, analysis, simulation, and optimization activities using those methods were gathered to enable the designer and boat builder to decide on the best design and manufacturing approach of a composite leisure boat.

Cong and Zhang (2012) studied CE for composites by focusing on manufacturability analysis. Manufacturability issues are the areas often being overlooked in CE studies. They pointed out that composite manufacturability analysis can be performed by identifying all of the manufacturing processes which can be used to develop the composite product, by analyzing the compatibility between the design and the manufacturing process, and by carrying out the manufacturability evaluation with the quantitative indices focusing on quality and cost. They stressed that a design can be manufactured if the sensitivities of the process parameters to the quality indices can be determined. They analyzed a few quality indices and process parameters in the autoclave curing process.

SOME COMMERCIALLY RELEVANT EXAMPLES OF RECENT DEVELOPMENT, TREND, AND ADVANCES IN THE AREA OF CE FOR COMPOSITES

Fifty composite-based companies around Europe, United States, Australia, and New Zealand had been contacted in order to obtain information regarding the implementation of CE in the composite industry but only a few companies had provided positive responses. From the survey conducted, it was discovered that industries may use CE or design for manufacturing (DFM) approach in their works but they were not aware of the terminology. Some of the companies had a nondisclosure policy due to their proprietary nature so information could not be obtained from them. From the survey, it was also found that there was a company in the Netherlands, which used the DFM method with the synthetic fiber reinforced polymer composites for their boats and building projects.

In spite of the lack of use of the official term for CE or DFM, the industries and researchers from the Western world and all around the world were practically using this method. Majority of the composite products had been commercialized and used for a long period of time. In the aerospace industry, for example, Andersson et al. (2014) from Linköping University, Sweden described the development of DFM strategy for SAAB aerostructures with Airbus and Boeing being their major customers. The structural parts, such as doors and ailerons for the commercial aircrafts were developed and manufactured using carbon fiber reinforced polymer composites.

In addition, a collaborative research by Bell Helicopter Textron in Canada and Texas with the Institute for Aerospace Research in Canada used the concept of DFM for a Bell Helicopter Model 407 aircraft tailboom using carbon-fiber composite structures. Some of the criteria taken into considerations were customer satisfaction, cost, weight, product readiness, and automated fiber placement (AFP) method for manufacturing (Marsden et al., 2011). The AFP tailboom prototype was used interchangeably with the existing metallic tailboom on the M407 commercial light helicopter.

Furthermore, Institute of Structural Mechanics, Germany was developing procedures for fiber composite airframe structures and they analyzed damage behavior, 3D temperature distribution, 3D stress distribution, and buckling behavior. A concurrent/integrated engineering (CIE) process was used, comprising conceptual and predesign tools, which allowed for a fast and precise analysis in product design and development (Rolfes et al., 2001).

Additionally, the European Commission had funded a cost effective small aircraft (CESAR) project by the National Aerospace Laboratory NLR, the Netherlands with their partners Piaggio Aero Industries, CIRA, MERL, HAI, and

FIGURE 4.5 CESAR composite wingbox (Klomp-de Boer, 2010).

IoA to develop and manufacture a semispan composite wingbox. They used the CE approach and remote team collaboration to shorten the design process in the development of the composite wingbox (Fig. 4.5) (Klomp-de Boer, 2010).

RTM technology is capable of making high quality, large, and complex 3D parts. Laurenzi and Marchetti (2012) discussed the use of RTM method using advanced composite materials for aerospace applications. The prototypes using CE approach presented a collaborative work of the School of Aerospace Engineering, Sapienza University, Rome, Italy, and Italian aerospace industries, that is, Agusta Westland and Aermacchi (Fig. 4.6).

Friction stir welding (FSW) is a welding process invented at The Welding Institute, United Kingdom in 1991 by Wayne Thomas, which allows the

FIGURE 4.6 Final aerospace component after demolding (Laurenzi and Marchetti, 2012).

joining or mixing of parts without fusion. Tavares (2013) from the Faculty of Engineering, University of Porto in his PhD dissertation explained the use of CE in aircraft industries, using the FSW method for composite products. There were also companies and industries that used the DFMA software for their products development. A famous DFMA software was developed by Boothroyd Dewhurst Inc. (BDI), Massachusetts, USA and Dynisco Inc. used the DFMA software for the development of their composite products. The software combined two important analysis tools, that is, design for assembly (DFA) and DFM and was proved to be cost efficient and time saving in product designing process (Assembly Magazine, 2013). Bell Helicopter had also used the same software as a trial but they were reluctant to use the tool in their composite product design as a standard procedure. The issues, such as software compatibility with Bell operation and, time and budget constraints were the major setbacks (Bergerson, 2010).

Marine industry had also shown interest in CE software tools. It was reported by Guillermin (2010) that concurrent composite design and analysis of the marine structures were performed using VISTAGY's FiberSIM software (Fig. 4.7). The software was used by naval shipyards worldwide for the design and manufacture of structural components, such as composite deck in vessels (Guillermin, 2010).

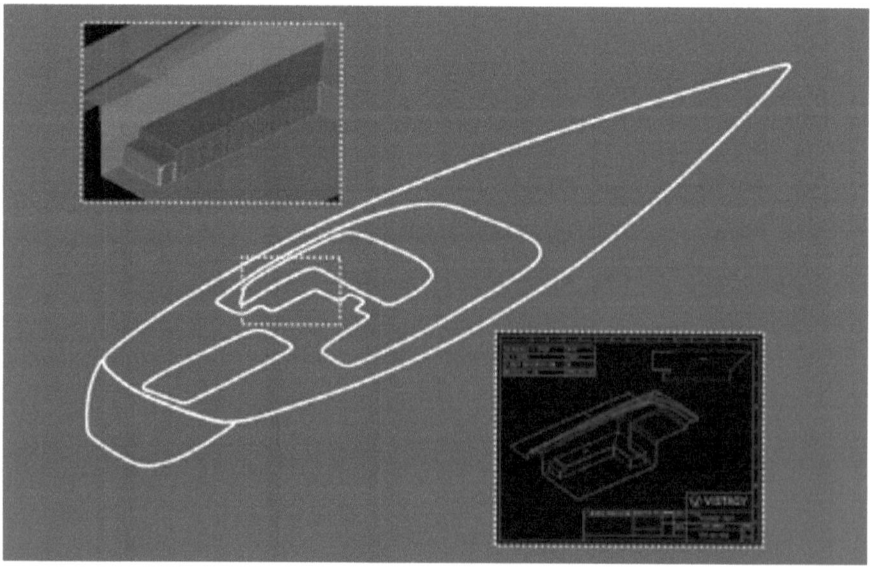

FIGURE 4.7 Producibility design of composite deck of luxury sailboat using FiberSIM software (Guillermin, 2010).

Many large companies, such as McDonnell Douglas, Boeing, AT&T, and IBM had implemented the concept of CE in their product development, including composite components where reduction of rework, part and materials used, time, defects, and assembly hour were reported (Das and Almonor, 2000). CE approach had also been used in the automotive industry around the world and Japanese companies, such as Toyota, Honda, Mazda, and Nissan had adopted CE for their product development, which may include composite components (Sobek et al., 1999). Although the exact term CE was not used because many people were still not familiar with the term at that time, the approach had been extensively applied in the areas of sales, engineering, and product development, which successfully boosted the sales of Japanese cars in the USA and the rest of the world.

Some case studies of CE practice within large companies had also been documented, such as Cincinnati Milacron, Xerox Corporation, Ford and Ingersoll Rand, Digital Equipment Corporation, Storage and Information Management Group, and Westinghouse Electronic Systems (Skalak et al., 1997).

In regulated products, such as medical devices, airplanes, and pharmaceuticals (some products may involve composites), the requirements for approval from authorities and agencies delays the product development and production (Das and Almonor, 2000).

All four major aircraft companies, that is, Boeing, Lockheed, Airbus, and McDonnell had used the CE approach with the composites materials in their aircraft development and construction. Boeing had documented experiences, data of service history, and feedback from other airplanes by arranging meetings with the CE teams. Development of Boeing 777 and Boeing 787 were some examples where CE was implemented and in those projects, concurrent product definition, focusing primarily on design and manufacturing integration was emphasized. For Boeing 777 well-known software CATIA (version 3) was used while for Boeing 787 CATIA (version 5) was used in the design (Arkell, 2005). Taking experiences from Boeing 777, which used 12% of composites in its construction, Boeing 787 was built using approximately 50% of composites (Hale, 2006) (Fig. 4.8).

Airbus had widely invested in the deployment and development of CE known as Airbus Concurrent Engineering (ACE) (Pardessus, 2004). The concept of design and product description called digital mock-up (DMU) was created with the support of CAD and managed by a product data management (PDM) system, which supported the industrial drawing process and configuration management. Fig. 4.9 shows the progress of the composite technology used in Airbus components from 1970 to 2013.

McDonnell Douglas Aerospace (MDA), a major aerospace manufacturing corporation and defense contractor established their communication interaction

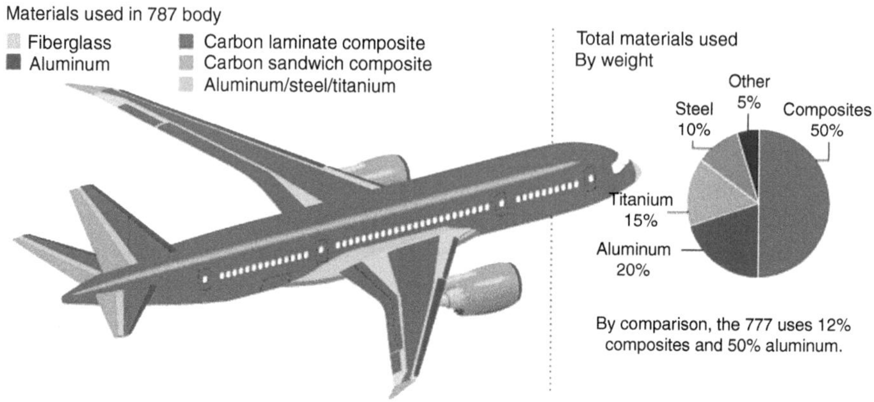

FIGURE 4.8 Materials used in Boeing 787 and its percentage (Hale, 2006).

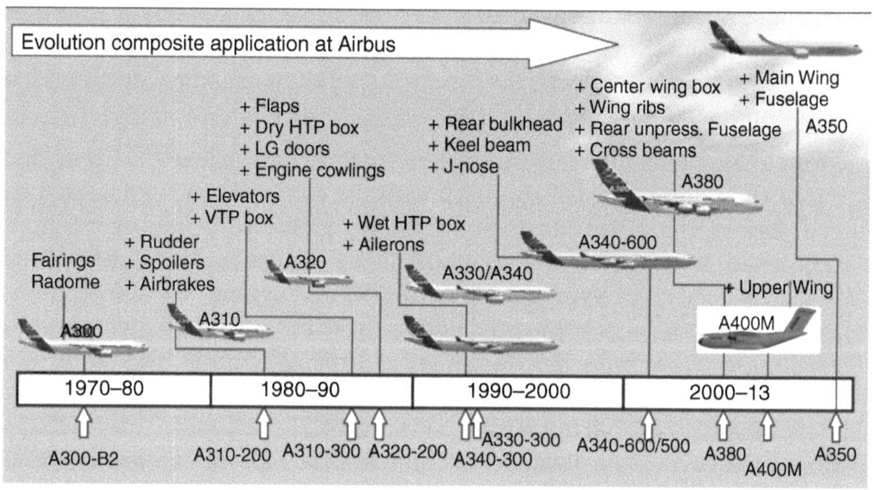

FIGURE 4.9 Evolution of application of composites at Airbus (Body, 2007).

for CE using the integrated product development (IPD) multifunctional teams (http://www.bmpcoe.org/bestpractices/internal/mdwes/mdwes_9.html). The teams were augmented with integrated design and manufacturing support system that create the environment for near real-time CE. The CE tactical and strategic plans also cover several Department of Defense (DoD) initiatives and instructions. Table 4.1 shows several military-type aircrafts using different types of composites and different compositions (Greene, 1999).

Table 4.1 Several Military-Type Aircrafts Using Different Types of Composites With Different Compositions (Greene, 1999)

Aircraft Type	Composites Composition	Composites Type
Advanced Tactical Fighter (ATF)	Approximately 50%	DuPont's Avimid K polyamide
Advanced Technology Bomber (B-2)	Each contains an estimated 40,000–50,000 lbs	Advanced composite materials
Second Generation British Harrier "Jump Jet" (AV-8B)	Approximately 26%	Bismaleimides (BMI's) used on the aircraft's underside and wing trailing edges
Navy Fighter Aircraft (F-18A)		wing skins used of graphite/epoxy
Osprey Tilt-Rotor (V-22)	Up to 40% of the airframe	AS-4 and IM-6 graphite fibers in 3501-6 epoxy

CE SEEMS TO BE AN OBVIOUS ROUTE TO ANY SORT OF NEW COMPOSITE PRODUCT DEVELOPMENT: HOW IT IS DIFFERENTIATED?

To answer this question, one should understand the definition of CE. The CE, also known as simultaneous engineering, integrates the cycle process of new product development from designing, manufacturing up to marketing. The design philosophy of CE managed to shorten the process of new product development, reduce the labor and repair costs, and improve the product quality. Development of a new product that does not run through the CE method will take more time to be developed and marketed because of cycle process of development. Some of the problems encountered include less improvement in product manufacturing and dissatisfaction or complaints from the customers or clients. The term and definition of CE sounds like a very obvious route for product development possibly because of some overlapping processes. Practically this methodology has been used in many large companies and a lot of small- and medium-sized companies are still unable to adopt and implement the methodology into their organizations, probably because of limited levels of management to implement the change (Skalak et al., 1997).

CE HAVE CHANGED THE WAYS INDUSTRIES WORK: INDUSTRIAL EXPERIENCE OF CE FOR COMPOSITES AT STEELCASE

The majority of the information presented in this section was obtained from Azim [Azim, A.A.R., 2016. Steelcase Manufacturing (M) Sdn Bhd, Puchong, Selangor, Malaysia, Personal Communication].

CE is not a quick fix for a company's problems and it is not just a way to improve engineering performance. It is a business strategy that addresses important company resources. The major objective that this business strategy aims to achieve is improved product development performance. CE is a long-term strategy, and it should be considered only by organizations willing to make up front investments and then wait several years for long-term benefits.

It involves major organizational and cultural changes. The problems with product development performance that CE aims to overcome are those of the traditional serial product development process in which people from different departments work one after the other on successive phases of development. In traditional serial development, the product is first completely defined by the design engineering department, after which the manufacturing process is defined by the manufacturing engineering department, etc. Usually this is a slow, costly, and low-quality approach, leading to a lot of engineering changes, production problems, product introduction delays, and a product that is less competitive than desired.

CE brings together multidisciplinary teams, in which product developers from different functions work together and in parallel from the start of a project with the intention of getting things right as quickly as possible, and as early as possible. Current CE that has been used widely is the parallel work during design stage, where design and engineering analysis at the same time engage all other related cross functions, for example, procurement (supply chain), material management, quality team, manufacturing and operation team, database team, designer, marketing team, and partnership supplier that as an agreement on nondisclosure (NDA). All these engagements are really beneficial for the engineering and design team to develop the criteria and design, and consider all the input from different cross function being taken into account during the very early stage (phase 1, a and b) (Fig. 4.11).

Anyway the complexity comes when a global product is developed that involves every cross-function mentioned earlier in all regions. For Steelcase (a company producing furniture particularly office chairs) there are three main product development centers for task seating; Grand Rapids (North America for American region), Sarrebourg (France for European region), and Kuala Lumpur (Malaysia for Asian region).

In the global product development (GPD) process, there are five phases as shown in Fig. 4.10 and all the CE takes places starting from Phase 1. In implementing GPD, the major difficulty occurs because the manufacturing locations are not located near the product development center (PDC). Steelcase has five factories for producing the task seating; Reynosa (Mexico), Sarrebourg (France), Puchong (Malaysia), Dongguan (China), and Pune (India). All these

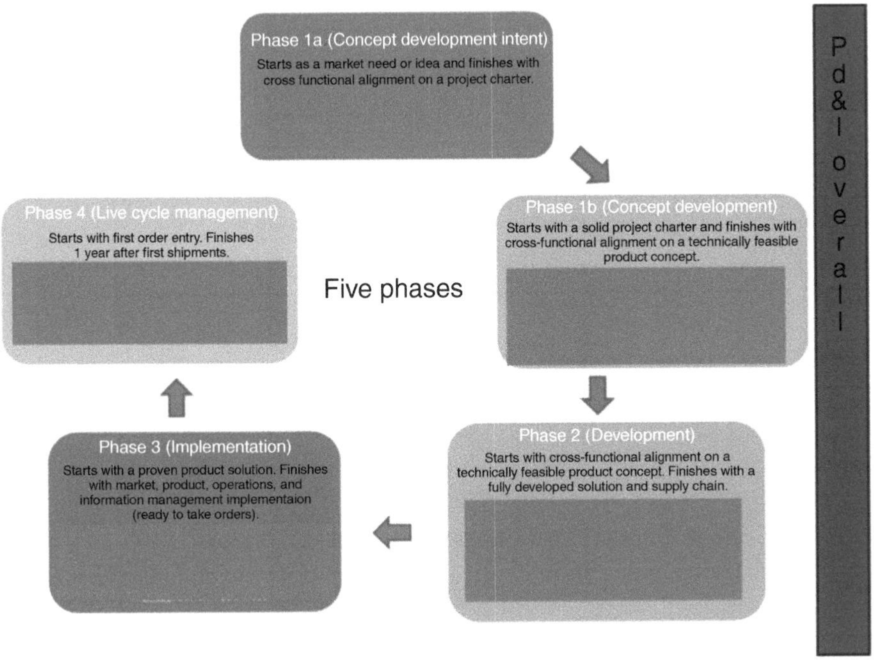

FIGURE 4.10 **Product development phases at Steelcase.**
Courtesy of Abdul Azim Abdul Rahman, Steelcase, Puchong, Malaysia.

different locations need good collaboration and planning in order to ensure all the activities for CE are monitored properly.

In implementing global product development, Steelcase have used a concept called "five phase cycles," and this concept is regarded as one of the company's best practices. Each phase in the product development cycle has been clearly defined for every cross function in terms of activities and deliverables. Each function plays its roles and parts throughout the product development process and is monitored by a global project manager. This is the high-level overview of the Global Product Development Concurrent Engineering.

In detail, the study on the postcondition process for glass fiber reinforced polyamide 6 (nylon) composites (PA6 + 30%GF) after injection molding process, provides better advantages using CE than the company currently experiences in terms of (Fig. 4.11):

- Cutting cost of capital (CAPEX) and expense (OPEX) of the project budget
- High quality
- Short development time

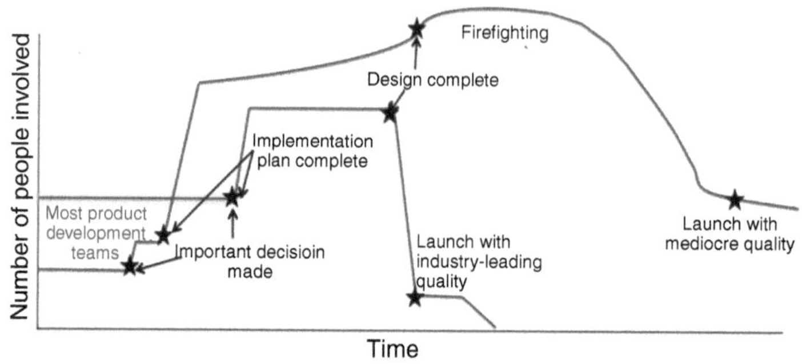

FIGURE 4.11 Major advantages of CE in Steelcase.
Courtesy of Abdul Azim Abdul Rahman, Steelcase, Puchong, Malaysia.

Advantages of CE at Steelcase

Cutting Cost of Capital and Expense of the Project Budget

By understanding the process control of the components and developing the proper testing condition, reduction of cost of CAPEX, as well as OPEX can be achieved. When dealing with nylon composite parts, the tooling can be developed anywhere, where the cost is reasonable and optimum, as there is no need to worry about the weather or climate that influences the performance of the nylon-based composites.

Testing, design evaluation and obtaining the compliance can be performed at any company's location in the world and there is no specific laboratory is assigned to handle a particular testing. As shown in Fig. 4.11 also in the early stage of the design, one might involve with more people than in the conventional way of product development because concurrent activities start very early in the product development process but the numbers reduce significantly as one moves along to other phases (later phases) because everything has been considered up front. This also saves in terms of labor cost throughout the project.

High Quality

With the CE activities, especially the latest study on the postcondition of nylon composite components after the injection process, the company is able to launch the product with high quality after everything has been considered at the upfront of the product development stage. No doubt that in CE design activities take longer time to define and lock the concept and design (refer to Fig. 4.11) due to input from all cross function in early stage and this enables the design and engineering team to have better concept and design consideration. Some of it is due to the input for better design for manufacturing, design for assembly, design for sustainability, and design for consumer.

Short Development Time

Many activities took place in the early phase of the design especially during phases 1 (a and b) and 2 as shown in Fig. 4.10. This makes a better flow of product development during the evaluation stage in phases 2 and 3. CE saving of about 40% of overall time of product development is reported and it includes time of troubleshooting product issue, before and after the product launch. There are also a lot of cross-functional activities among engineering, marketing, design, and operation in the early stage of the design process as shown in Fig. 4.12. "Fast failed" is the term used in Steelcase during the product development cycle to identify as quickly as possible the failure mode so that the company can understand better the overall performance of the product itself.

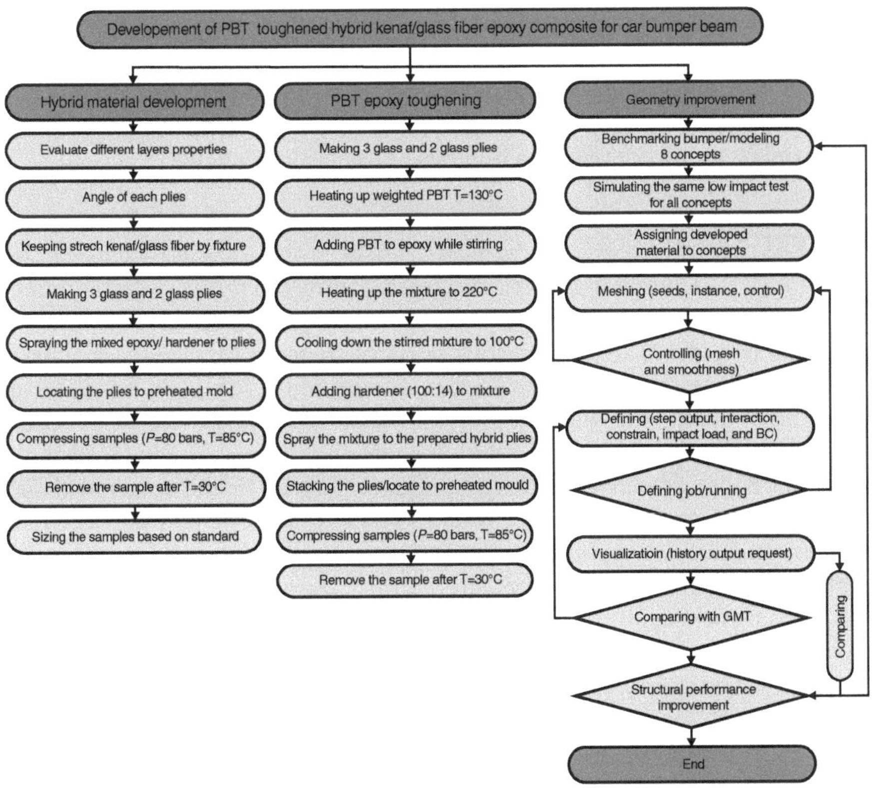

FIGURE 4.12 Development of kenaf/glass epoxy hybrid composite automotive bumper beam (Davoodi et al., 2011)

PREVIOUS WORKS BY THE AUTHOR AND HIS COWORKERS IN CE FOR COMPOSITES

Background

Selected works from the author's research group comprising his own PhD work, and his Master of Science and PhD students at Universiti Putra Malaysia (UPM) are reviewed here. Table 4.2 summarizes those activities, with materials and major tools used, as well as components studied. The major emphasis of individual projects include total design, concept development, concept selection, materials selection, manufacturing optimization, design for sustainability, mold flow analysis, design analysis, and rapid prototyping. The materials used range from hybrid composites, polymer composites with conventional fibers, such as glass fibers, biocomposites, CMC, and MMC. The major tools used include finite element analysis (FEA), Moldflow Plastic Insight (MPI), design methods, Taguchi method, SLA, 3D printing, AHP, ANP, TOPSIS, QFD, Java, ANN, TRIZ, and experimental work. The products being developed include automotive components, such as pedal box system (clutch, accelerator, and brake pedals), bumper system (fascia, bumper beam, and energy absorber), parking brake lever, anti-roll bar, disk brake, dashboard, and engine components (piston, connecting rod, and piston ring), household appliance and furniture, such as shoe shelf, table, and chair; boat, hovercraft hull base, and food packaging.

Bumper System

Davoodi (2012) simultaneously considered composite material development and geometry improvement in research related to composite automotive bumper beam (Fig. 4.12). While many work on composites focused on characterization and determination of properties of composites (e.g., Aridi et al., 2016; Jumaidin et al., 2016), Davoodi (2012) studied combination of characterization and product development simultaneously. Composite materials development is divided into two main components, that is, hybrid material development from kenaf/glass fiber reinforced epoxy composite and polybutylene terephalate (PBT) epoxy toughening. As far as product design is concerned, geometry improvement of composite bumper beam was carried out, which was focusing principally on design concept selection to fulfil the safety requirements of product design specification (PDS). In addition, impact test simulation was performed using FEA software (ABAQUS).

The major findings resulted from this investigation related to CE for composites include:

- Hybrid composite development (Davoodi et al., 2010)
  In this study, the thickness of composite plies for bumper beam, the dimensions of bumper beam overall geometry, hybrid materials

Table 4.2 CE Research Activities at UPM

Main Activities	Components	Tools	Materials	Researchers
Materials selection	Automotive	AHP	Biocomposites	Al-Oqla (2015)
Concept selection	Bumper beam	TOPSIS	Glass/kenaf composite	Davoodi (2012)
Concept design	Clutch pedal	AHP/TRIZ	Glass/kenaf composite	Rosli (2014)
Concept selection	Brake disk	AHP/ANP/QFD	MMC	Fatchurrohman (2013)
Materials selection	Dash board	AHP/ANN/Java	Biocomposites	Ahmed Ali (2015)
Materials selection	Boat component	Expert system	Polymer composites	Fairuz (2011)
Conceptual design	Brake lever	AHP/TRIZ	Natural fiber/glass composite	Mansor (2014)
Fabrication optimization	Pultruded rods	Taguchi (DoE)	Kenaf vinyl ester composite	Fairuz (2016)
Design analysis	Clutch pedal	FEA/MPI	PA 6,6/glass composite	Imihezri (2005)
Conceptual design	Bumper absorber	FEA	Glass/epoxy composite	Davoodi (2007)
Rapid prototyping	Pedal box system	SLA/3D printer	PA 6,6/glass composite	Lee (2003)
Conceptual design	Bumper fascia	Total design	Glass polymer composite	Suddin (2003)
Concept/material selection	Antiroll bar	TRIZ/AHP	Glass/sugar palm/PU	Mastura et al. (2016)
Concept/material selection	Bumper beam	AHP	Glass polymer composite	Hambali (2009)
Mold Flow Analysis	Plate	MPI	Glass/polymer composite	Zainudin (2002)
Total design	Telephone table	Design methods	Banana/epoxy composite	Sapuan and Maleque (2005)
Total design	Coffee table	Design methods	Banana/epoxy composite	Sapuan et al. (2007)
Materials selection	Engine components	Expert system	CMC	Jacob (2002)
Materials selection	General application	TOPSIS	Polymer composites	Jahan (2011)
Mold flow analysis	Plate	MPI	Polymer composites	Azaman (2015)
Design for sustainability	Food packaging	Experimental	Biopolymer composite	Sahari (2013)
Design for sustainability	Food packaging	Experimental	Biopolymer composite	Sanyang (2015)
Total design	Pedal box	Expert system	Glass/PA 6,6 composite	Sapuan (1998a)
Design analysis	Hovercraft	FEA	Glass polymer composite	Amirruddin (2007)
Total design	Shoe shelf	Design methods	Kenaf/epoxy composite	Norianti (2016)
Total design	Small boat	Design methods	Glass/sugar palm composite	Misri (2011)

(kenaf/glass epoxy composites), fiber orientation, and manufacturing process were determined. For fabrication of hybrid composites, fiber stretching fixture was designed and fabricated for the purpose of stretching kenaf fibers as these fibers were naturally twisted. The mold used to fabricate hybrid composites was simplified sheet molding compound (SMC). Specimens of hybrid composites were prepared and tested mechanically and the outcome was that hybrid composites were better in terms of tensile and flexural properties but inferior in terms of impact properties compared to materials generally used to make bumper beams, that is, glass mat thermoplastics (GMT). Therefore, the next stage was to improve the impact properties by incorporating toughening mechanism.

- Concept selection (Davoodi et al., 2011)
  Geometrical optimization of composite bumper beam was carried out to improve structural strength. Eight different bumper beam concepts were developed with the same frontal curvature, thickness, and overall dimensions. The major finding of this research, as far as the CE is concerned, is that from decision matrix and TOPSIS method, the selection of the best design geometry resulted in the concept called double hat profile for composite bumper beam out of eight concepts.
- Thermoplastic toughening (Davoodi et al., 2012a)
  The thermoplastic toughening (PBT toughening) method with modified SMC process was carried out to improve the impact properties of the bumper beam. From the investigation, improvement of over 50% was seen by using PBT toughening compared to nontoughened natural fiber hybrid composites.
- Low impact test (Davoodi et al., 2012b)

The addition of strengthening rib in bumper beam was very important through the simulation of low-speed impact using FEA (ABAQUS) software. The addition of strengthening ribs can improve strain energy and reduce deflection in bumper beam at the expense of weight increase and manufacturability (Davoodi et al., 2011; Davoodi, 2012).

In another study by Davoodi (2007), composite energy absorber for automotive bumper system was developed (Fig. 4.13). The focus of the study was on concept generation and development of bumper absorber. A design method called "systematic exploitation of proven ideas or of experience" introduced by Pahl et al. (2007) was adopted for concept generation in this study. This method was used to generate new ideas or to enhance the existing design by studying the earlier version of the same company's products, competitor's products or similar products with subfunctions can be used in the new design. Davoodi (2007) used this method to develop design concepts for bumper absorber from composites. Details of this development are discussed in

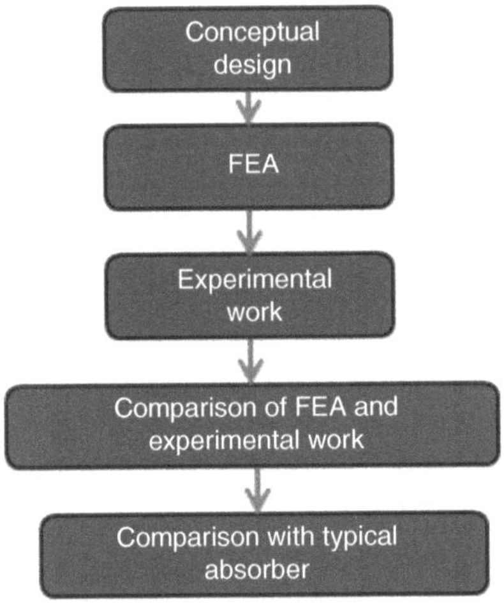

FIGURE 4.13 Flowchart of composite energy absorber (Davoodi, 2007).

Chapter 5. Experimental work was carried out mainly to determine the number of energy absorbers required to satisfy the energy of pedestrian leg-form. FEA was used for design analysis of fiber reinforced epoxy composite energy absorber (Davoodi et al., 2008).

Suddin (2003) performed DFM of polymer composite automotive bumper fascia (Fig. 4.14). Automotive bumper fascia is a nonstructural component, which contributes to vehicle crashworthiness during front or rear collisions

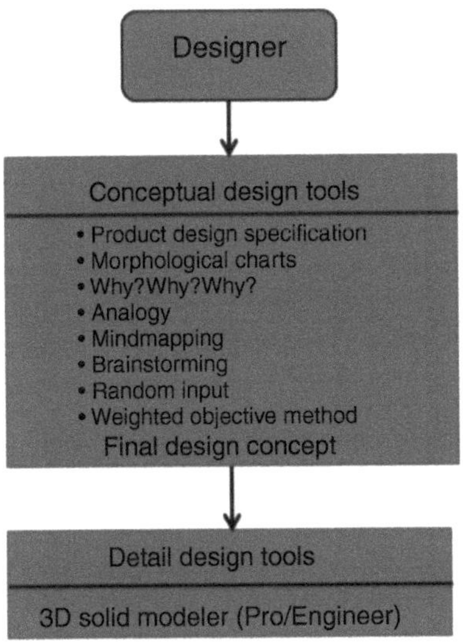

FIGURE 4.14 Development of composite bumper fascia (Sapuan et al., 2005; Suddin, 2003)

and it was generally made of plastics. In this research, the bumper fascia made of polymer composites was designed in solid modeling software. The work relied on creativity methods in product development focusing on conceptual design stage. Concept generation techniques, such as random input, product design specifications, brainstorming, mind-mapping, morphological chart, and analogy were used. Concept evaluation was performed using weighted objective method and the highest utility value appeared to be the best design concept (Sapuan et al., 2005).

Hambali (2009) used analytical hierarchy process (AHP) method to develop automotive bumper beam from composites under the umbrella of CE. Commercial AHP software, Expert Choice was used throughout the entire research (Fig. 4.15). Selection of manufacturing process, selection of materials, and

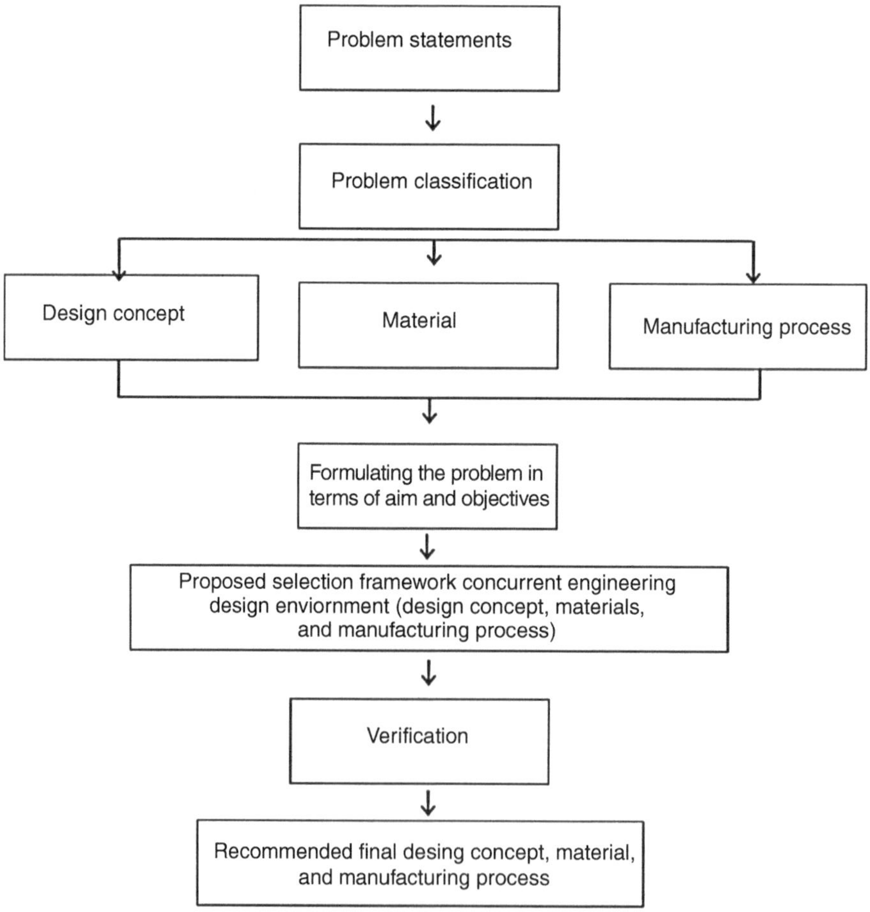

FIGURE 4.15 CE design system for composite bumper beam (Hambali, 2009).

selection of design concepts were performed in an integrated way (Hambali et al., 2010, 2011). A comprehensive product design specification (PDS) was prepared at the beginning of the work. The need to consider manufacturing issues in the early part of the design process had led to the development of an integrated selection system: materials, design concepts, and fabrication techniques.

The success of a product is based on appropriate decisions on material, design concept, and manufacturing process. Incorrect decisions on design concept, material and manufacturing process lead to huge cost involvement and premature component or product failure. CE is an approach, which enables designers to carry out early decision making for the components. The model developed is a framework that can guide designers to evaluate and determine the best design concept, material and manufacturing process simultaneously during conceptual design stage. The final decisions were verified by simulating various scenarios, either by increasing or decreasing the values of the priority vectors of the main criteria, to study the effect of different factors in making the best decision, which is called sensitivity analysis.

Pedal Box System

Sapuan (1998a) developed a CE design system for polymer composite automotive pedal box system. He designed a pedal box system (mounting bracket, clutch, and accelerator pedals) using Pro/Engineer CAD software and carried out a design analysis (FEA). He concurrently carried out materials selection and design to cost by means of a knowledge-based system shell called KEE (Sapuan and Abdalla, 1998). The materials selected for pedals include short glass fiber reinforced PA composites. Fig. 4.16 shows the methodology implemented in the research on CE of polymer composite automotive pedal box system and Fig. 4.17 shows architecture of the research of such components.

Lee (2003) extended the work of Sapuan (1998a) on CE design system for polymer composite automotive pedal box system. Sapuan's model of CE for composite was further extended to include several modules, such as shown in Fig. 4.18 (Lee, 2003). In his research, Lee (2003) performed design of composite pedal system for Proton car, a Malaysian-based car. Closed form or non-numerical data analysis of composite pedal box system was carried out. Calculations of bending stiffness and torsion strength to decide on the best pedal lever cross-section were performed. This research includes measuring process with a coordinate measuring machine of steel pedal box system so that its composite counterpart can be designed conveniently and followed by modeling the design using solid modeling system. Rapid prototyping models using stereolithography (SLA) and 3D printer were produced. SLA and 3D printing were compared in terms of cost, speed of process, user friendliness, and model

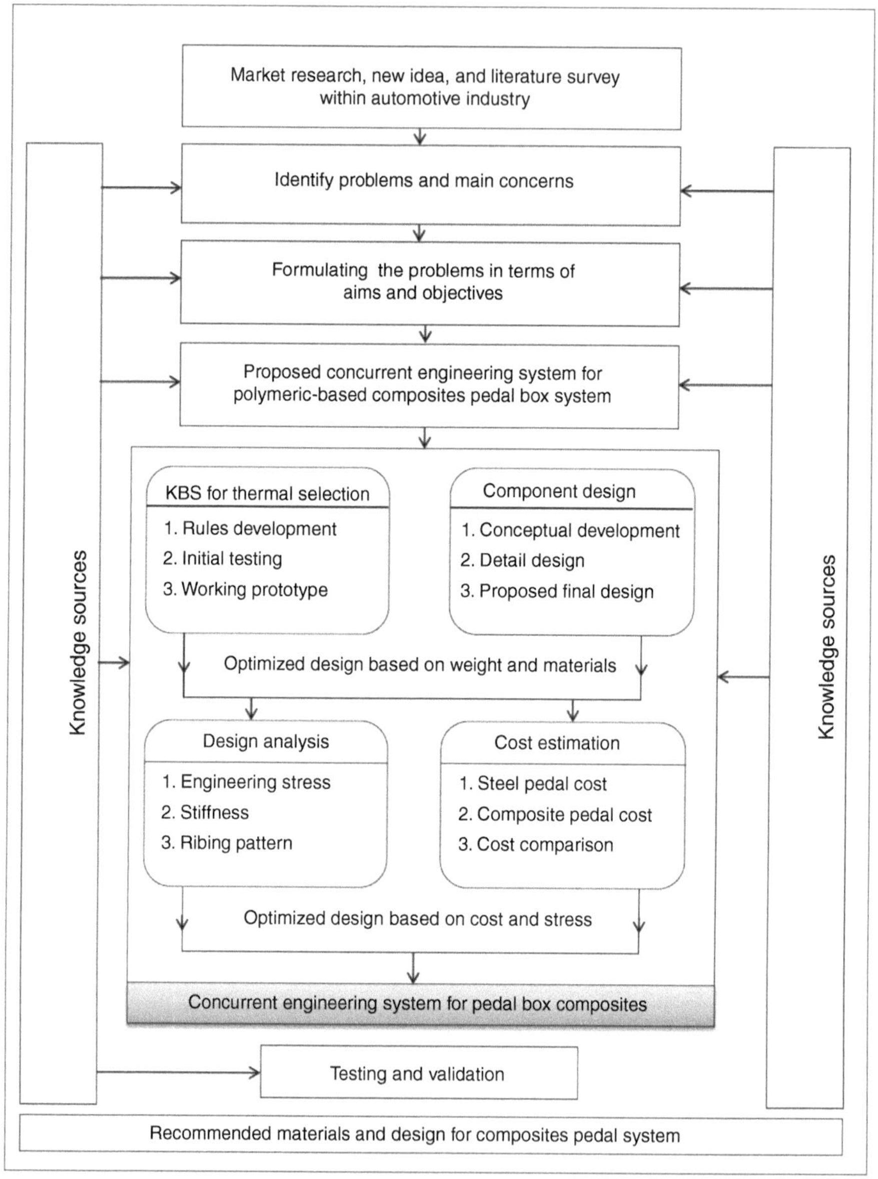

FIGURE 4.16 The methodology implemented in composite pedal box system research (Sapuan, 1998a).

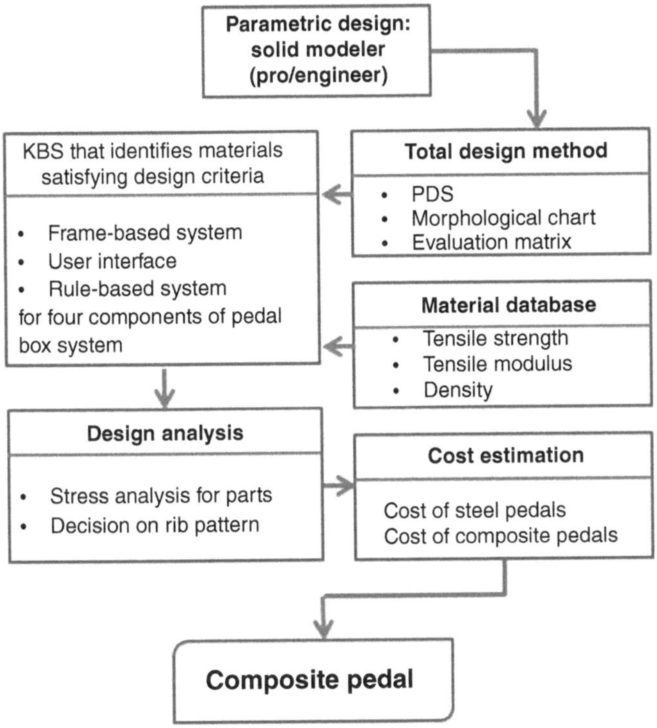

FIGURE 4.17 The architecture of the research of composite automotive pedal box system (Sapuan 1998a).

quality. Rapid prototyping models were developed to facilitate easy communication of products to the customers. The pedal box system was designed to meet the final objective of weight reduction using injection molding process.

Imihezri (2005) further extended the work of Sapuan (1998a) and Lee (2003) by looking at the design analysis, mold flow analysis, and fabrication of composite automotive clutch pedals. She analyzed and simulated the effects increasing the number of gates of injection molding process for a clutch pedal using short glass fiber reinforced polyamide 6,6 composites using FEA and Moldflow Plastic Insight (MPI) software packages. She compared the effects of single-gated and double-gated molds (Imihezri et al., 2005, 2006). MPI software was used to investigate the effects of mold design on fill time, pressure, temperature weld line, air traps, and fiber orientation and distribution. The earlier work using MPI software for composite products was performed by Zainudin (2002) in the development of a simple composite plate. Imihezri (2005) used FEA, LUSAS to analyze stress distribution for pedal profiles and rib patterns. The work of Imihezri (2005) on MPI of composite structures was

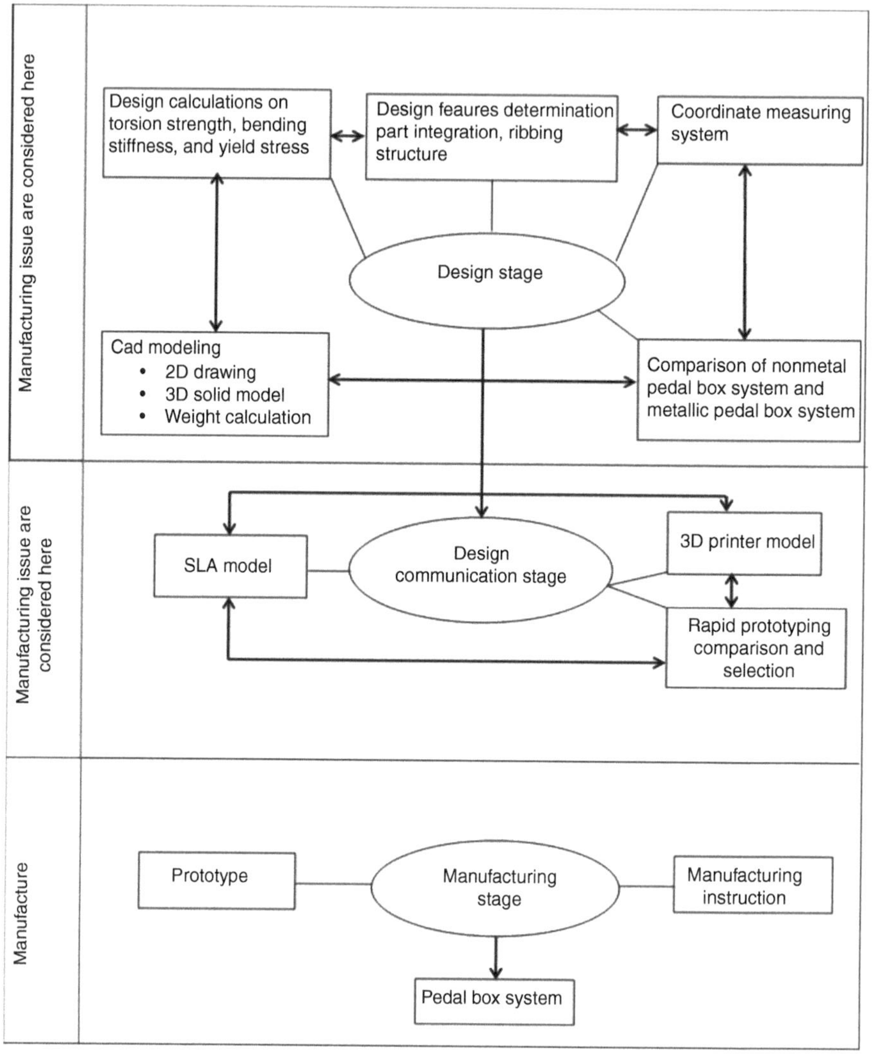

FIGURE 4.18 CE approach of composite pedal box system developed by Lee (2003).

extended by Azaman (2015) on wood-filled polypropylene thin wall composite parts and it has resulted in many interesting findings. The manufacturing process used was injection molding (Azaman et al., 2013a,b). The overview of the structure of research on mold flow analysis and FEA of composite clutch pedal is shown in Fig. 4.19 and CE design system architecture for the analysis of composite clutch pedal is shown in Fig. 4.20.

Rosli (2014) developed the conceptual design support system using integrated TRIZ and AHP methods (Fig. 4.21) and had used this system in the conceptual

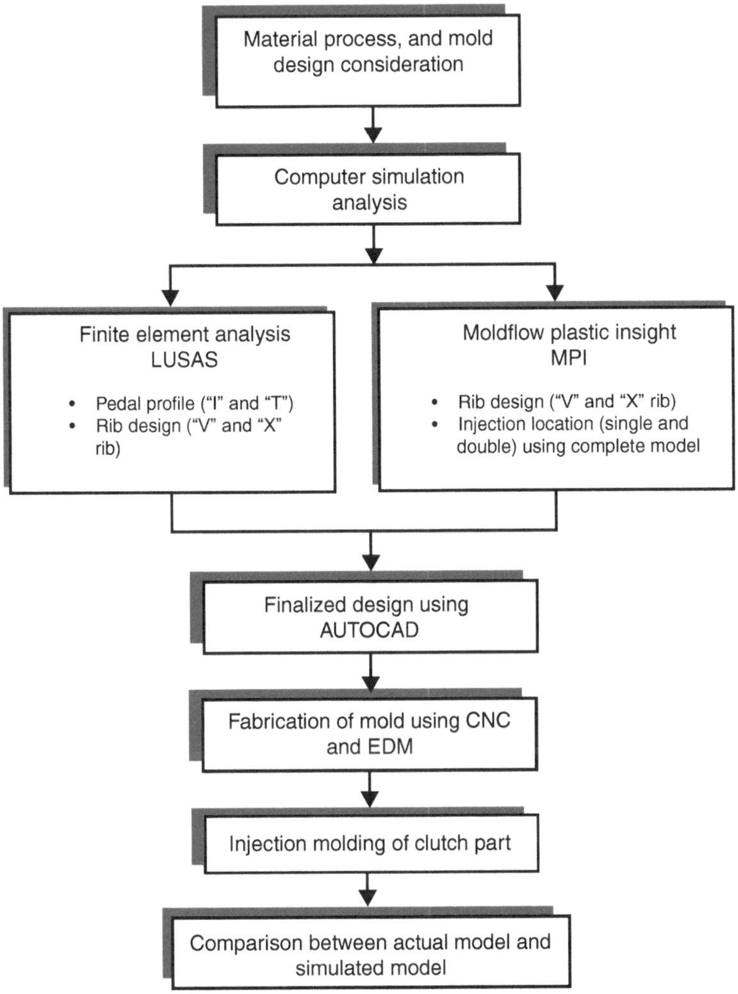

FIGURE 4.19 Overview of the structure of research on analysis of composite clutch pedal (Imihezri, 2005).

design of various engineering products, and one of the products was an automotive clutch pedal made from glass/kenaf hybrid composites. A framework for concept generation was established using the TRIZ contradiction matrix and 40 inventive principle solution tools and AHP to provide weightage to design concept criteria. The framework was formulated in a system called computer-based conceptual design support system. For composite clutch pedal, the final concept sketch was generated with ribbing and layering the pedal with fire and water resistant thin films. Significant improvement was obtained in terms of ecoproduct evaluation compared to the existing product.

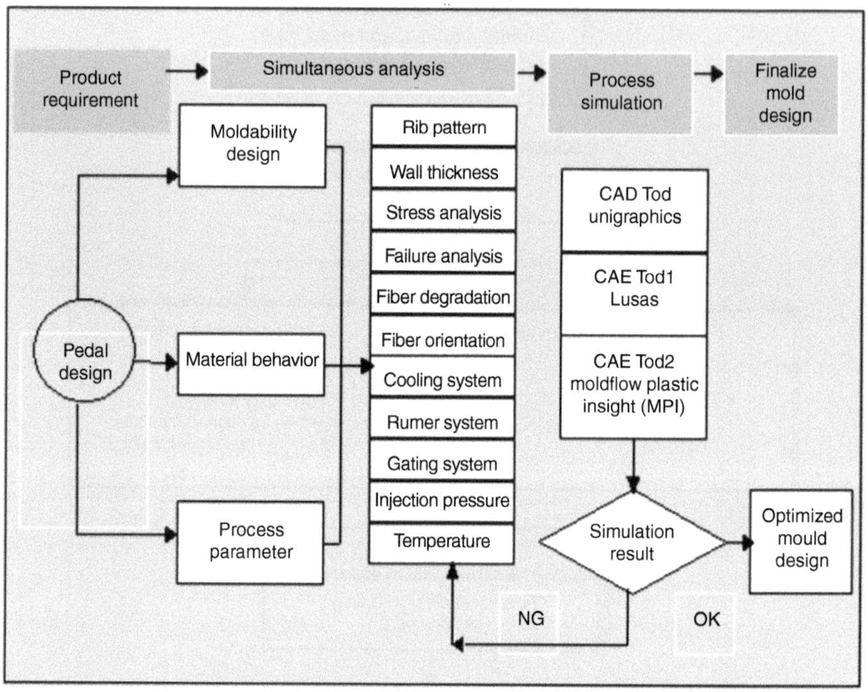

FIGURE 4.20 CE design system architecture for the analysis of composite clutch pedal.
Reproduced from Imihezri, S.S.S., Sapuan, S.M., Ahmad, M.M.H.M., Sulaiman, S., 2004. Design of composite clutch pedal, Sudan Eng. Soc. J. 50(42), 19–23.

Other Automotive Components

CE for composites was carried out by Jacob (2002) and Sapuan et al. (2002) but the emphasis was on materials selection. In their work, materials selection process of ceramic matrix composites (CMC) for automotive engine components (piston ring and connecting rod) was carried out using expert system shell Kappa PC. Similar work was performed by Jahan (2011), Ahmed Ali (2015), and Al-Oqla (2015). Jahan (2011) and Jahan et al. (2011) reported the development of materials selection work utilizing various tools and techniques, such as TOPSIS, ordinance data and had been applied to select composite materials. Al-Oqla (2015) and Al-Oqla et al. (2015a,b) utilized AHP as the major tool in their studies. However, the work of Al-Oqla (2015) focused mainly on materials selection of natural fiber composites for automotive components. Similarly, Ahmed Ali (2015) used AHP in materials selection of natural fiber composites for automotive components. This was also reported by Ahmed Ali et al. (2015). The work of Ahmed Ali (2015) also utilized other software-based materials selection tools, such as ANN, AHP, and Java for automotive composite dash board.

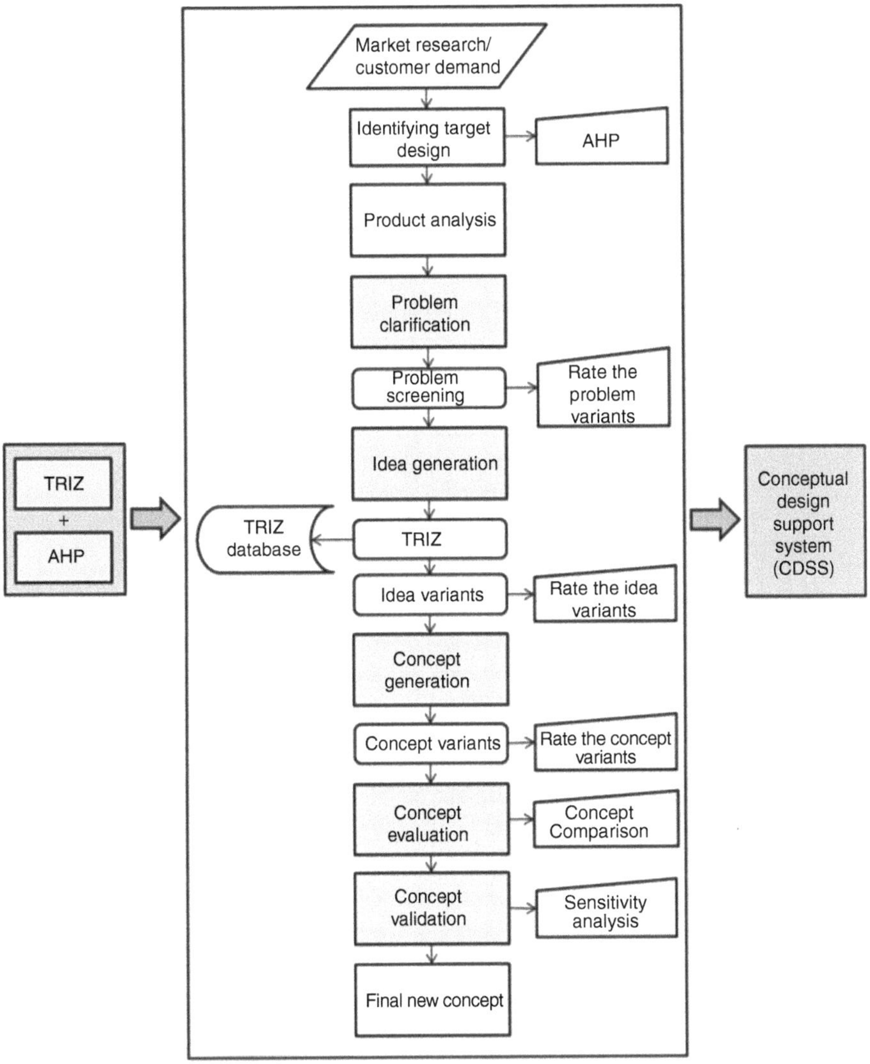

FIGURE 4.21 Integration of TRIZ and AHP methods in composite product development (Rosli, 2014).

Mansor (2014) performed concurrent conceptual design of hybrid natural/glass fiber reinforced thermoplastic composites for automotive parking brake lever. It started with market investigation, PDS, and conceptual design (Fig. 4.22). In product design, the first step to consider after the design brief is market investigation. The activities under this stage include information gathering, competitors' study, marketing research, and studying the voice of customers. Design information from books, journals, trade literature, internet sources,

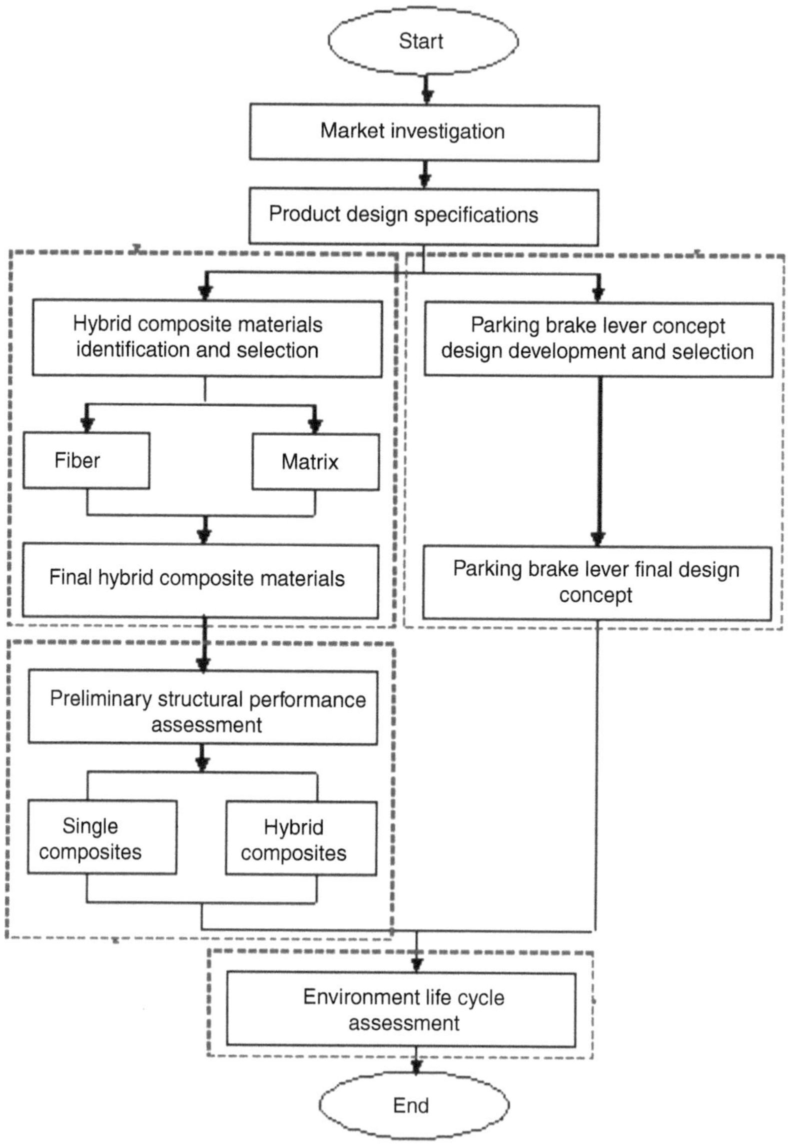

FIGURE 4.22 CE design system for composite automotive parking brake lever (Mansor, 2014).

magazines, conference proceedings, reports, standards, patents, and company brochure can be gathered using many sources but the most important source of information can be obtained through internet sources and computerized materials and design databases (Sapuan, 1998b), which are now mostly available online at some or no cost. Conceptual design was divided into four modules, namely, hybrid composite materials identification and selection, parking brake lever concept design development and selection, preliminary structural performance assessment, and environmental life cycle assessment (Mansor et al., 2013, 2014).

A recent work of Mastura et al. (2016) is concerned with the use of CE for composites for automotive antiroll bar. The CE design system for sugar palm/glass fiber reinforced polymer hybrid composite for automotive antiroll was developed comprising several modulus, such as materials selection, manufacturing process selection, and concept generation and evaluation using various tools, such as TRIZ, AHP, fuzzy systems, strategy canvas, and environmental-based QFD. The work was carried out in collaboration with a leading Malaysian automobile manufacturer, PROTON.

Automotive brake disk was designed using CE and the materials studied were metal matrix composites (MMC) (Fatchurrohman, 2013). The work comprised three stages: product investigation, product design specification, and conceptual selection. All these three activities were considered simultaneously. The model was finally validated using sensitivity analysis. Fig. 4.23 shows the CE model proposed by Fatchurrohman (2013) and he termed it as Integrated Conceptual Selection (ICS) framework. He initially stated the development by carrying out market investigation or product investigation followed by product design specification. In conceptual selection stage, he focused mainly on the manufacturing process and material selection. He proposed a

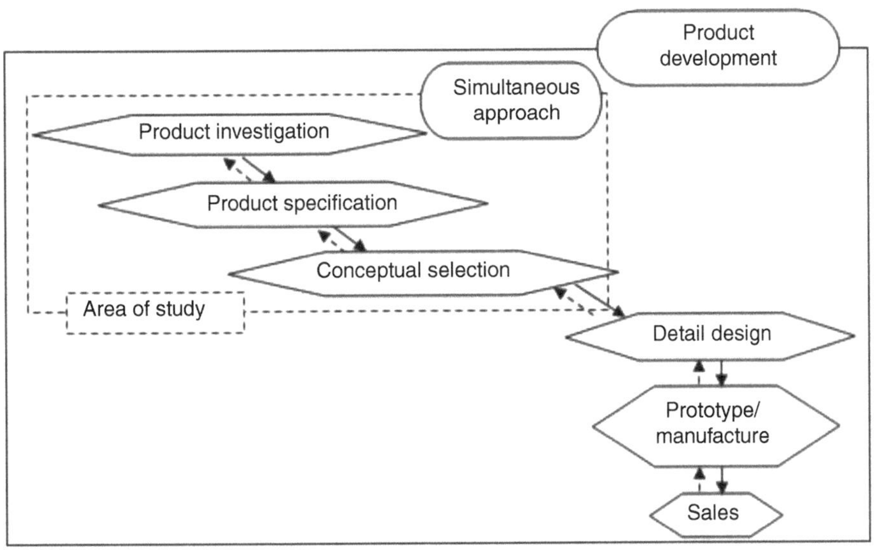

FIGURE 4.23 CE model for automotive brake disk from MMC by Fatchurrohman (2013).

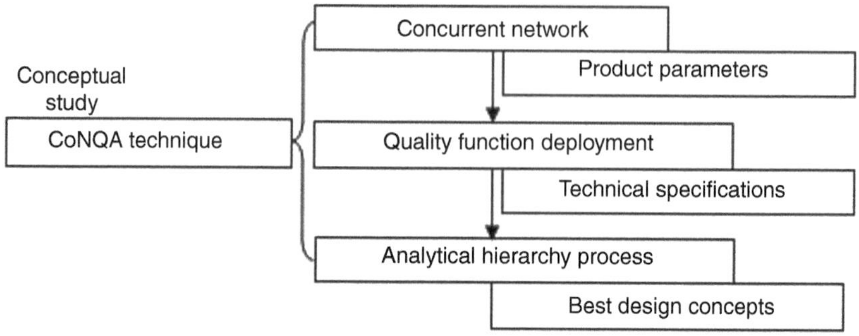

FIGURE 4.24 CoNQA technique in automotive brake disk from MMC (Fatchurrohman, 2013).

new technique based on multicriteria decision making (MCDM) called CoNQA; a combination of several activities like concurrent network (ConN), quality function deployment (QFD), and analytical hierarchy process (AHP) (Fig. 4.24).

ConN is the combination of concurrent approach (CA) and analytical network process (ANP). CA was represented by product investigation and specifications and from CA, product performance indicators and product conceptual parameters (CP) were established. The functions of ANP were to determine the influence and interdependence of parameters and to limit priorities of parameters. QFD was used to establish relationships between CPs with technical specifications and to calculate the priority weights of technical specifications.

In the manufacturing process selection, employing AHP, the goal set was to determine the best manufacturing process for MMC brake disk. There are 14 technical specifications that must be satisfied by a manufacturing process for MMC brake disk. Examples include shape complexity, internal metal soundness, dimensional accuracy rating, production rate, and ease of engineering changes. Manufacturing processes listed as candidate processes included sand casting, permanent mold casting, pressure die casting, investment casting, shell casting, plaster mold casting, and evaporative pattern casting (Fatchurrohman et al., 2012, 2013).

For material selection, the criteria set for MMC brake disk were tensile strength, yield strength, Young's modulus, ductility, density, maximum operating temperature, thermal conductivity, fracture strength, corrosion resistance, machining quality, price of materials recyclability, and wear resistance. Candidate materials comprised 21 MMCs with different matrix and reinforcement systems that must satisfy all the requirements to be selected as the suitable material for brake disk (Fatchurrohman et al., 2012, 2013).

Marine Application

Fairuz's (2011) work was performed using expert system shell Exsys Corvid to select the suitable materials for three components of boat structures, namely, stanchion, rib, and beam using polymer-based composites. The proposed work comprised materials selection framework and expert system approach using If-Then logic rules (Fairuz et al., 2012). The work of Amirruddin (2007) was concerned with design, analysis, and fabrication of polymer composite hovercraft hull base. Configuration structural analysis was carried out on glass fiber reinforced composite hovercraft hull base using NASTRAN/PATRAN software. The development of composite hovercraft hull base was carried out following the development of aluminum hovercraft hull base with slight modification (Amiruddin et al., 2008).

Misri (2011) used total design in the development of a small boat made from hybrid composites. During market investigation stage, survey questionnaires were sent to 50 fishermen and boat manufacturers in the areas of Sekinchan, Morib, Bagan Lalang, and Port Dickson, in Malaysia. The information from the survey was used to prepare a product design specification (PDS). Concept generation and development were then performed and a small composite was fabricated using the hand lay-up process. The performance of sugar palm/glass fiber reinforced unsaturated polyester composites were determined mechanically. Design and development process for glass/sugar palm unsaturated polyester hybrid composite small boat is shown in Fig. 4.25.

Other Applications

Other products developed with the CE for composites research group, UPM include furniture, such as chair, coffee tables, telephone stand, and shoe shelf, biopolymer-based biocomposite packaging, and pultruded rods normally used as materials for making outdoor domestic products, such as drain cover and house gate. Sapuan and Maleque (2005) used total design approach to design and fabricate a household telephone stand made from banana pseudo-stem fiber reinforced epoxy composites. Sapuan et al. (2007) also used total design approach in the design and development of a multipurpose small table made from banana pseudo-stem fiber reinforced epoxy composites.

Norianti (2016) applied total design in the development of kenaf fiber reinforced polymer composite household shoe shelf similar to the work of Misri (2011). She used several idea generation techniques in developing design concepts for composite shoe shelf, such as mind mapping, braining storming, and TRIZ. Fig. 4.26 shows the design and development process of kenaf fiber reinforced composite shoe shelf. Sanyang (2015) and Sahari (2013) focused on the development and characterization of sugar palm filled sugar palm starch biocomposite films for food packaging. The ultimate aim of their effort is to

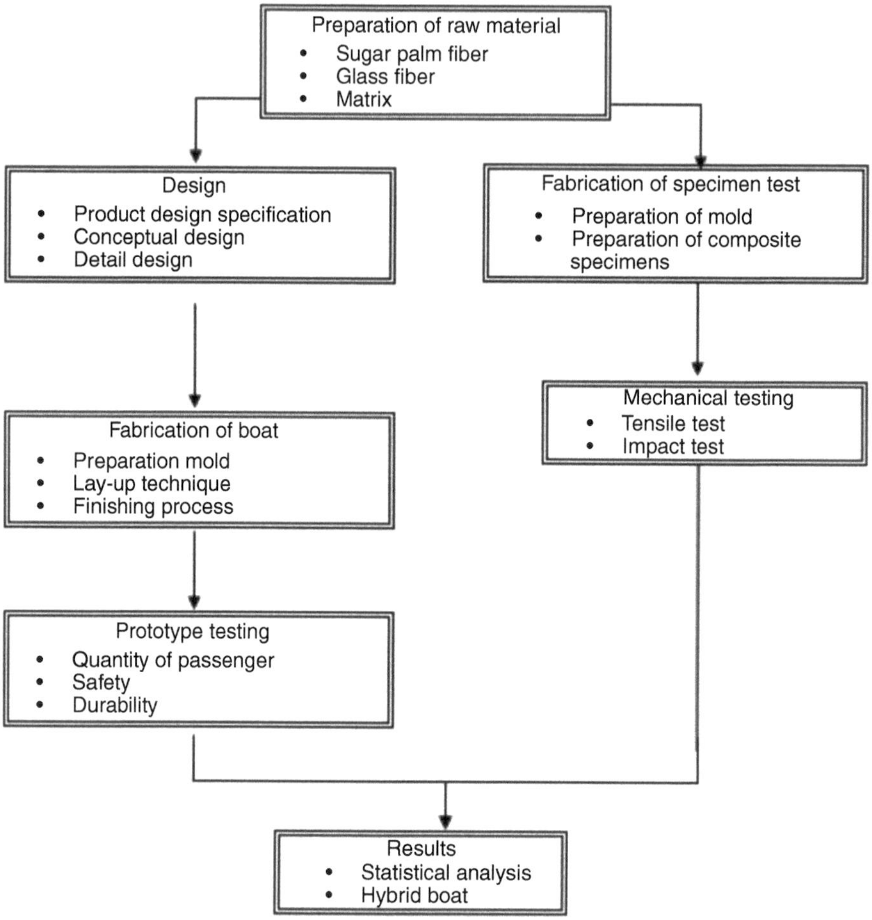

FIGURE 4.25 Design and development process for glass/sugar palm unsaturated polyester hybrid composite small boat (Misri, 2011).

design packaging products with environmental issues in mind, such as biodegradability and recyclability (Sanyang et al., 2016a,b; Sahari et al., 2013). It should fit very well with the notion of design for sustainability; an important so-called subset of CE. According to Karbhari et al. (1991), in the study of CE for composites, many tools and techniques were required to support its successful implementation. Statistical tools like the Taguchi method, in the areas of parametric design and tolerance design. The recent work by Fairuz (2016) is related to the Taguchi method emphasizing on design of experiment (DoE) in the pultrusion process using kenaf fiber composites. Through the DoE, the best manufacturing parameters, such as kenaf filler loading, pultrusion speed, and temperatures are obtained (Fairuz et al., 2015a). The results of this study have enabled the determination of mechanical properties at different parameters

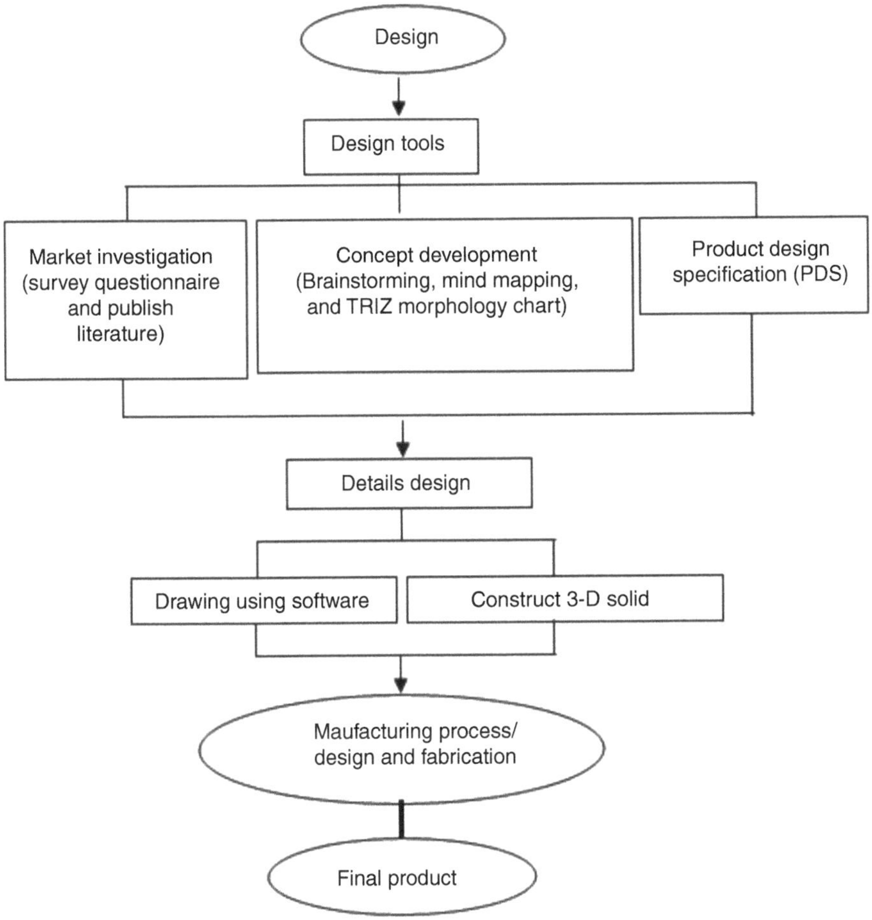

FIGURE 4.26 Design and development of kenaf fiber composite shoe shelf (Norianti, 2016).

that can be done without undergoing a lengthy and expensive process (Fairuz et al., 2015b).

SUMMARY

In this chapter, the implementation of CE technique in composite product development is reported based on the work carried out in research and academic institutions, as well as in industries. From the review, it can be concluded that CE had been performed very effectively both in industries and academics; in majority of the cases, they were successfully achieved particularly in terms of quality improvement, and cost and time reduction. Surprisingly, the aerospace industry has been reported to be deriving tremendous benefits from its implementation. Even though CE has been around for almost three decades, it is still very relevant as far as composite industry is concerned. A case study on the use of CE in the development of composite-based furniture product in an international furniture firm is also presented. Finally, to appreciate the use of CE in composite product development, previous works of the author based on his 23-year involvement in research and development of composite products were reviewed.

References

Ahmed Ali, B.A., 2015. Web-Based Expert System for Material Selection of Natural Fiber-Reinforced Polymer Composites, Universiti Putra Malaysia, PhD Thesis.

Ahmed Ali, B.A., Sapuan, S.M., Zainudin, E.S., Othman, M., 2015. Implementation of expert decision system for environmental assessment in composite material selection. J. Cleaner Prod. 107, 557–567.

Albiñana, J.C., Vila, C., 2012. A framework for concurrent material and process selection during conceptual product design stages. Mater. Des. 41, 433–446.

Al-Oqla, F.M., 2015. Enhancement of Evaluation Methodologies for Natural Fiber Composites Material Selection System, Universiti Putra Malaysia, PhD Thesis.

Al-Oqla, F.M., Sapuan, S.M., Ishak, M.R., Nuraini, A.A., 2015a. Predicting the potential of agro waste fibers for sustainable automotive industry using a decision making model. Comput. Electr. Agric. 113, 116–127.

Al-Oqla, F.M., Sapuan, S.M., Ishak, M.R., Nuraini, A.A., 2015b. A decision making model for optimal reinforcement condition of natural fiber composites. Fibers Polym. 16, 153–163.

Amirruddin, A.K., 2007. Design, Analysis and Fabrication of A Hovercraft Hull Base Prototype, Universiti Putra Malaysia, Master of Science.

Amiruddin, A.K., Sapuan, S.M., Jaafar, A.A., 2008. Analysis of glass fibre reinforced epoxy composite hovercraft hull base. Mater. Des. 29, 1453–1458.

Andersson, F., Hagqvist, A., Sundin, E., Björkman, M., 2014. Design for manufacturing of composite structures for commercial aircraft—the development of a DFM strategy at SAAB aerostructures. Proc. CIRP 17, 362–367.

Aridi, N.A.M., Sapuan, S.M., Zainudin, E.S., Al-Oqla, F.M., 2016. Mechanical and morphological properties of injection-molded rice husk polypropylene composites. Int. J. Polym. Anal. Charact. 21, 305–313.

Arkell, D., 2005. Similarities between 777, 787 help airlines, passengers and Boeing. Boeing Frontiers 4 (5), www.boeing.com/news/frontiers/archive/september/i_ca1.html.

Assembly Magazine, 2013. DFMA software generates cost-efficient product design. Available from: http://www.assemblymag.com/articles/91090-dfma-software-generates-cost-efficient-product-design

Azaman, M.D., 2015. Injection Moulding Simulation of Wood-Filled Polypropylene Thin-Walled Composite Parts, Universiti Putra Malaysia, PhD Thesis.

Azaman, M.D., Sapuan, S.M., Sulaiman, S., Zainudin, E.S., Abdan, K., 2013a. An investigation of the processability of natural fibre reinforced polymer composites on shallow and flat thin-walled parts by injection moulding process. Mater. Des. 50, 451–456.

Azaman, M.D., Sapuan, S.M., Sulaiman, S., Zainudin, E.S., Khalina, A., 2013b. Shrinkages and warpage in the processability of wood-filled polypropylene composite thin-walled parts formed by injection molding. Mater. Des. 52, 1018–1026.

Babu, B.J.C., Prabhakaran, R.T.D., Agrawal, V.P., 2008. DFX analysis applied to composite products. J. Reinf. Plast. Compos. 27, 287–312.

Barbosa, G.F., Carvalho, J., 2013. Analytical model for aircraft design based on Design for Excellence (DFX) concepts and use of composite material oriented to automated processes. Int. J. Adv. Manuf. Technol. 69, 2333–2342.

Belloy, P., Foucard, E., 2000. Roughness and materials in concurrent engineering systems. Comput. Mater. Sci. 19, 166–169.

Bergerson, A., 2010. DFMA Challenges and Success at Bell Helicopter, Proceedings of the 2010 International Forum on Design for Manufacture and Assembly (DFMA), 14–16 June, Providence, Rhode Island, 8 p.

Body, J., 2007. Airbus Future Composite Wing, SE Composite Seminar, Airbus S.A.S. Available from: www.gla.ac.uk

Bralla, J., 1996. Design for Excellence. McGraw-Hill, New York.

Choi, J.W., Kelly, D., Raju, J., 2007. A knowledge-based engineering tool to estimate cost and weight of composite aerospace structures at the conceptual stage of the design process. Aircr. Eng. Aerosp. Technol. 79, 459–468.

Cong, J., Zhang, B., 2012. Methodology for evaluating manufacturability of composite materials. Appl. Compos. Mater. 19, 189–201.

Das, S.K., Almonor, J.B., 2000. A concurrent engineering approach for the development of medical devices. Int. J. Comput. Integr. Manuf. 13, 139–147.

Davoodi, M.M., 2007. Development of Fiber-Reinforced Epoxy Composite Energy Absorber for Automotive Bumper System, Universiti Putra Malaysia, Master of Science Thesis.

Davoodi, M.M., 2012. Development of Thermoplastic Toughened Hybrid Kenaf/Glass Fibre Reinforced Epoxy Composite for Automotive Bumper Beam, Universiti Putra Malaysia, PhD Thesis.

Davoodi, M.M., Sapuan, S.M., Ahmad, D., Ali, A., Khalina, A., Jonoobi, M., 2010. Mechanical properties of hybrid kenaf/glass reinforced epoxy composite for passenger car bumper beam. Mater. Des. 31, 4927–4932.

Davoodi, M.M., Sapuan, S.M., Ahmad, D., Aidy, A., Khalina, A., Jonoobi, M., 2011. Concept selection of car bumper beam with developed hybrid bio-composite material. Mater. Des. 32, 4857–4865.

Davoodi, M.M., Sapuan, S.M., Ahmad, D., Ali, A., Khalina, A., Jonoobi, M., 2012a. Effect of polybutylene terephthalate (PBT) on impact property improvement of hybrid kenaf/glass epoxy composite. Mater. Lett. 67, 5–7.

Davoodi, M.M., Sapuan, S.M., Ali, A., Ahmad, D., 2012b. Effect of the strengthed ribs in hybrid toughened kenaf/glass epoxy composites bumper beam. Life Sci. J. 9, 210–213.

Davoodi, M.M., Sapuan, S.M., Robiah, Y., 2008. Conceptual design of a polymer composite automotive bumper energy absorber. Mater. Des. 29, 1447–1452.

Edwards, K.L., 2004. Exploiting new materials and processes for higher productivity: use of advanced composite technologies. Mater. Des. 25, 565–571.

Fairuz, A.M., 2011. Development of Material Selection Expert System for Polymer-Based Composite Materials, Universiti Putra Malaysia, Master of Science Thesis.

Fairuz, A.M., 2016. Mechanical Properties of Pultruded Kenaf Reinforced Vinyl Ester Composites, Universiti Putra Malaysia, PhD Thesis.

Fairuz, A.M., Sapuan, S.M., Zainudin, E.S., 2012. Prototype expert system for material selection of polymeric-based composites for small fishing boat components. J. Food Agric. Environ. 10, 1543–1549.

Fairuz, A.M., Sapuan, S.M., Zainudin, E.S., Jaafar, C.N.A., 2015a. Optimization of pultrusion process for kenaf fibre reinforced vinyl ester composites. Appl. Mech. Mater. 761, 499–503.

Fairuz, A.M., Sapuan, S.M., Zainudin, E.S., Jaafar, C.N.A., 2015b. The effect of gelation and curing temperatures on mechanical properties of pultruded kenaf fibre reinforced vinyl ester composites. Fibers Polym. 16, 2645–2651.

Fatchurrohman, N., 2013. Manufacturing Process and Material Selection Using Modified Multicriteria Decision Making for Brake Disc, Universiti Putra Malaysia, PhD Thesis.

Fatchurrohman, N., Sulaiman, S., Sapuan, S.M., Ariffin, M.K.A., Baharuddin, B.T.H.T., 2012. A new concurrent engineering—multi criteria decision making technique for conceptual design selection. Appl. Mech. Mater. 225, 293–298.

Fatchurrohman, N., Sulaiman, S., Sapuan, S.M., Ariffin, M.K.A., Baharuddin, B.T.H.T., 2013. Generic framework for conceptual design using concurrent engineering strategy: a case study: advanced material application product development of metal matrix composite component. Adv. Mater. Res. 789, 82–86.

Gandi, M.V., Thomson, B.S., 1990. manufacturing process driven design methodologies for components fabricated in composite materials. Mater. Des. 11, 235–242.

Greene, E., 1999. Marine Composites. Eric Greene Associates, Inc, Annapolis, pp. 285–299.

Guillermin, O., 2010. Composites put wind in the sails of all kinds of vessels. Available from: http://m.plm.automation.siemens.com/en_us/Images/reinforced-plastics-marine-composites-201008_tcm1224-197893.pdf

Hale, J., 2006. Boeing 787 from the ground up. Aero Quart. 4, 17–24.

Hale, R.D., Vasey-Glandon, M., 2001. PACKS: an affordable knowledge-driven composite design for manufacturing process, Proceedings of SAMPE Conference 2001, 6–10th May, Long Beach, CA, pp. 1–14.

Hambali, A., 2009. Selection of Conceptual Design using Analytical Hierarchy Process for Automotive Bumper Beam under Concurrent Engineering Environment, Universiti Putra Malaysia, PhD Thesis.

Hambali, A., Sapuan, S.M., Ismail, N., Nukman, Y., 2010. Materials selection of polymer based composite bumper beam using analytical hierarchy process. J. Cent. South Univ. Technol. 17, 244–256.

Hambali, A., Sapuan, S.M., Ismail, N., Nukman, Y., 2011. Concurrent decisions on design concept and material selection using analytical hierarchy process at the conceptual design stage. Concurrent Eng. Res. Appl. 19, 111–121.

Imihezri, S.S.S., 2005. Design, Analysis and Fabrication of Glass Fiber-Reinforced PA 6,6 Composite Automotive Clutch Pedals, Universiti Putra Malaysia, Master of Science Thesis.

Imihezri, S.S.S., Sapuan, S.M., Hamdan, M., Sulaiman, S., 2005. A study on comparison of V and X ribbings for polymeric based composite automotive clutch pedal using mould flow analysis software. Mater. Des. 26, 157–166.

Imihezri, S.S.S., Sapuan, S.M., Sulaiman, S., Hamdan, M.M., Zainudin, E.S., Osman, M.R., Rahman, M.Z.A., 2006. Mould flow and component design analysis of polymeric based composite automotive clutch pedals. J. Mater. Process. Technol. 171, 358–365.

Jacob, M.S.D., 2002. A Prototype Knowledge-Based System for Ceramic Matrix Composites Material Selection of Automotive Engine Components, Universiti Putra Malaysia, Master of Science Thesis.

Jahan, A., 2011. Improving Multi-Criteria Decision Making Algorithm for Material Selection, Universiti Putra Malaysia, PhD Thesis.

Jahan, A., Mustapha, F., Ismail, M.Y., Sapuan, S.M., Bahraminasab, M.A., 2011. Comprehensive VIKOR method for material selection. Mater. Des. 32, 1215–1221.

Johnson, A.F., Marchant, A., 1990. Design and analysis of composite structures. In: Hollaway, L. (Ed.), Polymers and Polymer Composites in Construction. Thomas Telford, London.

Jones, M.D.C., 1995. Towards a concurrent engineering solution for RTM manufacture of complex high performance polymer composite components, 4th International Conference on Automated Composites (ICAC95): Conference Pre-prints, 6–7 September, Nottingham, pp. 15–22.

Jumaidin, R., Sapuan, S.M., Jawaid, M., Ishak, M.R., Sahari, J., 2016. Characteristics of thermoplastic sugar palm starch/agar blend: thermal, tensile, and physical properties. Int. J. Biol. Macromol. 89, 575–581.

Karandikar, H., Mistree, F., 1992. Designing a composite material pressure vessel for manufacture: a case study in concurrent engineering. Eng. Optimiz. 18, 235–262.

Karbhari, V.M., Wilkins, D.J., Henshaw, J.M., 1991. Concurrent Engineering for Composites, Final Report, U.S. Army Research Office, Research Triangle Park, NC.

Kim, J., Kim, C., Hong, C., Hahn, H.T., 2000. Development of concurrent engineering system for design of composite structures. Compos. Struct. 50, 29–309.

Klomp-de Boer, R., 2010. Development of a cost effective composite wingbox for small aircraft. NLR-TP-2010-047. Tech Report, National Aerospace Laboratory NLR, The Netherlands.

Laurenzi, S., Marchetti, M., 2012. In: Hu, N. (Ed.), Advanced Composite Materials by Resin Transfer Molding for Aerospace Applications, Composites and Their Properties. InTech Open Access Publisher, Rijeka, Croazia, pp. 127–226.

Lee, H.B., 2003. Comparison of Cost, Surface Roughness and Time Using Stereolithography and 3D Printer for the Design of Composite Pedal, Universiti Putra Malaysia, Master of Science Thesis.

Lee, R.S., Chen, Y.M., Lee, C.Z., 1997. Development of a concurrent mold design system: a knowledge-based approach. Comput. Integr. Manuf. Syst. 10, 287–307.

Mansor, M.R., 2014. Concurrent Conceptual Design of Hybrid Natural/Glass Fiber Reinforced Thermoplastic Composites for Automotive Parking Brake Lever, Universiti Putra Malaysia, PhD Thesis.

Mansor, M.R., Sapuan, S.M., Zainudin, E.S., Nuraini, A.A., Hambali, A., 2013. Hybrid natural fiber/glass fiber reinforced polymer composites material selection using Analytical Hierarchy Process for automotive gear lever design. Mater. Des. 51, 484–492.

Mansor, M.R., Sapuan, S.M., Zainudin, E.S., Nuraini, A.A., Hambali, A., 2014. Conceptual design of kenaf fiber polymer composite automotive parking brake lever using integrated TRIZ-morphological chart-AHP method. Mater. Des. 54, 473–482.

Marchant, A., 1994. Methodology and management of a design project. In: Hollaway, L. (Ed.), Handbook of Polymer Composites for Engineers. Woodhead Publishing Ltd, Cambridge.

Marsden, C., Fews, R., Oldroyd, P., Yousefpour, A., 2011. Design for manufacturing—one-piece, fibre-placed composite helicopter tailboom, IOP Conference Series: Materials Science and Engineering, vol. 26, paper no. 012016.

Mastura, M.T., Sapuan, S.M., Mansor, M.R., Nuraini, A.A., 2016. Environmentally conscious hybrid bio-composite material selection for automotive anti-roll bar. Int. J. Adv. Manuf. Technol., DOI 10 1007/s00170-016-9217-9.

Mayer, G., 1991. Advanced Materials Aspects of Concurrent Engineering, October, Institute for Defense Analyses, Alexandria, Virginia, IDA Document D-1033.

Misri, S., 2011. Design and Fabrication of a Boat using Sugar Palm Glass Fibre Reinforced Unsaturated Polyester Composite, Master of Science, Universiti Putra Malaysia, Master of Science Thesis.

Norianti, M., 2016. Design and Fabrication of Kenaf (*Hibiscus cannabinus* L.) Fiber Reinforced Unsaturated Polyester Composite Shoe Shelf, Universiti Putra Malaysia, Master of Science Thesis.

Pahl, G., Beitz, W., Feldhusen, W., Grote, K.H., 2007. Engineering Design: A Systematic Approach, third ed. Springer-Verlag London Limited, London.

Pardessus, T., 2004. Concurrent engineering development and practices for aircraft design at Airbus, Proceedings of the 24th International Council of the Aeronautical Sciences Conference, 29th Aug–3rd Sept, Yokohama, Japan, 9 p.

Perry, N., Bernard, A., Laroche, F., Pompidou, S., 2012. Improving design for recycling—application to composites. CIRP Ann. Manuf. Technol. 61, 151–154.

Prabhakaran, R.T.D., Babu, B.J.C., Agrawal, V.P., 2006. Design for 'X'-abilities of RTM products—a graph theoretic approach. Concurrent Eng. Res. Appl. 14 (2), 151–161.

Prasad, B., 1996. Concurrent Engineering Fundamentals, Integrated Product and Process OrganizationPrentice-Hall, Inc, Upper Saddle River, New Jersey.

Pugh, S., 1991. Total Design: Integrated Methods for Successful Product Engineering. Addison Wesley Publishers Ltd, Wokingham, England.

Quinn, J.A., 1999. Composites: Design Manual. Technomic Publishing Co., Inc, Lancaster, USA.

Reichert, S., Schwinn, T., Magna, R.L., Waimer, F., Knippers, J., Menges, A., 2014. Fibrous structures: an integrative approach to design computation, simulation and fabrication for lightweight, glass and carbon fibre composite structures in architecture based on biomimetic design principles. Comput. Aided Des. 52, 27–39.

Rolfes, R., Baaran, J., Juhasz, J., Nolden, P., Tessmer, J., Topfer, G., Zimmermann, R., 2000. Advanced analysis tools for concurrent/integrated engineering of composite airframe structures, Proceedings of the 1st Stade Composite Colloquium, 7–8th September, Stade, Germany, pp. 225–244.

Rolfes, R., Noack, J., Ruiz-Valdepenas, F, Taeschner, M., Zimmermann, R., 2001. Fast analysis tools for concurrent/integrated engineering of composite airframe structures, Optimization in Industry II: Proceedings of the Conference, 6–11 June 1999, Banff, pp. 331–338, American Society of Mechanical Engineers (ASME).

Rosli, M.U.M., 2014. Development of Conceptual Design Support System Based on Integrated Theory of Inventive Problem Solving and Analytical Hierarchy Process, Universiti Putra Malaysia, PhD Thesis.

Sahari, J., 2013. Characterization and Development of Biopolymer Derived from Arenga pinnata and their Biocomposites, Universiti Putra Malaysia, PhD Thesis.

Sahari, J., Sapuan, S.M., Zainudin, E.S., Maleque, M.A., 2013. Thermo-mechanical behaviours of thermoplastic starch derived from sugar palm tree (*Arenga pinnata*). Carbohydr. Polym. 91, 1711–1716.

Sanyang, M.L., 2015. Biodegradable Composite Films from Modified Sugar Palm (*Arenga pinnata* (Wurmb) Merr.) Starch for Food Packaging Applications, Universiti Putra Malaysia, PhD Thesis.

Sanyang, M.L., Sapuan, S.M., Jawaid, M., Ishak, M.R., Sahari, J., 2016a. Development and characterization of sugar palm starch and poly(lactic acid) bilayer films. Carbohydr. Polym. 146, 36–45.

Sanyang, M.L., Sapuan, S.M., Jawaid, M., Ishak, M.R., Sahari, J., 2016b. Recent developments in sugar palm (*Arenga pinnata*) based biocomposites and their potential industrial applications: a review. Renew. Sust. Energy Rev. 54, 533–549.

Sapuan, S.M., 1998a. A Concurrent Engineering Design System for Polymeric Based Composite Automotive Components, De Montfort University, UK, PhD Thesis.

Sapuan, S.M., 1998b. Computerized information retrieval system for the postgraduate research degree in mechanical engineering design. AEESEAP J. Eng. Educ. 28, 27–36.

Sapuan, S.M., Abdalla, H.S., 1998. A prototype knowledge-based system for the material selection of polymeric-based composites for automotive components. Compos. Part A Appl. Sci. Manuf. 29A, 731–742.

Sapuan, S.M., Jacob, M.S.D., Mustapha, F., Ismail, N., 2002. A prototype knowledge based system of material selection for ceramic matrix composites of automotive engine components. Mater. Des. 23, 701–708.

Sapuan, S.M., Harun, N., Abbas, K.A., 2007. Design and fabrication of a multi-purpose table using composite of epoxy and pseudo-stem banana fibres. J. Trop. Agric. 45, 66–68.

Sapuan, S.M., Maleque, M.A., 2005. Design and fabrication of natural woven fabric reinforced epoxy composite for household telephone stand. Mater. Des. 26, 65–71.

Sapuan, S.M., Maleque, M.A., Hameedullah, M., Suddin, M.N., Ismail, N., 2005. A note on the conceptual design of polymeric composite automotive bumper system. J. Mater. Process. Technol. 159, 145–151.

Skalak, S.C., Kemser, H.P., Ter-Minassian, N., 1997. Defining a product development methodology with concurrent engineering for small manufacturing companies. J. Eng. Des. 8, 305–328.

Sobek, D.K., Ward, A.C., Liker, J.K., 1999. Toyota's principles of set-based concurrent engineering. Sloan Manag. Rev. 40, 67–84.

Sobey, A.J., 2010. Concurrent Engineering in the Context of the Composite Leisure Boatbuilding Industry, University of Southampton, UK, PhD Thesis.

Srinivasan, R., Karandikar, H.M., Mistree, F., 1991. Understanding design-manufacture interaction using compromise decision support system—I. A formulation for composite pressure vessels. Comp. Struct. 40, 679–692.

Strong, A.B., 2006. Plastics: Materials and Processing, third ed. Pearson Prentice Hall, Upper Saddle River, New Jersey.

Suddin, M.N., 2003. Design of Polymeric Based Composite Automotive bumper Fascia, Universiti Putra Malaysia, Master of Science Thesis.

Swanstrom, F.M., Hawke, T., 2000. Design for manufacturing and assembly: a case study in reduction for composite wing tip structures. SAMPE J. 36, 9–16.

Tavares, S.M.O., 2013. Design and Advanced Manufacturing of Aircraft Structures Using Friction Stir Welding, University of Porto, Portugal, PhD Thesis.

Wall, K.O., Crawford, R.J., 1995. The 'design for manufacture' of continuous fibre-reinforced thermoplastic products in primary aircraft structure. Compos. Manuf. 6, 245–254.

Zainudin, E.S., 2002. Numerical Investigation of Fibre Orientation in Short Glass Fibre Reinforced Injection-Moulded Thermoplastic Composites, Universiti Putra Malaysia, Master of Science Thesis.

CHAPTER 5

Conceptual Design in Concurrent Engineering for Composites

INTRODUCTION

Composite materials have replaced metals in many structural and nonstructural applications due to desirable attributes such as light weight, corrosion resistance, comparable specific strength and stiffness properties to metals, part integration, ease of processing, and good damping properties. Composites are applied in aerospace, marine, automotive, building and construction (Fig. 5.1), furniture, sport and recreation, and telecommunication industries in the recent years. Manufacturing processes to fabricate composite components are commercially available made from polymer composites, metal matrix composites (MMC), and ceramic matrix composites (CMC). Characterization, product development, design, and fabrication of composites have been the subjects of great interest in the research and development of the materials either by research institutions or industries.

Designing with composites is a complex process and this is particularly true when we are designing composite laminates. Decision on the appropriate types of fiber, and matrix and the correct fiber volume fraction is important. Similarly, determination of the optimum fiber orientation in each layer and the stacking sequence of the layers are also critical. Finally, the number of plies required in each direction must also be found (Uusitalo, 2013).

In design with fiber reinforced composites, the approach is different from isotropic materials like metals and polymers as composites are orthotropic materials. Design of such composite materials can be found in standard composite textbooks (Barbero, 1999; Mallick, 2008). Damage and fracture of composites such as delamination, fiber de-bonding, and fiber breaking can be included in the design using several failure criteria like maximum stress criterion, Tsai-Hill failure criterion, and Tsai-Wu failure criterion (Barbero, 1999).

Detail design of composites is very important in composite product development. It involves using various failure criteria, understanding of various laminated theories, and consideration of design for strength. In a nutshell, design

CONTENTS

Introduction 141

What Is Conceptual Design? 142

Conceptual Design in CE Environment.............. 144

Why Conceptual Design Is an Important Element of CE? *144*

Design and Development of Composite Products Emphasizing on Conceptual Design.... 145

A Literature Review on Conceptual Design of Composite Product Development *145*

Conceptual Design Methods for Composites 149

"Preconcept Generation" Methods for the Design of Composite Products *150*

Concept Generation Methods for Composites *160*

Concept Evaluation and Selection Methods for Composites *195*

Conclusions 203

References 204

Composite Materials. http://dx.doi.org/10.1016/B978-0-12-802507-9.00005-2

FIGURE 5.1 Structural component from kenaf fiber reinforced composites.
Courtesy Institute of Forestry and Forest Products (INTROP), Universiti Putra Malaysia.

with composites demands that within predetermined service life, a component or structure must not fail and this is basic philosophy in composite design (Mallick, 2008).

During detail design of composites, numerical design analysis and simulation plays an important role and numerous works had been carried out in this area of research. For instance, finite element analysis has been extensively used in the design analysis of composites. Similarly, manufacturing cost consideration should also be extensively studied.

In composites, engineering designers are designing a material as well as a component, and in this regard, the understanding on how the part is fabricated is essential. In the absence or in the lack of the understanding of fabrication process, it is very difficult for the design engineer to meet design criteria. Therefore, early and simultaneous consideration of all life cycle issues (Sapuan et al., 2006) is important in composite product development (Fig. 5.2). Therefore study of conceptual design of composites should be given high priority and yet there are limited publications dealing with this topic. In this chapter conceptual design of fiber reinforced composites is presented by paying particular attention on CE approach.

WHAT IS CONCEPTUAL DESIGN?

It is generally understood that conceptual design can be broadly divided into concept generation and concept evaluation (in this book, a so-called pre-concept generation is also introduced) (Fig. 5.3). Before going into detail on

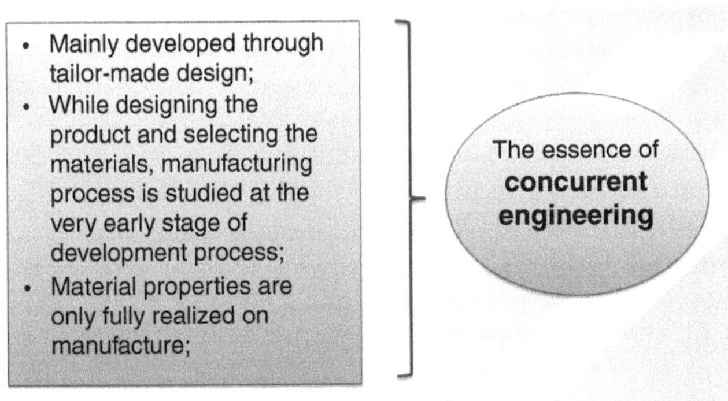

FIGURE 5.2 Early and simultaneous consideration of all life cycle issues in composite product development.

conceptual design of composite materials, this section discusses conceptual design in general. The conceptual design is performed to consider several alternative solutions, and in this stage creativity and lateral thinking play vital roles. Conceptual design activities that are related to providing design ideas are sometimes referred to as design concept, conceptualization, concept creation, and concept initiation while selecting the most suitable concept is referred to as concept evaluation or selection. During conceptual design stage, design engineers use creativity and imagination to develop, envisage, and generate the several design approaches or alternatives to fulfil product design specification (PDS) and design needs by having creative, unconfined, and uninhibited mindset and it is different from the mindset of engineers, who are normally analytical in nature (Hyman, 1998).

French (1999) refers conceptual design as the early stage of design after recognizing the design need and the outcome of the conceptual design stage is termed "schemes." Wright (1998) categorized creativity methods and evaluation and

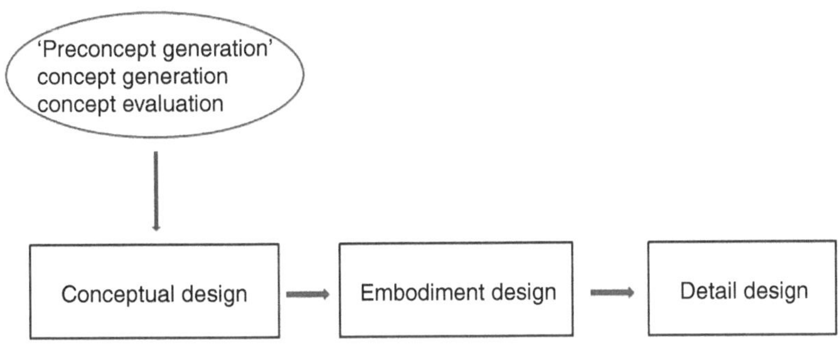

FIGURE 5.3 Engineering design process with conceptual design encompassing "preconcept generation".

selection as important elements of conceptual design. Creativity methods are meant for concept generation; and evaluation and selection are meant for concept evaluation.

Pugh (1991) did not include embodiment design in his total design model. Therefore, what can be learnt from his model is that, any activities before detail design is still included in conceptual design. Dieter and Schmidt (2009) differed from Pugh (1991) as they regarded market investigation and PDS as parts of conceptual design. The final output of conceptual design is a final concept. Andreasen et al. (2015) defined concept as a design proposal that is detailed enough to justify if it is a good answer to the task and intention, and show a high probability of realization and success. Pahl et al. (2007) provided very detailed procedures on how to perform effective conceptual design and among others, are clarification of the design task and setting up the list of design requirement, abstraction to identify the problems (formulating the design problems), and formulating the overall function structures.

CONCEPTUAL DESIGN IN CE ENVIRONMENT

In this chapter, conceptual design of composite products is presented with the emphasis on its use in CE environment. Conceptualization and design are the cores of product development and these are the predominant activities in launching competitive products (Andreasen et al., 2015). Making the best decision on design concept is not an easy task as selection process involves many different criteria and constraints and inaccurate decision may lead to the need of redesigning or remanufacturing of the product (Hambali, 2009).

Why Conceptual Design Is an Important Element of CE?

In this chapter, conceptual design is considered an important activity in CE and the statements of many product design and development experts supported the claim very well. The CE seeks early involvement of other downstream functions during the early stage of the design process (Sapuan and Mansor, 2014) and the most important and distinctive activity during the early stage of design process is mainly conceptual design.

Despite conceptual design used minimum amount of budget of a company as revealed by Ulrich and Eppinger (2004) by stating that *"concept generation had typically consumed less than 5 percent of the budget and 15 percent of development time"* but a number of experts agreed that this stage is very crucial in design because important decision has to be made during this stage. A survey carried by Loiter (1986) reveals that 75% of the product cost is decided in the early stage of the design process and Corbett (1986) also mentioned almost similar percentage. Munroe (1995) is reported to make this important statement:

"after all, 70% of a product's total cost is determined by its design, and that cost includes material, facilities, tooling, labor, and other support costs." Barton et al. (2001) stated in their work by posing question to the readers *"design determines 70% of cost?"*.

Other experts such as Suh (1990) and Boothroyd (1988) also affirmed that conceptual design is important because 70% cost of manufacture is committed during this stage. Seo et al. (2002) emphasized, after knowing that 70% of the total cost of a product is decided at the outset of the design process, design engineers should be able to significantly shorten the product life cycle through life cycle implications of design decisions. Excellent detail design cannot compensate poor design concept (Hsu and Woon, 1998), so appropriate tools and techniques play a major role in this regard and investment of tools and technology must be evenly distributed for concept and detail design. Fig. 5.4 is an artist impression of the importance of conceptual design in product development process.

DESIGN AND DEVELOPMENT OF COMPOSITE PRODUCTS EMPHASIZING ON CONCEPTUAL DESIGN

It is important to note that conceptual design publications on composite materials mainly describe conceptual design as part of complete composite product design and development. Very limited publications are dealing with conceptual design by emphasizing tools or methods in composite product development and publications of this nature are mainly written by academics whose research interests are on design tools and methods. However, both types of publications are considered in this chapter (Fig. 5.5).

A Literature Review on Conceptual Design of Composite Product Development

Markov and Cheng (1996) developed novel conceptual design of robotizing the asymmetrical part in filament winding process for continuous fiber reinforced polymer composites. Robotic manipulators in filament winding used mechatronic approach in their design. The manipulators were for winding helical and polar patterns in filament winding process. The paper focused mainly on the conceptual design of robotic manipulators for filament winding machine and was not concerned with the products from filament wound composites.

Imihezri et al. (2006) carried out research in the design of glass fiber reinforced polyamide composite automotive clutch pedal. The design concept was started by deciding the section profile of the clutch pedal, that is, "I," "T," "U," and "□" profiles. The determination of stress and flow behavior of composite clutch

FIGURE 5.4 The importance of conceptual design in product development process.

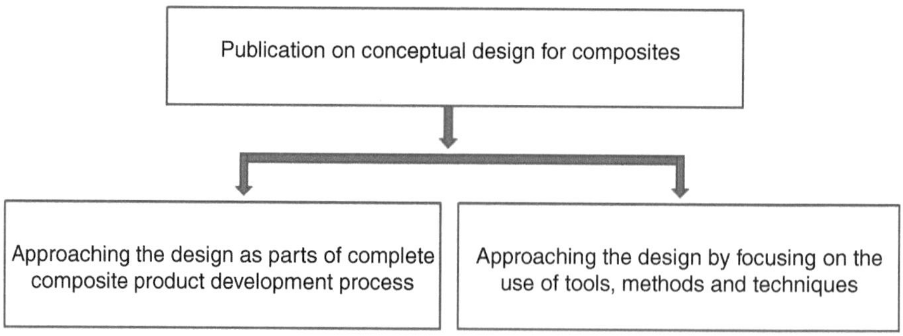

FIGURE 5.5 Two approaches of the publication in conceptual design for composites.

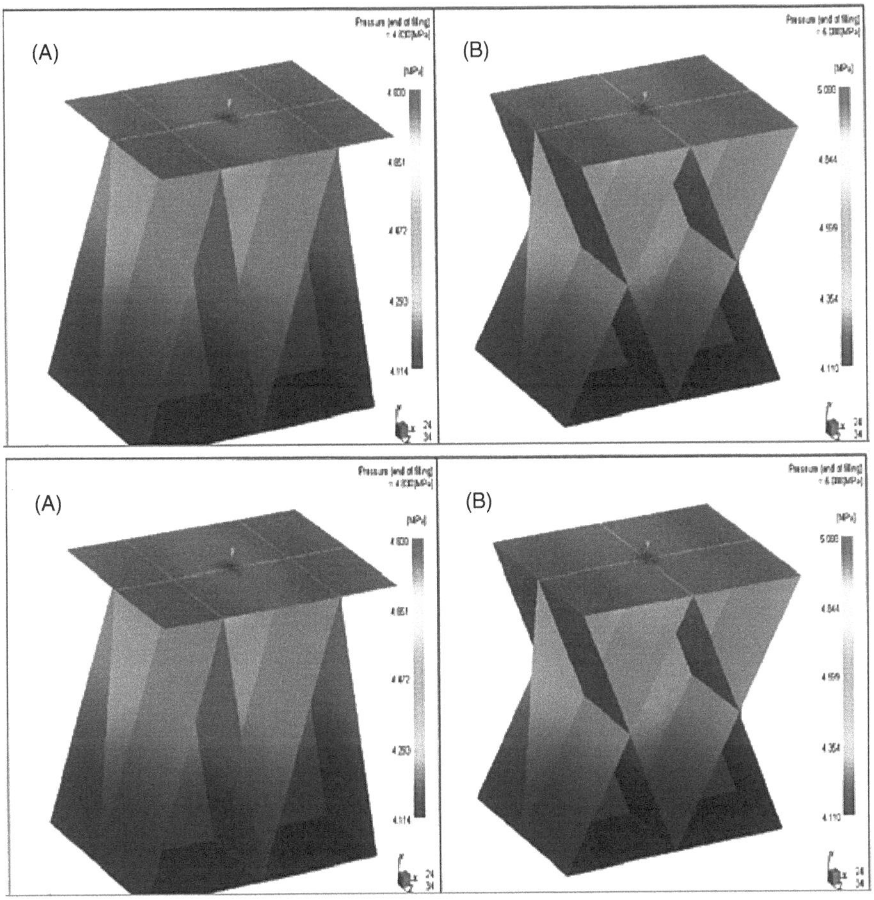

FIGURE 5.6 FEA to determine the best pattern of rib in composite clutch pedal (Imihezri et al., 2006).

pedals were concurrently carried out to determine which rib structure ("X" or "V" pattern) is more suitable to be incorporated in composite clutch pedal lever. The analysis was performed using finite element and mold flow analyses. Fig. 5.6 shows FEA models to study stress analysis in "X" and "V" ribs (rib pattern selection) for composite pedal levers.

Kapidžić et al. (2014) investigated and compared the performance of two design concepts of hybrid composite–aluminum generic wing box for aircraft components. The baseline concept was made from aluminum and the concepts 1 and 2 were made from hybrid of composite–aluminum but they are different in their configurations. Requirements that formed the basis of comparison of the two concepts include weight, thermal properties, bolted joints, buckling, and failures. Additional requirements include the composite in the hybrid materials should have the plies being laid in 0, 90, and ±45 degrees and brittle damage should be considered.

Amirruddin (2007) performed conceptual and detail design study of polymer composite hovercraft hull base. In the design, the length of the hovercraft was made longer than the width with the tail section stretched wide for aerodynamic reason. In the body design, the function needed for positioning the lifting and thrust engine, drive units, and passenger sections are combined into a system together with the aerodynamic requirements. All designs are applied with varied number of seats (till four seats) and detail design was performed with the help of FEA. Composite materials have good potential to replace metals such as aluminum particularly in the hull base.

Netten and Vingerhoeds (1997) divided conceptual design into three phases, that is, prototype selection, concept selection, and concept modification applied to fiber reinforced composite panels for aerospace structure. They defined conceptual design as configuration (prototype) of concept and parametric design of design concept. Prototype is the configuration of design. Examples of prototypes include sandwich, L-stiffened, I-stiffened, Z-stiffened, or hat-stiffened for panel type prototype and carbon, aramid, glass fiber reinforced plastics, unidirectional, cross-ply materials, and aluminum alloys for material prototypes. Both the prototype and design concept must satisfy functional requirements such as stiffness, strength and impact strengths, fatigue resistance, weight, maintenance, and production cost. All these requirements have been documented in PDSs. Concept generation and selection were mentioned in this paper; prototype selection and concept selection are regarded the same as concept evaluation. This fits within the proposed Pugh total design model where conceptual design is divided into concept generation and concept evaluation (selection).

In aerospace industry, generally design process is carried out with extensive materials qualification program, even during conceptual design. Bergers (2016) carried out research on conceptual design of main fitting of a cantilever type nose landing gear of a helicopter from composites. "Adoption of proven solution" was used to generate the design concepts. Four design concepts were developed on the basis of the most critical load. Sketches of the design were made using pencil and papers and further developed using CAD software. The work was based on the existing design of helicopter landing gear by Fokker Landing Gear BV, the Netherlands.

In a study by Turner and Grande (1978), several design concepts were developed for aircraft wing structure such as skin stiffener, stiffened skin sandwich, and conventional sandwich design made of borsic aluminum. Conventional sandwich panel was used for fuselage skin panel and upper and lower surface panels of the aero wing structure used high strength graphite/polyimide composites. Encouraging results were obtained from the study and it opened up the possibility of using such materials in aircraft components.

According to Dominy (2000), the basic principle of conceptual design with composites is that composite and metal structures are different as far design is concerned. Composite designers can benefit from the inclusion of features such as ribs, bosses, cores, insets, and sandwiches during fabrication. These can help improve the properties of composites. Dominy (2000) further highlighted that conceptual design of laminated composites for a cantilever beam, two alternative designs are possible, that is, cantilever beam from unidirectional composites and with unidirectional skins and a low density core. The latter retains majority of strength and modulus of the former. For the latter, it demonstrates excellent specific strength and stiffness compared to the former.

BioComposites Centre (BC) Bangor University (2014) listed market study and technology study as two main activities during concept or idea evaluation stage of biocomposite product development, with no detailed account of the activities. BioComposites Centre (BC) Bangor University (2014) provided different approach in idea generation techniques particularly in terms of activities in development of biocomposite products. In addition to conventional brainstorming technique, they offered patent search, benchmarking study, technical literature search, collaborative study, and funding research as their activities (BioComposites Centre (BC) Bangor University, 2014). However, in Pugh model, technical literature search, patent search, and benchmarking study are done during market investigation stage.

Lee (2004) performed conceptual design study of fiber reinforced polymer composite automotive pedal box system. He carried out closed form data analysis of pedal box system. He performed bending stiffness and torsion strength calculation to decide on the best pedal lever cross-section. He also made rapid prototyping models of the composite pedal box system using stereolithography (SLA) and 3D printer; their performances were compared in terms of cost, speed of process, user friendliness, and model quality. Fig. 5.7 shows rapid prototyping models of composite-based automotive clutch pedals using SLA and 3D printer.

CONCEPTUAL DESIGN METHODS FOR COMPOSITES

In this chapter, conceptual design methods can be divided into three components, that is, "preconcept generation," concept generation, and concept evaluation. These methods can successfully be implemented for the design of composite and noncomposite products but with certain exceptions that are unique only to composites.

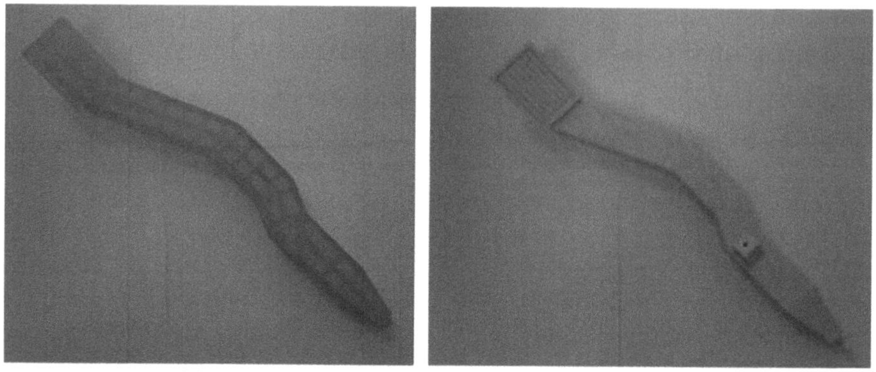

(A) (B)

FIGURE 5.7
Rapid prototyping models of automotive composite clutch pedals (A) stereolithography (B) 3D printer (Lee, 2004).

"Preconcept Generation" Methods for the Design of Composite Products

The activities or methods during conceptual design stage but prior to concept generation are called "preconcept generation" in this book, which include problem formulation and information gathering.

Problem Formulation

The first step in design is to recognize the design need. In composite design, as in any other designs, design engineers should know their client, that is who they are working for. Therefore the design engineers must be very clear with the design brief. The successful implementation of the design project depends largely on the agreement between the client and the design engineers. However, for some products, whose need may not be originated from a client, particularly for consumer products, company/design engineers need to identify the customer needs and determine the customer requirements by carrying out appropriate surveys on customer need (Wright, 1998).

Problem Definition

Then design engineer should be able to define the design problem as normally design problem is considered ill-structured or ill-defined, that is, it is not clearly defined (Cross, 1994). It is well known that designing products from composites is different from designing with metal counterpart. There are several properties that should be considered in the design of composite products (Fig. 5.8). Examples of these properties include low density, high strength, high stiffness, good thermal performance, corrosion resistance, wear resistance, low heat transmission, good environmental performance, good fatigue performance, good electrical insulation, and low sound transmission (Mayer, 1993; Sapuan and

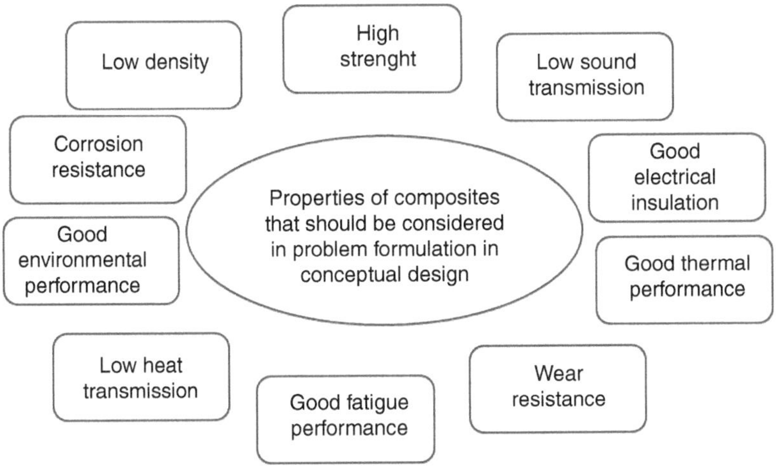

FIGURE 5.8 Properties of composites to be considered in the problem formulation in conceptual design.

Yusoff, 2015) should be taken into consideration in the design. While designing with composites, design engineers must be aware of those properties that some of them may not be found in metals and should consider some of them in their designs. In the initial stage of the design of product from composites, the design engineer has to realize that there is no resource to a clearly defined range of proven materials. Even during the design phase, the question of manufacturability has to be borne in mind and the potential manufacturing techniques have to be determined (Sapuan et al., 1995).

The next step is to define the design problems in terms of goal, objectives, and constraints as emphasized by Hyman (1998). Consider the example of the development of an orthopedic device to assist patients to walk (Dominy, 2000). The intention is to replace the existing design from metal to composites. While carrying out this task, the design engineers must consider the important properties of composites (as stated by Mayer, 1993) in their design. Mayer (1993) further emphasized on the advantages of composites that the design engineers should consider in their design such as textured surface, self-coloring, part integrations, economy of scale, molding direct to final dimensions, efficient use of materials, durability, and lifetime costing attractive. The customer/client's need is to develop an orthopedic device to help paraplegic patients to walk. The goal is to design a light weight hip cradle. The objectives of the new product should

- be low cost
- have attractive surface finish
- provide increased stiffness to allow the device to be used also for heavier patients
- be light weight for reducing effort in walking
- be aesthetically pleasing

- not break that can cause injury
- be able to prevent progressive collapse

In the new design, the design constraints are to avoid heavy structure and if the structure is caused to collapse, say by a fall, it must not break which might cause another injury. Based on the needs, goals, objectives, and constraints, an extensive PDS can be prepared and concept generation and other downstream activities can be performed.

Another example of problem formulation in composite product development is in the development of photo frames from natural fiber composites. Top management of a university approached a group of designers to develop frames from natural fiber composites to be sold to the graduates during their convocation ceremony. The university top management felt that awarding contract to outside companies to supply and sell frames made from wood (Fig. 5.9) plastics, or metals for degree certificates and photographs was to their disadvantage as the university staffs in some selected schools have the capability to provide the same service.

The need of this project is to support the "green campus" policy, and to provide income for poor students the university has considered the staff to be involved in developing "green" products. The goal is to utilize expertise within the university to develop natural fiber composite frame; the objectives are as follows:

- low cost
- light weight
- aesthetically pleasing
- "green material"
- sufficient strength
- sufficient stiffness
- the frame can be used to display certificates and photographs.

The constraints are the products to be produced in house facilities such as injection molding, twin screw extruder, and hot press, and the task to be performed during summer vacation by a group of students from low income family, and the maximum quantity is 5000 units per year. Based on the aforementioned problem formulation, the design engineers are expected to perform design task and finally fabricate the frames with the help of the students. The materials being considered are locally available natural fibers, such as kenaf and oil palm fibers and thermoplastic polymers like high density polyethylene (HDPE).

Product Design Specification

For composites, Mayer (1993) listed few principal design considerations such as new or improved product, functional requirements, product cost, product life, material type, shape, colors and texture, operating environment, performance, environmental impact, and production volume and these design considerations can be included in the PDS. Some authors considered PDS as

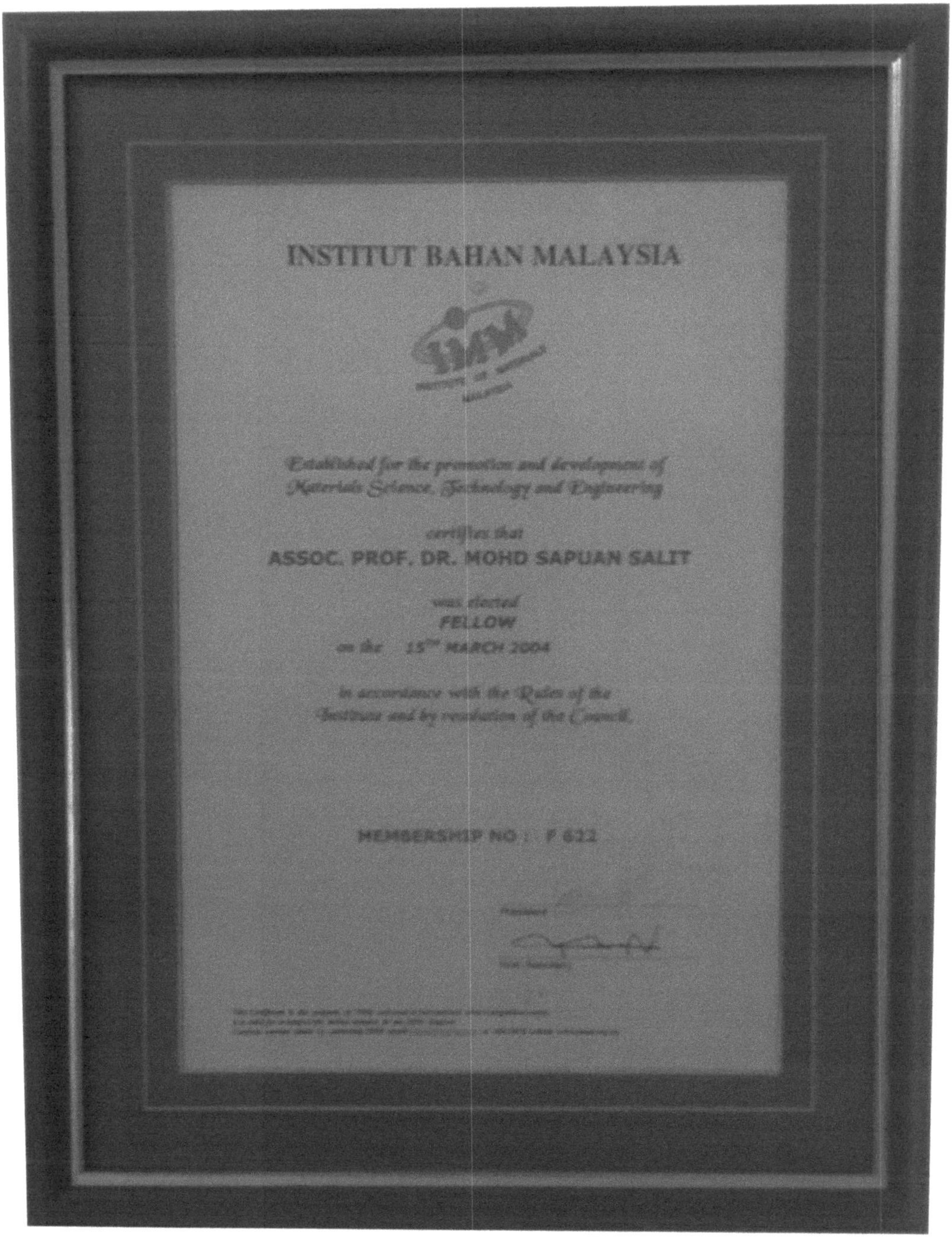

FIGURE 5.9 A wooden frame to be replaced with natural fiber composite frame.

Table 5.1 Elements of PDS

1.	Performance
2.	Environment
3.	Life in service
4.	Maintenance
5.	Target product cost
6.	Competition
7.	Packing
8.	Shipping/transport
9.	Quantity
10.	Manufacturing facility
11.	Size
12.	Weight
13.	Aesthetics
14.	Materials
15.	Product life span
16.	Standards/specifications
17.	Ergonomics
18.	Customer
19.	Quality and reliability
20.	Shelf life storage
21.	Processes
22.	Time-scale
23.	Testing
24.	Safety
25.	Company constraints
26.	Market constraints
27.	Patents
28.	Political and social implications
29.	Disposal
30.	Installation
31.	Documentation
32.	Legal

Adopted from Pugh, S, 1991. Total Design: Integrated Methods for Successful Product Engineering. Addison Wesley Publishers Ltd, Wokingham, England.

separate activity after market investigation and determination of customer requirements (Pugh, 1991; Wright, 1998) but authors like Dieter and Schmidt (2009) considered PDS as one of the activities in problem formation and definition. Pugh (1991) listed 32 elements of PDS as shown in Table 5.1. However, many of the elements of PDS may not be applicable to composite products.

The following are some important attributes of a PDS (Pahl et al., 2007; Pugh, 1991; Wright, 1998):

- PDS is a dynamic document and its contents can be changed at any stage of design process.

- PDS acts as a control document, which can be referred by relevant parties at any stage of design process.
- Although Pugh (1991) proposed that 32 elements of PDS should be considered in the design, taking selected important elements of PDS is necessary and it is termed partial PDS. Each product will have its unique set of PDS elements, that is elements of PDS selected for a chair should be different from an automotive component.
- In the author's opinion, design engineers are free to add other elements that may not be covered by Pugh (1991) especially in the design for composites. Issues like tailormade design, the manner the fibers and matrices are incorporated (Mayer, 1993) to form composites (1) by having separate fiber and matrices added together by the user, (2) semicured composites (also called intermediate materials or molding compounds), and (3) finished composite products like pultruded rods or filament wound tubes, the use of closed or open molds, and any other issues that are only applicable to composites.
- Design engineers should provide range of values when specifying the quantitative data like cost or mechanical properties as precise values are not normally practical.
- Some experts suggested that "Demand" and "Wish" should be included in the elements of PDS to ensure which elements are compulsory to be satisfied or just a matter of option (Wright, 1998).
- The contents of PDS should represent what to be achieved and should not report what had been achieved.
- PDS should be written clearly, can be understood, and it is actually a user document.
- PDS should be prepared in point form and not in the form of an essay for easy referral.
- Each time an amendment is made on PDS, it should be well documented and the date of the changes made should also be recorded.

Two PDS documents related to the design of products from composites are reviewed in this subsection. The first PDS is concerned with the design of glass–sugar palm fiber reinforced unsaturated polyester hybrid composite boat (Misri, 2011) and the PDS of this product is shown in Appendix A. The second PDS is related to the design of glass–kenaf fiber reinforced polypropylene hybrid composite automotive hand brake lever (Mansor, 2014) and it is presented in Appendix B. It should be noted in both documents, only partial PDSs were prepared for the obvious reason that only selected elements of PDS are applicable to respective components (boat and automotive component). For the former, PDS was prepared in a most straightforward manner as the product is meant to be used in small rivers. In this case, safety issue related to operating in the river was considered. In the

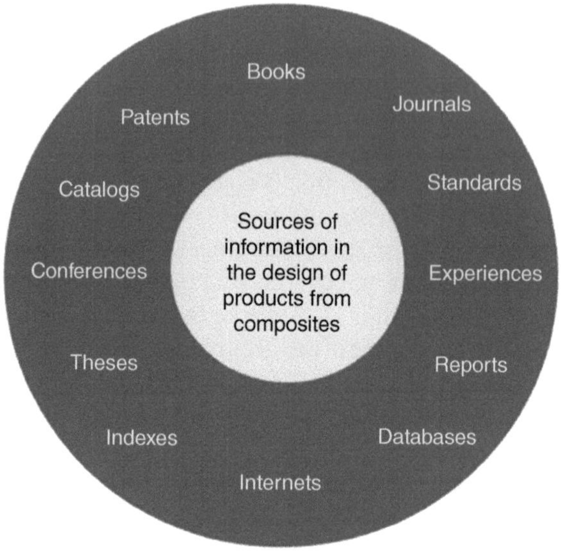

FIGURE 5.10 Sources of information in the design of products from composites.

latter, as far as possible, the information of the existing component is included in the PDS. Many of the elements of the PDS were done by referring to some relevant safety standard.

Information Gathering

Actually, information gathering is not really a concept generation method but it is rather an important activity during conceptual design stage. Dieter and Schmidt (2009) listed information gathering as an important activity in conceptual design along with problem definition, concept generation, and concept evaluation. Among the activities that were included in information gathering in steps of design process of Dieter and Schmidt (2009) are Internet, patents, technical articles, trade journals, and consultants. However, in total design model by Pugh (1991), information gathering is not included in conceptual design but it is an activity during market investigation stage. In composite conceptual design, the information required are normally from patents, databases, brochures, Internet sources, text books (Sapuan, 1998a) in engineering design and design with composites. Fig. 5.10 provides the sources of information in the design of products from composites and they are discussed in the subsections.

Books

Design engineer should explore any research progress in both product area and analogous areas that are related to his product area. Books are good sources of references and he should search product information in related area and in our case in the design of products from composites, and should refresh fundamental knowledge as well through textbooks. Books on research and state-of-the-art reviews in the area of design of composite products gave exhaustive and detailed coverage of the topic. It could be in the form of authored or edited books. Other types of books such as design handbooks, materials handbooks, composite handbooks, data books, yearbooks, encyclopedia, and dictionaries in the area of design for composite products can also be consulted appropriately (Anthony, 1986).

Composite Material Databases

There has been a huge rise in the amount of data and published information about design in general, and in the design with composites, in particular. Databases have now become important tool for design engineer. Data are known facts used as a basis for reasoning, and a data bank is a store of data. A database is a structured set of data held in computer (Maier, 1997). Computer helped in making rapid advances in the improvement of materials' property data (Rumble, 1986). Materials' selection is one of the important activities in product design. Therefore, design requires advanced tools to help design engineers to select the materials. Design engineers could use available material databases such as AGATE, CMH-17, NCAMP, IDES Composites, MaterialUniverse (Marsden and Warde, 2016), PLASFIND, CAMPUS, FUNDUS, CES, PERITUS, WIAM, Knovel Polymer Matrix Composites Database, Helius Composite, the Hub, NIMS Materials Database, CompoSIDE CMDB, MatWeb, MatDat, Prospector:Composites and Worldwide Composites Search Engine (WWC Search Engine) and majority of these databases are related to composite materials and they are useful information in the design of composite products.

Indexes and Abstracts

Indexes and abstracts stored in computerized data searches and modern information retrieval systems can be found in many large libraries and in the recent years are available online. The systems maintain databases and the information can be used by users to find the contents. They guide users to the contents of peer-reviewed journals, books, articles, reports, conferences, and theses. The systems contain subject keywords and references that have information regarding where the publications can be found and provide abstracts of these items that are obtained from publishers and other organizations. Most of these systems enable users to view or purchase the full-text materials by visiting the publisher's website. Some examples of these indexes include Current Technology Index, Science Citation Index, Material Science Citation Index, INSPEC, Index to Thesis, SCOPUS, Science Citation Index Expanded (SciSearch), Journal Citation Reports/Science Edition, Chemical Abstracts Service (CAS), Google Scholar, CSA, Academic OneFile, EI-Compendex, Expanded Academic, OCLC, Polymer Library, SCImago, Summon by ProQuest Technical Index, and New Scientist Index.

Journal Papers

Technical journals normally report new research findings on certain topics and reviews of some important topics in the areas. High impact journals are normally indexed and abstracted in well-known databases such as Science Citation Index by Thomson Reuters, and SCOPUS and journals are ranked according to their positions in quartiles, which can be found in databases such as Journal Citation Reports also by Thomson Reuters. In 2016, Composites Science and Technology, published by Elsevier is ranked the first in the area of composites. In this journal and other related journals, a huge amount of information related to the design of composites can be found, and the sources are considered reliable as the papers are thoroughly reviewed by anonymous experts in the relevant areas.

Conference/Seminar Papers and Proceedings

Two different sources of design information can be obtained under this subsection. The first is conference or seminar presentations and the second is conference proceedings. Although both are associated with conference the latter is considered to be more reliable. In general, papers presented in seminar or conference without being published in proceedings are not vetted but most of the papers presented in conferences and published in proceedings are reviewed but sometimes it may not be necessarily the case.

Technical Reports

Technical reports form very important documents during the conceptual design of composite products. Many technical reports had been released to the public and made available online and they contain very valuable information for the designers. However, in some cases the designers had to contact the respective companies and agencies in the cases when the reports are not available on the webs. Some reports contain confidential information, which can be accessed by the staff of the companies or the agencies alone.

Standards

The purposes of a standard are to enhance quality of products and services in terms of safety, reliability, efficiency, and compatibility. It also enables the products or services being offered to be cost-effective and sustainable. The most important attribute of the standard is to ensure that all products or services provided by the manufacturers and service providers are consistent irrespective of who makes the products and where they are made. For instance, in composite design, a designer engineer should design his products by following standards such as ASTM International, British Standards Institute, (BS), International Organization for Standardization (ISO), European Committee for Standardization (EN), International Electrotechnical Commission (IEC) in China, and TAPPI Standards. A certain standard is only meant for specific purposes or experiments, for example, ASTM D5467 is the standard for the in-plane compressive properties of polymer matrix composite materials reinforced by high-modulus fibers in a sandwich beam configuration.

Intellectual Property Rights (Patent and Industrial Design)

A patent, in the context of design of a composite product, is an exclusive right for the product or invention, which technically solves the design problem. Patent document is generally assigned with a patent number and information such as the technical field, background of the invention, summary of the invention, brief description of the drawing, references and claims made for the invention are included in the document. When a design engineer designs a product, he should perform patent search in his product area and analogous product areas so that he is aware of any legislation and byelaws that may affect the product. At the same time, studying the existing patents enables design engineer to generate ideas in designing the product. A design often focuses on aesthetic value of a product, in the form of 3D representation like shape and configuration or 2D representation like ornamentation and pattern and this type of invention is called industrial design. A design engineer should be aware of any industrial design related to his design and avoid infringing them.

Company Brochures and Product Catalogs

Company brochures and catalogs obtained from other companies are important sources of information for design engineers to help them in generating ideas and to know about their competitors. Nowadays, company brochures and product catalogs are made available in a computerized form, and are published online. Trade literatures like trade magazines are available in hard copy or electronically and they provide good information on competitor's products and the design engineer should use these types of information in developing his product.

Internet Sources

The Internet or the World Wide Web literally linking millions of computers around the world currently constitutes the global phenomenon. There are so many sites on the Internet, which provide useful information on the design of a composite product. However, they have to be filtered thoroughly as not all of them come from reliable and reputable sources (Sapuan et al., 2001). Without any doubt, this source of information of design of composite products is the most powerful, important, and the most encompassing because almost all of the aforementioned sources of information can be found in Internet. They include conference proceedings, journal papers, e-books, research reports, company brochures and pamphlets, trade literature, competitors' products, standards, and patents. Some of the information can be obtained at no cost while majority of them required substantial or minimum amount of subscription fees.

Theses

Theses (and dissertations) of Master and PhD degrees related to research in the design of composite products are another good source of information in the design of composite products. Examples of the theses related to the design of composite products were written by Sobey (2010), Mansor (2014), Hambali (2009), Davoodi (2012), and Sapuan (1998b). Even the final reports of the final year projects of bachelor degree students can provide a lot of information of the design of composite products. These theses can be borrowed from respective university libraries or can be borrowed using interlibrary loan facility. Some databases like Index to Thesis provide information related to the theses written by different scholars.

Information Gathering Based on Expert Knowledge and Experiences

Design information from design experts based on their expertise and experiences can be very valuable information in design. Design engineers dealing with products from composites have knowledge that sometimes cannot be found in open literature and yet it is the deciding factor in the development of a product (Fig. 5.11). They can be reached through personal communication. For instance, designers with experience in design of pultruded kenaf composites (biocomposites) (Fig. 5.12) are in the best position to provide the detailed process description and very limited information on the design and fabrication process can be found in the open literature.

Concept Generation Methods for Composites

Wright (1998) refers to concept generation stage as "initiation of concept solutions to the design problem" and this stage involved the use of concept generation tools and techniques. It should be noted that majority of the methods are not developed to cater for concept generation of composite products. It is the author's initiatives in this chapter to formalize the existing methods in different domains to be used in composites. This is done by initially introducing the methods, and wherever possible to discuss brief history or origin of the methods and trying to discuss them in the contexts of composites and to present some examples of the development that the author and his team had involved and from other relevant studies by different authors.

Design concepts need to be expressed in graphical, diagrammatical, or modeling form. Software can be used, especially 3D images, or the design engineers can use conventional sketches. As design concepts are normally shared with others, it is important for the design engineer to have sufficient clarity so that his ideas can be understood by others. Table 5.2 shows the methods of concept generation discussed in detail in this chapter. The table also provides information on the inventor, date of invention, and the reference sources.

FIGURE 5.11 Consultation with experienced personnel help solving design problem.

Method 1: Brainstorming

Brainstorming was developed by Alex F. Osborne in the 1950s, wherein the participants use their brain to storm a problem or to generate ideas. The aim of brainstorming is to spontaneously stimulate the ideas by generating as many potential solutions as possible quickly (Osborne, 1953). It is possibly the most established and widely used idea generation technique, and it is applied in various different fields. Wright (1998) discussed two different techniques of brainstorming, that is, the controlled input method and the 6.3.5 method. The two important elements in brainstorming are suspending judgment until ideas are exhausted and generation of multiple concepts. The procedures to conduct a brainstorming session are setting up a multidisciplinary group, no criticisms are allowed and to avoid nonhierarchical structure, and election of a leader or coordinator to facilitate the session, provide suitable working environment, and to propose the topic to be brainstorming and it should focus on specific

FIGURE 5.12 Pultrusion process for kenaf fiber reinforced composite product.
Courtesy Dr Mohd Fairuz Abdul Manab, University College Linton, Malaysia.

product context. Each member of the group will propose solutions (can be multiple solutions) and the ideas are consolidated (Wright, 1998).

Brainstorming for Idea Generation of a Natural Fiber Composite Product

Brainstorming was used in the development of a product from composites at Universiti Putra Malaysia (UPM). Eight experts with different backgrounds such as mechanical engineering, chemical engineering, material engineering, agricultural science, petroleum engineering, manufacturing engineering, and design engineering sat down to generate a list of solutions during brainstorming session (Fig. 5.13). All the participants have postgraduate qualifications

Table 5.2 Inventors of Concept Generation Methods

Methods	Introduced by	Date	Reference
Brainstorming	Osborne	1950s	Osborne (1953)
Biomimetics	Otto Schmitt	1950s	Vincent et al. (2006)
Cross-industry innovation	NA	NA	Vullings and Heleven (2015)
Analysis of existing system	Pahl and Beitz	1984	Pahl et al. (2007)
Why? Why? Why?	NA	NA	Cross (1994)
Gallery method	Hellfritz	1978	Pahl et al. (2007)
Morphological chart	Zwicky	1948	Zwicky (1948)
Blue ocean strategy	Kim and Mauborgne	2005	Kim and Mauborgne (2005)
Mind mapping	Buzan	1970s	Buzan (2005)
TRIZ	Genrich Altshuller	1940s	Altshuller (1998)

FIGURE 5.13 Brainstorming session to generate solutions for natural fiber composite flower vase.

either at master or PhD in natural fiber composites. The task given to them was to design a flower vase to be fabricated from natural fiber reinforced polymer composites. The vase should be unique and be fabricated from locally available natural fibers such as oil palm, and kenaf as reinforcements for composites. In this exercise, a coordinator was elected to facilitate the session and he initially presents the design brief to the group members (Fig. 5.14). Another member of the team was assigned to write all the suggestions made by each member of the team and ideas were consolidated and important and practical ideas are further strengthened as potential candidate solutions.

The results of the brainstorming are listed subsequently.

Problem: to design a flower vase from natural fiber composites.

Light weight product
"Green" product
Low cost
Variety of shapes
Unbreakable
Strong
Biodegradable
Aesthetically pleasing
Easy to handle

FIGURE 5.14 The coordinator presents the design brief to the grosp members.

Recyclable
Stable base
Easy to fabricate
Good modulus of elasticity
Ease of storage
Vase for souvenir
Can produce many units
What fibers to use?
Can resist humid environment
Biopolymer composites

After the concepts are generated, the design teams should further combine and refine the design concepts and suggest only few selected concepts (say four concepts) to be further developed and finally the best concept will be selected.

Method 2: Biomimicry/Biomimetics (Analysis of Natural System)

Analogical thinking is also known as synectics if it is done formally and synectics is normally carried out in a group in the same way as brainstorming is conducted (Cross, 1994). In this chapter, synectics (analogy), it is divided into two major classifications, that is biomimetics and cross-industry innovation.

Biomimetic or biomimicry can be categorized as a direct analogy method by seeking biological solution to a similar problem. It is also known as an analysis of natural system (Pahl et al., 2007) and it involved an innovation inspired by and imitating the nature. Biomimetics was coined by Otto Schmitt in 1950s (Vincent et al., 2006) and since then, the topic had increasingly been studied by materials scientists and it is also known as bionics. The concept of biomimicry in design can be traced from the time of Leonardo da Vinci, where his design of flying contraption was based on the wings of a bat (Benyus, 1997). Although the terms biomimicry and biomimetics are used interchangeably as they are considered synonymous, Pawlyn (2011) provides a slight difference in the meaning of biomimicry and biomimetics and he further stated *"biomimicry is specifically focused on developing sustainable solutions while biomimetics can be applied to fields of endeavor such as military technology."*

Helms et al. (2009) used biomimetic principles in product design and they proposed steps in product development based on this method, that is, problem formulation, reframing the problem, searching for biomimetic solutions in design, defining the solutions, and incorporating the idea found from biomimetic concepts into the actual design concept.

Biomimetics in the Design of Composite Products

In composite design, biomimetics or biomimicry can be a powerful innovation technique to generate design concepts. Easterling (1990) reported that the idea of the design of composite sandwich panels is mimicked from the scanning electron micrograph of a leaf of a lily tree as shown in Fig. 5.15. Pahl et al. (2007) reported the design work that mimicked natural principle in products like honeycomb structures in sandwich composite construction used in aircraft components (Fig. 5.16).

Pawlyn (2011) studied the concept of biomimicry in the field of architecture and he presented a work on expedition boat made from polymer composites. The boat hulls were made from large bundles of bottles and the frame for the bottle was made from a sheet material known as self-reinforced polyethylene terephthalate (PET), where PET acted as both fiber and matrix (composites). The biomimetic concept applied in this design was related to the idea of pressurizing bottles with air by mimicking the pomegranates, which had a number of individual segments packed together tightly inside a strong skin.

Masselter and Speck (2011) carried out extensive research on biomimetics and reported research work on composites made from braiding process to make branched braiding composites. Braiding is a composite manufacturing technique for automatically interlacing fibers into complex shapes to provide high strength three-dimensional reinforcement. The concept of branched composites mimicks the branch of some plants like cactus tree. The works of Burns

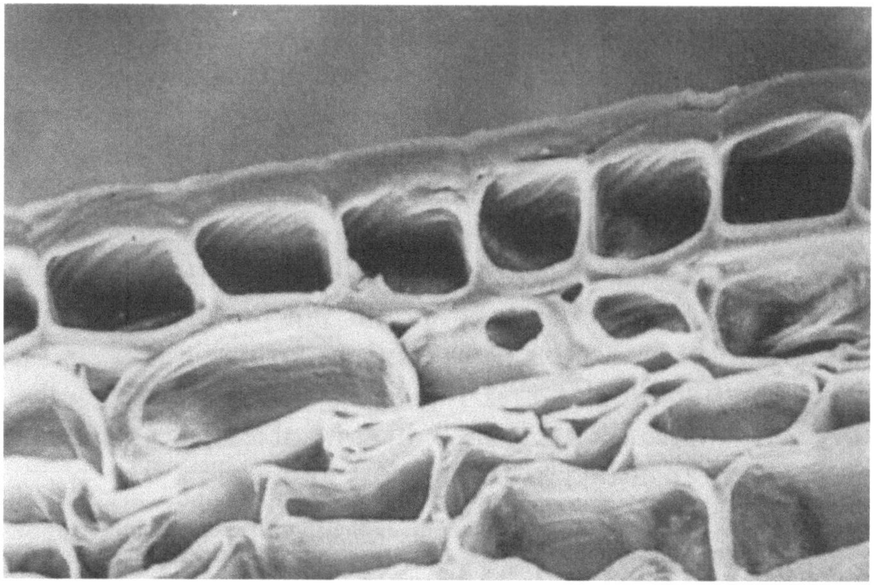

FIGURE 5.15 Scanning electron micrograph of the cross-section of the leaf of a lily tree (Easterling, 1990).

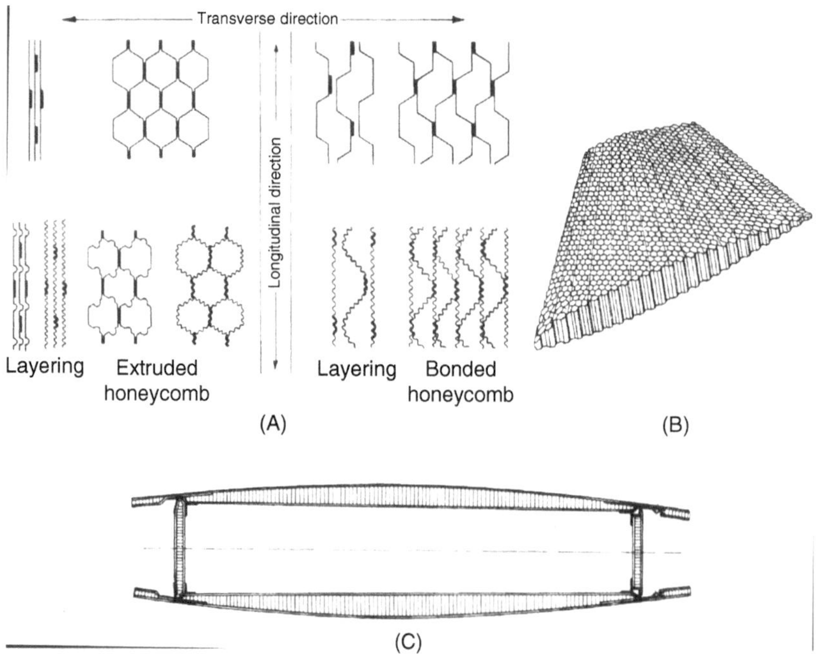

FIGURE 5.16 Honeycomb structures in sandwich composite construction in aircraft components (Pahl et al., 2007).

et al. (2012a,b) were related to joining strength improvement of aerospace structural composite components through the application of bio-inspired T-joint adapted from process of adaptive growth of wood branch–trunk joint. Li et al. (1995) carried out a developmental research on a double helical structural design using biomimetic principle. The design was inspired by bamboo bast fiber structure and this led to significant improvement in the compressive strength (15% improvement compared to original structures) of composite laminates in structural components.

Method 3: Cross-Industry Innovation

Cross-industry innovation is an analogy method that systematically transfers the idea, solution, or technique from outside one's industry, domain, sector, or field. Examples of cross-industry innovation include copying the idea of the baggage carousel system in the airports for transferring food for the customers in Shushi bar and copying the concept of airplane's retractable landing gear in the development of foldable lightweight baby pram (Vullings and Heleven, 2015).

Sapuan (1994) applied cross-industry innovation in the design of a stacking system for straw bales. One of the concepts that he developed, the method of "jack block" in building construction was imitated (Anon, 1962). In this system, the entire construction was carried out at or close to the ground level. The top floor of the building was jacked up to allow the level underneath to be constructed below. Then this floor was jacked again and so on.

Cross-Industry Innovation in Design of Composite Products

In composite product development, the following are examples of cross-industry innovation (Table 5.3). In the author's opinion, many terms used in composite industry are actually borrowed from other established industries. However, since some of the terms are already well accepted among the composite community, it is very difficult to claim that those terms are not originally meant for composites. In the limited knowledge that the author has, it is believed the terms presented in the left column of Table 5.3 are actually adopted from different industries given in the right column of the table.

A sandwich is a food item comprising one or more types of food, like egg, tuna, meat, cheese, vegetables, and pickles that are placed between slices of bread (Fig. 5.17A). In composite industry, sandwich structures are developed to take the idea of making sandwich in food industry. Sandwich construction (Fig. 5.17B) is described by Dorworth et al. (2009) in the following manner: *"sandwich panel laminates are designed with an integral lightweight core material to provide web space between laminate skins, increasing the stiffness and strength-to-weight ratio of the panel."*

In Fig. 5.17B, sandwich structure comprises two thin skins separated by thick, lightweight core material. Sandwich construction demonstrates lightweight

Table 5.3 Cross Industry Innovation for Composites

Composite Terms	Original Industry
Sandwich structure	Food industry
Tailor made	Clothing industry
Plain weave	Textile industry
Satin weave	Textile industry
Twill weave	Textile industry
Needle punching	Textile industry
Stitching	Textile industry
Weaving	Textile industry
Knitting	Textile industry
Needling	Textile industry
Yarn	Textile industry
Crimp	Textile industry
Blow molding—thermoplastics	Glass industry

and yet high strength and stiffness. The core can be made from balsa wood, foam cores, and honeycomb cores. Sandwich structures are normally used in advanced structures like composite kayak canoes and interiors of private and business jet aircraft (Dorworth et al., 2009).

Tailor is an individual whose job is making clothes such as suits, shirts, pants, and jackets based on the requirement of the customers after the size of the final products is measured (Fig. 5.18A). For composites an adjective "tailor" is used to emphasize how the materials are worked in the manner same as how a tailor works on his cloth. He has complete freedom to work with cloth and making it into final dress depends on his skill. To emphasize on the skill of the operator, and the similarity of the process, working with composites is compared with working with cloth. However, it is only true mainly for selected processes, which require high skill operator to fabricate the final composite

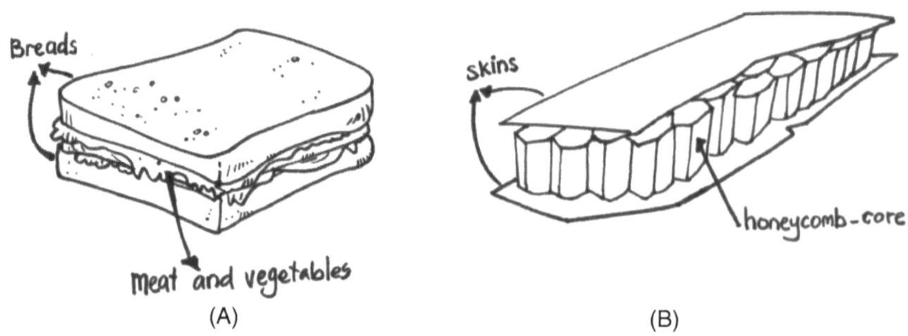

FIGURE 5.17 (A) Sandwich (B) sandwich construction.

FIGURE 5.18 (A) A tailor (B) a tailor-made process (hand lay-up).

products such as hand lay-up, spray up, vacuum bagging, and autoclave molding. As shown in Fig. 5.18B, a skillful operator fabricated a small composite boat using hand lay-up process. The boat is tailormade according to the need of the client or customer and the operator designs his material and product in a manner similar to how a tailor designs the cloth to produce a finished item. A mold is required to produce the boat and initially he applied mold release agent on the surface of the mold to facilitate easy detachment of the boat from the mold surface. Then, he used a gel coat made from polymer to ensure smooth and strong surface of the product, and it is also pigmented to provide desired color without going through the painting process. He works on the materials using a roller to consolidate the composite layers to the required thickness by alternately placing fibers and resin on the mold surface. For the highly automated processes like injection molding, and compression molding, tailormade processing may not be so relevant.

The terms used in fiber reinforcement technology, such as plain weave, satin weave, twill weave, yarn, crimp, needle punching, stitching, knitting, weaving, and needling are mainly taken from textile industry counterpart. Due to the similarity in the terms used, people working with textile industry often confused whether the terms used in fiber technology are referred to their area of expertise.

Enkel and Gassmann (2010) reported work on cross-industry innovation for replacing steel cable in elevator with the composite of aramid fiber/carbon fiber rope. The application of aramid rope in mountain climbing activity (and other nonelevator applications) is adopted in elevator manufacturing company (Schindler Group). Similar work reported by Enkel and Gassmann (2010) based on cross-industry innovation is in the development of frames of chaise longue bicycles from carbon fiber composites as found in Formula One monocoques to replace currently heavy material, that is, steel.

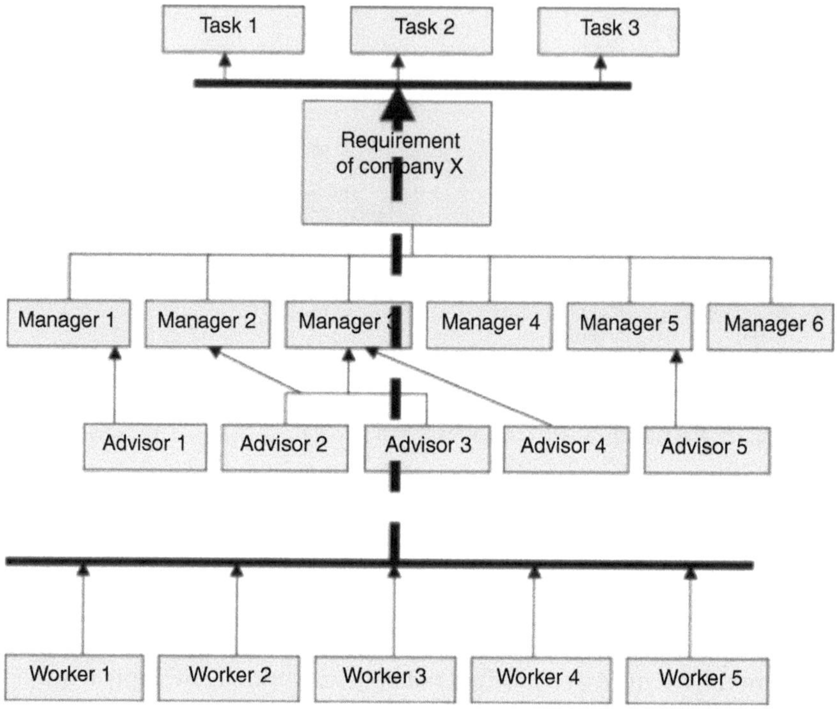

FIGURE 5.19 The management style of company "X" in executing three tasks (Sapuan et al., 2005).

Sapuan et al. (2005) involved in the design of automotive bumper system from composites. They used total design model to design automotive bumper fascia. Selected design methods such as morphological chart, mind mapping, and analogy were used in conceptual design. Analogy method based on cross-industry innovation was used to select the most suitable material for composite bumper system. The management method of a company "X" was studied (Fig. 5.19) and its management style is imitated for the materials selection (Fig. 5.20) of composite bumper system.

Method 4: Analysis of Existing Technical Systems

Pahl et al. (2007) introduced a method for idea generation called analysis of existing technical systems. It involves dissection of existing product and acquiring subfunctions from the current arrangement. After some analysis, results can be identified from these subfunctions and potential solutions can be suggested. Existing technical system includes competitors' products, earlier version of products from within the organization or, similar products with similar subfunctions. The solutions can be obtained by systematically exploiting proven ideas, or of experience.

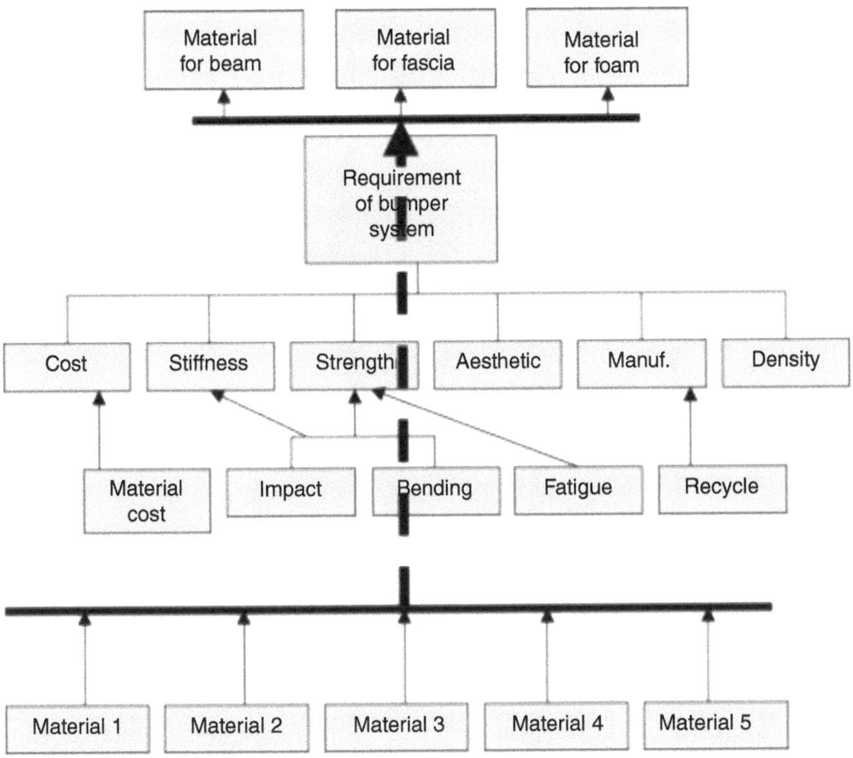

FIGURE 5.20 Cross-industry innovation: materials selection for composite bumper system (Sapuan et al., 2005).

Analysis of Existing Technical Systems in the Design of Composite Products

A new composite product can be developed based on the existing products, which may be in the forms of products or methods of fabrication from the competitors, earlier version of products and similar products (with some of their parts being same as those with new products to be developed) (Pahl et al., 2007). The following example on the development of a composite bumper component (Davoodi et al., 2008) can illustrate this idea.

The works of Neopolen® (2006) and the AISI (2004) were used as guides with some modification in Neopolen® (2006) work. In Neopolen® (2006) a series of absorbers made from expanded polypropylene (EPP) was placed in between bumper fascia and reinforcing beam. The concept of torsion bar (AISI, 2004) was adapted in the design of composite bumper along with concept from the work of Neopolen® (2006). By combining these two concepts developed by AISI (2004) and Neopolen® (2006), an elliptical energy absorber was proposed (Fig. 5.21). After performing some experimental work and design calculation, the new design

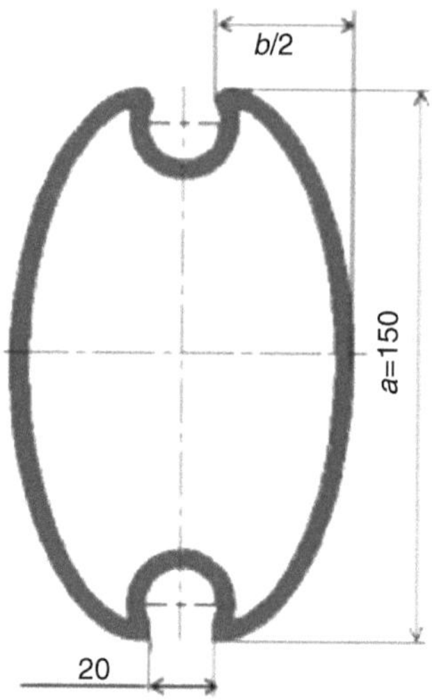

FIGURE 5.21 Elliptical energy absorber concept; unit in mm (Davoodi et al., 2008).

of bumper beam required 11 energy absorbers to satisfy the energy of pedestrian leg form. The final design of composite bumper beam is shown in Fig. 5.22 developed based on the method called analysis of existing technical systems.

Analysis of existing technical systems led to another new development of composite product. By analyzing existing steel drain cover (Fig. 5.23), with the drawbacks that it posed, a new drain cover was developed from composites (Fig. 5.24). The problems with the existing design of steel drain cover include

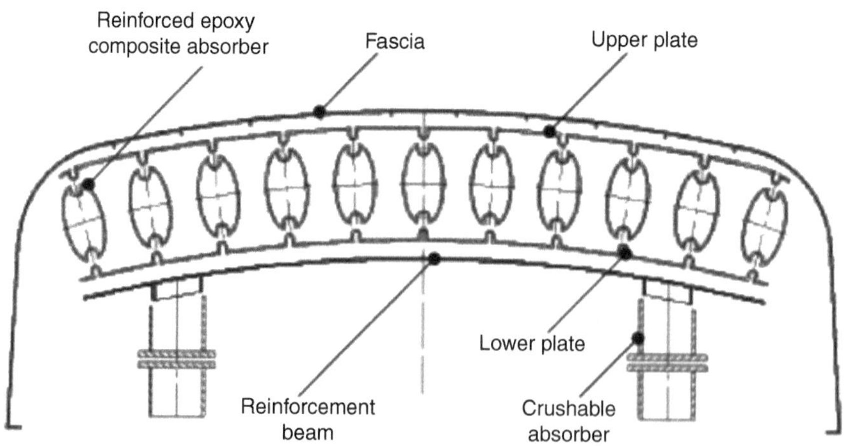

FIGURE 5.22 Schematic view of the new concept of bumper absorber (Davoodi et al., 2008).

FIGURE 5.23 Existing steel drain cover.

FIGURE 5.24 New composite drain cover.
Courtesy Rasidin Senawi, Innovative Pultrusion, Sdn. Bhd., Seramban, Malaysia.

problem with theft, heavy, and not aesthetically pleasing. According to Ashby and Johnson (2002), design is adaptive; taking existing concept and seeking incremental advance in performance through refinement. The major refinement made in this design is on the use of different and more appropriate material (composites). The new composite drain cover includes less attractive to thieves due to unfamiliar nature of the materials, more aesthetically pleasing, lower cost, and lighter weight. In addition, the strength of composite drain cover is sufficient for the intended use.

Method 5: Why? Why? Why?

Why? Why? Why? is one of the methods of enlarging or extending the search space. It is done by asking a question and then once the answer is found, and the next question is asked like a persistent child (Fig. 5.25) until no more questions are asked or unexpected answer is found (Cross, 1994). The process is stopped as currently no more questions can be asked. Fig. 5.26 shows the use of Why?Why?Why? method in the development of graphane-based composites.

For a design engineer working in the design team to develop automotive components from composite materials such as pedals, he may ask questions like

FIGURE 5.25 A persistent child with inquisitive mind.

Why do we need to design a transparent graphane polyurethane composite body armor?

Why do we need to use graphane as fiber?

Why do we need to invest so much money on this material?

Why do we need to use polyurethane as matrix?

Why can't we use Kevlar as fiber?

Why the current testing and fabrication methods for this product are still unavailable?

Why don't buy the existing product with this material?

Why do we need to use transparent graphane?

Why other researchers have not venture into this technology?

Why do we need this product at all?

FIGURE 5.26 An exercise of the use of Why?Why?Why? in development of graphane based composites.

"why do we need a clutch pedal?" "Why can't bracket be eliminated?" "Why do we need clutch pedal from composites." Each answer must be followed up, with another "Why" until a conclusion is reached or an unexpected answer prompts an idea for a solution.

Method 6: Gallery Method

The gallery method is an idea generation method carried out by a group of design engineers by displaying a number of concepts simultaneously. The group is arranged in the same manner as brainstorming and a coordinator is elected to lead the session. Generally, there are no formal formats or rules to be followed. The coordinator initially introduces the problem to be solved to the group members. In the idea generation stage 1, each group member works toward solving the design problem and in this stage, the individual contribution can be assessed (Pahl et al., 2007). The individual design concept should be developed in the forms of diagrams (sketches) or a mixture of sketches and texts and it should preferably be presented one concept per sheet of paper.

The design concepts generated by individual members of the team are wall mounted by means of pins or tapes. Thereafter, all members look at the design at the same time, discuss, and move past the wall mounted concepts, as in an art gallery. This stage is called association of ideas and this is from where the name of the method is originated. In this stage, the member can compare the design and can produce hybrid design solution (Wright, 1998). Then the designer engineers sat down again to generate combined concept and finally the selection of the most suitable concept is made. In this method, individual and group efforts can be merged and the most appropriate solution is achieved (Ulrich and Eppinger, 2004).

Gallery Method in Composite Product Development

A gallery method was used to generate idea for a composite product, namely ceiling fan blades to be made from natural fiber composites. Initially, the coordinator introduced the problem to be solved to the group. Then, each member was given 15–20 min to generate the concept individually (Fig. 5.27A). The association of idea stage was then performed by preparing the concept in the form of text and sketches on a piece of paper. The concepts were displayed by mounting them on the wall using suitable tapes. In this gallery method, the designers walked around, looked closely at the concepts, and discussed the concepts (Fig. 5.27B and Fig. 5.28). They sat down again, but this time in a collective manner to refine the concepts by looking at the strengths and weaknesses of each concept to get the best combination of concepts (Fig. 5.27C). Finally the most promising idea was selected (Fig. 5.27D). The sketch and text of the final concept is shown in Fig. 5.29. It comprised a ceiling fan with three blades, to be made from locally available natural fibers like kenaf or oil palm fibers, with color being "woody" color, and for making prototype, thermoset materials were fabricated using hand

(A) (B) (C) (D)

FIGURE 5.27 Gallery method in the design of ceiling fan blades from natural fiber composites (A) individual concept generation (B) problem association in a "gallery" (C) combined concepts are developed, and (D) the most suitable concept is selected.

FIGURE 5.28 A gallery method: problem association for composite fan blades.

lay-up process but for mass production, thermoplastic materials should be used as the matrices and to be fabricated using injection or compression molding process.

Method 7: Morphological Chart Method

Morphology is a field of study related to structure of form of things. Morphological chart is also known as morphology chart or morphology box and its origin can be traced from an early work of Zwicky (1948) and it is originated from morphological analysis. A morphological analysis is a systematic attempt to analyze the form that a product might take and a morphological chart is a summary of this analysis. The purpose of this method is to generate different arrangements and to enable design engineer to choose new combinations of elements. Morphological chart method is an idea generation method employing a chart to consider all possible subsolutions for various functions of a product being developed. This chart is a grid of empty squares. Then, subsolutions are combined into different complete solutions. In this method, the design engineer considers all of different ways that can meet the various function requirements of the problem (Wright, 1998).

This method is appropriate for products with large functional requirements or features. All the possible alternatives are considered and by doing this, the

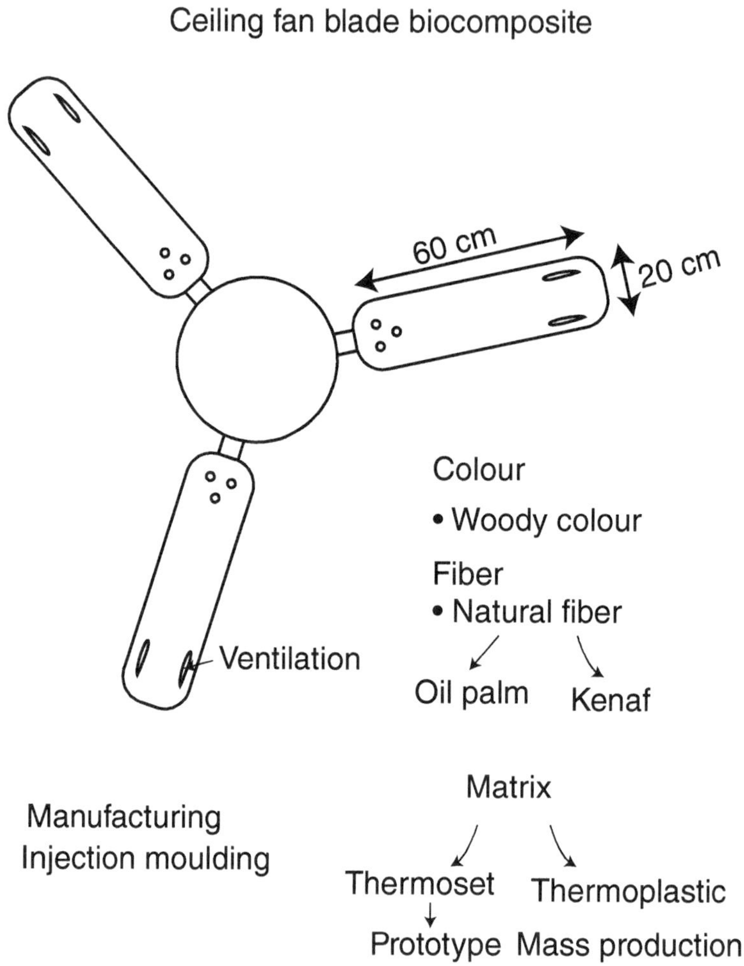

FIGURE 5.29 A concept of natural fiber composite ceiling fan blades developed from gallery method.

tendency of overlooking particular concepts is avoided. The solutions and the subfunctions can be prepared in the form of sketch or pictorially, in a written description or the mixture of them in the chart. Generally, the several subfunctions of a product are listed on the left-hand side. Then for each row of the chart possible solutions to achieve these subfunctions are provided, one solution per grid. Each solution is a stand-alone solution and it has connection or link with solution in the next grids. We may have, for instance, three means of achieving subfunction 1, four solutions of achieving subfunction 2, and so on. Then, feasible combinations of subsolutions can be identified (Cross, 1994; Dieter and Schmidt, 2009; Pahl et al., 2007).

Morphological Chart in Design of Composite Products

Sapuan (2006) presented the use of morphological chart method for the design of polymer composite automotive pedals. He developed four charts for each component of automotive pedals namely mounting bracket, accelerator, clutch, and brake pedals from composites. After the morphological chart was developed, the best combination of all subsolutions is decided and the concept was further refined toward achieving the design solution (Sapuan, 2005).

Fig. 5.30 shows the morphological chart for the composite boat. It has five solution options to its six subfunctions listed on the left column of the figure from where the solutions are generated. This concept of the boat can be applied on river, so it needs the capability of four structural ribs. The safety for this boat requires three buoys. The fabrication process is hand lay-up and the number of passengers on the boat is 4. The glass–sugar palm fiber reinforced unsaturated polyester hybrid composite boat is shown in Fig. 5.31.

Sub function	Solution				
	1	2	3	4	5
Capability on boat structure	no rib	2 ribs	3 ribs	4 ribs	5 ribs
Shape of boat	simple	medium	complex		
Boat fabrication process	hand lay-up	resin transfer moulding (RTM)	resin infusion moulding		
Number of passengers	1	2	3	4	5
Safety	1 buoy	2 buoys	3 buoys	4 buoys	5 buoys
Boat application	lake and small river	river	Sea		

FIGURE 5.30 Morphological chart for a glass/sugar palm reinforced unsaturated polyester hybrid composite small boat (Misri et al., 2014).

FIGURE 5.31 Glass–sugar palm fiber reinforced unsaturated polyester hybrid composite boat.

Researchers have made some modifications on the actual morphological chart by integrating this method with other methods such as TRIZ method as reported by Mansor et al. (2014). TRIZ is used as idea generation and morphological chart as idea refinement. Based on the combination of solutions in morphological chart and TRIZ, the concepts are further developed. In the traditional morphological chart, subfunctions, are normally placed on the left column of the chart, but in this integrated method, subfunctions are developed as a result of TRIZ solution principles and design strategy. For instance, for subfunction number of sections, it is not a stand-alone subsection but it has been derived from TRIZ design solution #1, that is segmentation (Fig. 5.32). In fact, under

TRIZ solution principles and design strategy	Design features	Solutions				
		1	2	3	4	5
#1. Segmentation i) Product the component in different sections and the sections should be asymmetric as possible to simplify and ease the design and manufacturing process ii) Joining the sections using pin-and-boss method for easy assembly and disassembly process	**Cross-section profile**	Double-T	Square	U-type		
	No. of sections	2	3	4		
	Sectioning type	Symmetrical	Non-symmetrical			
	Part section assembly method	Screw-boss	Circular pin-boss	Square pin-boss	Adhesive	
#3. Local quality i) Vary the thickness of the component according to the stress concentration value. Thicker component at higher stress location points. ii) Brake lever body casing designed with ribs to reinforce and strengthened the structure as well as with pin-and-boss features to provide quick and easy assembly method. Joint different functioning feature together to the same component	**Body type**	Solid	Shell			
	Ribbing pattern	I	V	X	I+V	I+X

Note: *Example for Concept Design 1 = Square + 2-sections + symmetrical + circular pin-boss + shell + I-rib*

FIGURE 5.32 Integrated morphological chart-TRIZ method in the design of –glass kenaf fiber reinforced polypropylene hybrid composites (Mansor et al., 2014).

FIGURE 5.33 Blue ocean strategy (*left*) and red ocean strategy (*right*).

the TRIZ design solution #1 (segmentation), four subfunctions have been generated, that is, section profile, number of sections, sectioning type, and part section assembly method.

Method 8: Blue Ocean Strategy

Kim and Mauborgne (2005) developed a systematic approach in business called blue ocean strategy (BOS), with the major aim being to make the competition irrelevant by creating uncontested market space and placing so much emphasis on value innovation. BOS (Fig. 5.33, left) was introduced to counter the widely known business practice called red oceans (Fig. 5.33, right), which focuses on the existing market share. The market share is always very competitive and overcrowded, which may appear to be *"cutthroat competition leading to the red ocean bloody"* (Kim and Mauborgne, 2005). Many tools within BOS practice such as the strategy canvas (value curve) and the four actions framework (the eliminate–reduce–raise–create grid) are very valuable tools that help achieve the business success. However, the implementation should be extended beyond business and in this chapter BOS is implemented in the design of composite products.

Concept Generation Using BOS Tools for the Design of Composite Product

1. A four actions framework
   Engineering Composites Research Group at UPM utilizes the advantages and effectiveness of BOS to help in generating ideas during conceptual design stage of a hybrid natural fiber composite automotive component (Sapuan et al., 2016). Two major tools within the BOS are mainly used and they are the four actions framework and strategy canvas. The product under consideration is an automotive antiroll bar (Fig. 5.34). The aim is to replace the use of an existing metal automotive antiroll bar with hybrid composites.

FIGURE 5.34 Basic shape of an automotive antiroll bar.
Courtesy Mastura Mohamad Taha, UPM, Malaysia.

The first tool of BOS to generate the design concepts is the four actions framework. In this proposed approach, four actions framework of BOS, which is referred to as "eliminate–reduce–raise–create grid" is used for concept generation. The four actions framework of BOS (Kim and Mauborgne, 2005) for conceptual design application is described as follows:

- Eliminate—which part of the design that is dysfunctional and should be eliminated?
- Raise—which parts should be raised well above the design standard?
- Reduce—which parts should be reduced well below the design standard?
- Create—which parts should be created that the design has never featured?

In Table 5.4, based on the four actions framework of BOS, the specific solution strategy for the design concept of automotive antiroll bar (ARB) is presented. From the specific solution strategy that is generated in Table 5.4, modification of the conventional automotive ARB is performed to satisfy the constraints from the material properties where the number of fiber orientations could be raised for better mechanical

Table 5.4 BOS Four Actions Framework

Eliminate	Critical point that gained the highest stress concentration should be eliminated
Raise	The number of fiber orientations could be increased for better mechanical properties of the composite materials
Reduce	Cross-section of the ARB could be in a hollow shape to reduce the weight while maintaining the performance of the component
Create	The bar could be designed with reinforced feature to strengthen the structure

Courtesy Mastura Mohamad Taha, UPM, Malaysia.

properties. From the framework, the eliminated action suggested to eliminate the stress concentration point based on the failure mode evaluation. Generally, the design concepts of hybridized natural fiber composite automotive ARB could be comparable with the conventional steel automotive ARB as the design considered all the technical requirements as well as nontechnical requirements.

2. Strategy canvas

   Another method in BOS that is used in the development of product from composites is strategy canvas. The strategy canvas of ARB that is based on steel (S), carbon fiber reinforced composite (CFRC) (C), and natural fiber reinforced composite (NFRC) (N). The design requirements of automotive ARB are shown in Table 5.5. The results of the comparison (strategy canvas) are shown in Fig. 5.35.

   The score is assigned for each material based on how well it satisfies the design requirements. For instance, for the best achievement, the score of 5 is assigned, for the worst achievement, it is assigned 1, and for the intermediate performance it is given 3.

   From Fig. 5.35, it is clearly shown that NFRC has better performance compared to CFRC and steel. NFRC has the lowest raw material cost, the most readily available and the lowest low price compared to CFRC and steel. Cost is the main setback of using CFRC in automotive components to replace steel. The main reason for selecting NFRC is its environmentally friendliness, and this is where it got the highest score

Table 5.5 Design Requirements of ARB and Their Abbreviations

Design Requirement	Abbreviation
Price	P
Easy to reuse	ERU
Easy to recycle	ERC
Less transportation	LT
Easy to manufacture	EMF
Durability	D
Lightweight	LW
Easy to maintain	EMT
Reliability	R
Long lifetime	LL
Endure to impact	EI
Not easy to break	NB
Free from hazardous materials	FH
Less material	LM
Environmental safety	ES

Courtesy of Mastura Mohamad Taha, UPM, Malaysia.

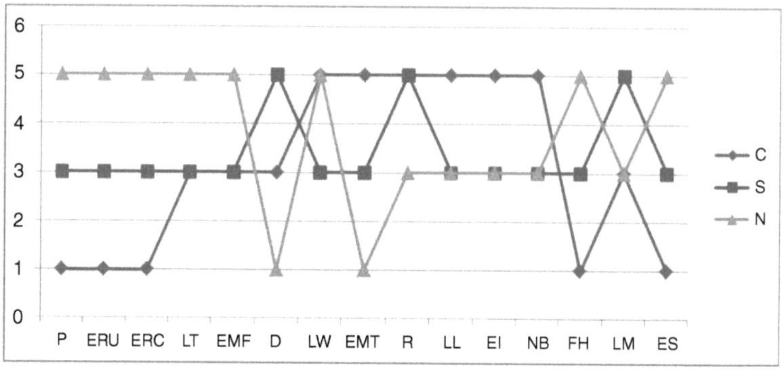

FIGURE 5.35 Strategy canvas of automotive antiroll bar.
Courtesy Mastura Mohamad Taha, UPM, Malaysia.

in the strategy canvas. NFRC is known to be bio-degradable, recyclable, renewable, and consumes low amount of energy consumption. With all these merits, NFRC can be an excellent contender in replacing traditional materials in an automobile.

3. National Blue Ocean Strategy (NBOS)
   National Blue Ocean Strategy (NBOS) is an approach in programs and services to public implemented nationwide by Malaysian government and sufficient funding was allocated to carry out the programs or to execute the services to public (NBOS, 2006). The programs or services to be delivered to the public should have high impact, low cost, and should be rapidly executed.

A research project funded by NBOS initiative was carried out to develop products from sugar palm (*Arenga pinnata Wurmb Merr.*) tree. The objectives of the project include implementing the concept of design for sustainability, proper utilization of agricultural waste, knowledge transfer from university to publics, offering employment and business opportunities, improving socio-economic level of low income group living in rural areas and involvement of younger generation in product development.

The sugar palm trees were selected in the project as they were not seriously being used to make products except for making sugar block in the past. The trees are rich with sources of fibers and starches and their utilization as commercialized products is still minimal. The concept adapted is to design and develop new materials for natural fiber composites (biopolymer and natural fibers) to produce biodegradable polymer composites to be used in biodegradable plastic bags and disposable food containers. Fig. 5.36 shows the utilization of sugar palm fibers for simple applications like making roof, ropes, brushes, and brooms. Figs. 5.37 and 5.38 show the sugar palm fibers and sugar palm starch ready for commercialization as raw materials for natural fiber composites.

FIGURE 5.36 Utilization of sugar palm fibers into various products.

The project was completed within 1 year, and the project team managed to train younger generation to start small business, to help low income group living in rural areas to sell their products, and to transfer knowledge from the university to the general public on how to make products from sugar palm tree. In addition, this project offered some solutions in managing agricultural wastes like fibers, where normally these fibers were not fully utilized and disposed. Through this project, the concept of design for sustainability (DFS) was implemented. All three national agendas are fulfiled such as project has very high impact especially to the low income group living in the rural areas, low cost of implementation as all technologies adopted are appropriate, and the raw materials are obtained free of charge and it was executed fairly rapidly, that is within 1 year. As far as the principle of BOS is concerned, the project has been able to create uncontested market as the group (Sanyang et al. 2016) had the expertise and experience in dealing with extracting, utilizing, designing, researching, and commercializing the sugar palm fibers, sugar palm starches, and sugar palm–based biopolymer composites.

Method 9: Mind Mapping

Mind map or mind mapping is one of the established idea or concept generation techniques and it was coined by Tony Buzan in the 1970s (Buzan, 2002).

FIGURE 5.37 Sugar palm fibers used as reinforcement in biopolymer composite ready to be commercialized.

It is also called concept map by Dieter and Schmidt (2009) and it is used for organizing information. Development of mind mapping requires certain steps to follow such as the use of color image and the need to produce branched curves (Buzan, 2005). In this method, the major tasks of a design engineer are to generate and to record the ideas. The method depends largely on the visual presentation of the concept. Each branch emanated from the central image is actually any activities or issues related to the design intent, that is, the main aim of the design. Any activities and issues related to branches can be written close

FIGURE 5.38 Sugar palm starch for biopolymer in biocomposite ready to be commercialized.

to these branches, and ideally the text written should be as brief as possible; perhaps one word is sufficient. Mind mapping or more accurately concept mapping is a means for rapidly externalizing the ideas generated in the brain. This method enables rapid and uncritical recording of the results of concept generation to be carried out. Dieter and Schmidt (2009) further stated that "*because it requires the mapping of associations between ideas it stimulates creative though*t". This method can be carried either in a group or individually. The design engineer should organize the map by connecting any related branches of activities or issues. Then the design engineer should tidy-up the results. Finally, the outcome of the concept generation session should be concluded by the preparation of a report.

Buzan (2005) listed steps in performing mind mapping task and they are as follows:

- Start in the center of a blank page
- Use image or picture for the central idea
- Use color throughout
- Connect the main branches to the central image and so on
- Make the branches curved and not straight-lined.

- Use one keyword per line
- Use images throughout

Some authors did not adhere strictly to the aforementioned steps, and yet they still call the activity as mind mapping (Wright, 1998) as the main issue that matters is the major framework of the method without entertaining so much detailed jargons. In author's opinion, the main issue is that the method should comprise topics or problems to be solved and it should be sketched or written at the center of the map, then writing down the subproblems and identifying issues or activities related to subtopics and representing them in brief by words or images.

Mind Mapping in Concept Generation of Composite Products

A mind mapping is prepared as a proposed procedure to perform conceptual design activities in developing products from composites (Fig. 5.39). The branched activities include problem formulation, information gathering, concept generation, concept evaluation, and interface with embodiment design. Except for interface with embodiment design, this mind mapping is actually guiding the readers to read this chapter in addition to providing information on procedures of how to design components from composites.

Sapuan et al. (2005) and Misri et al. (2014) attempted to use mind mapping in their research of composite bumper fascia and composite boat, respectively, but the effectiveness of method was not clearly evident as the maps were prepared in a most simplistic way. The work of Norianti (2016) seemed to be quite comprehensive and it is very clearly presented (Fig. 5.40). The central issue is

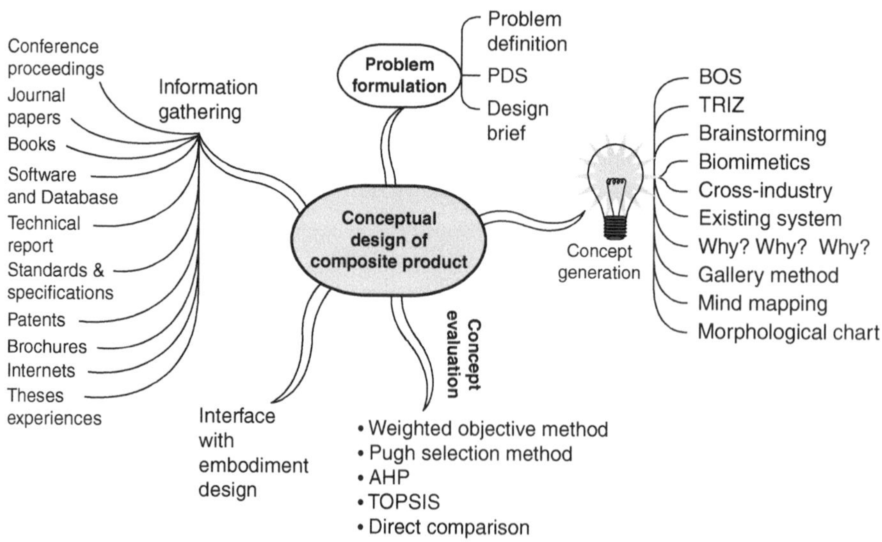

FIGURE 5.39 Mind mapping of conceptual design of composite product.

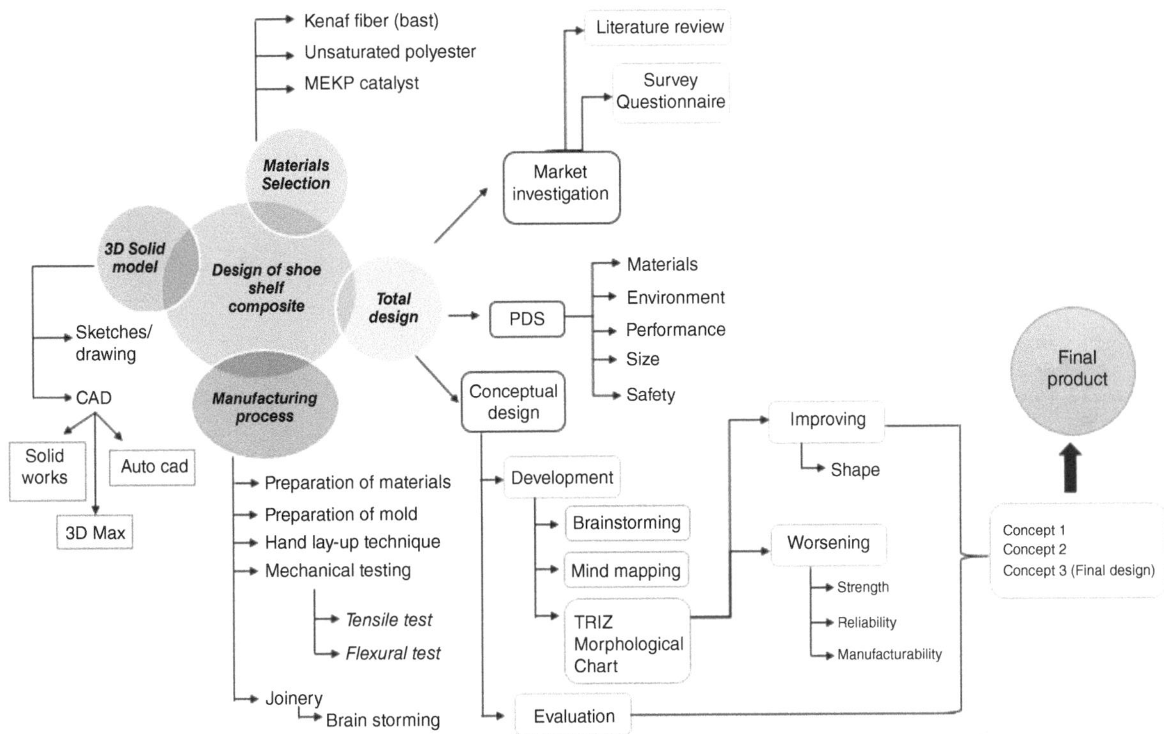

FIGURE 5.40 Concept mapping of design of kenaf fiber reinforced unsaturated polyester composite shoe shelf (Norianti, 2016).

the design of composite shoe shelf. The branches include materials selection, total design, manufacturing process, and 3D solid modeling. The branches are further subdivided into more specific issues or activities like under 3D solid model, there are conventional sketches and CAD model and CAD model is further divided into the actual software used such as 3D max, SolidWork, and CATIA. Fig. 5.41 shows kenaf fiber reinforced unsaturated polyester composite shoe shelf developed by Norianti (2016).

Method 10: TRIZ

A concept generation technique called TRIZ, an acronym for "Teoriya Resheniya Izobretatelskikh Zadatch" in Russian, which is translated in English as Theory of Inventive Problem Solving had attracted the attention of many designers in the recent years. TRIZ was founded by a Russian patent officer, Genrich Altshuller, in 1940s, where the method was developed through his early works on analyzing author certificates (in most of the countries they are known as patents). Out of 200,000 author certificates (patents) that he and his coworkers studied, they discovered majority of patents shared

FIGURE 5.41 Kenaf fiber reinforced unsaturated polyester composite shoe shelf (Norianti, 2016).

common design solutions (i.e., had similar group of solution principles) and these solutions were common even though the fields of studies were completely different.

The main aim of TRIZ is to design a product toward achieving ideality, by overcoming technical contradictions (TRIZ 39 system/engineering parameters) (Table 5.6) and by introducing inventive solutions synthesized from the patents and proposed in 40 inventive principles of TRIZ (Table 5.7), so that the design can be improved and the weakness in the design can be removed. TRIZ concept is different from the generally accepted design norm, that is, by allowing the trade-off or compromise of design requirements in solving a design problem (Altshuller, 1998).

TRIZ in the Design of Composite Products

In the area of composite materials, TRIZ complements composite materials very well as it is clearly stated in TRIZ 40 inventive principles; the last principle, that is, composite materials, is the solution proposed to solve the design problem (Sapuan and Mansor, 2016).

Table 5.6 A 39 System/Engineering Parameters Used in Explaining the Contradictions in TRIZ

#1:	Weight of moving object
#2:	Weight of stationary object
#3:	Length (angle) of moving object
#4:	Length or angle) of stationary object
#5:	Area of moving object
#6:	Area of stationary object
#7:	Volume of moving object
#8:	Volume of stationary object
#9:	Speed
#10:	Force
#11:	Stress/Pressure
#12:	Shape
#13:	Stability of object
#14:	Strength
#15:	Durability of moving object
#16:	Durability of stationary object
#17:	Temperature
#18:	Brightness
#19:	Energy spent by moving object
#20:	Energy spent by stationary object
#21:	Power
#22:	Waste of energy
#23:	Waste of substance
#24:	Loss of information
#25:	Waste of time
#26:	Amount of substance
#27:	Reliability
#28:	Accuracy of measurement
#29:	Accuracy of manufacturing
#30:	Harmful factors acting on object
#31:	Harmful side effects
#32:	Manufacturability
#33:	Convenience of use
#34:	Repairability
#35:	Adaptability
#36:	Complexity of device
#37:	Complexity of control
#38:	Level of automation
#39:	Productivity

Despite existence of the specific principle related to composites, the actual procedure to perform concept generation with TRIZ is currently presented. A case of the development of an automotive parking brake lever (Fig. 5.42) made from glass–kenaf fiber reinforced polypropylene hybrid composites is presented (Mansor et al., 2014).

Table 5.7 The 40 Inventive Principles of TRIZ to Solve the Problems of Technical Contradictions

#1:	Segmentation
#2:	Extraction
#3:	Local quality
#4:	Asymmetry
#5:	Combining
#6:	Universality
#7:	Nesting
#8:	Counterweight
#9:	Prior counteraction
#10:	Prior action
#11:	Cushion in advance
#12:	Equipotentiality
#13:	Inversion
#14:	Spheroidality—curvature
#15:	Dynamically
#16:	Partial or overdone action
#17:	Moving to a new dimension
#18:	Mechanical vibration
#19:	Periodic action
#20:	Continuity of useful action
#21:	Rushing through
#22:	Convert harm into benefit
#23:	Feedback
#24:	Mediator
#25:	Self-service
#26:	Copying
#27:	An inexpensive short-lived object instead of expensive durable one
#28:	Replacement of a mechanical system
#29:	Use of a pneumatic or hydraulic construction
#30:	Flexible film or thin membranes
#31:	Use of porous material
#32:	Change the color
#33:	Homogeneity
#34:	Rejecting and regenerating parts
#35:	Transformation of physical and chemical states of an object
#36:	Phase transition
#37:	Thermal expansion
#38:	Use strong oxidizers
#39:	Inert environment
#40:	Composite materials

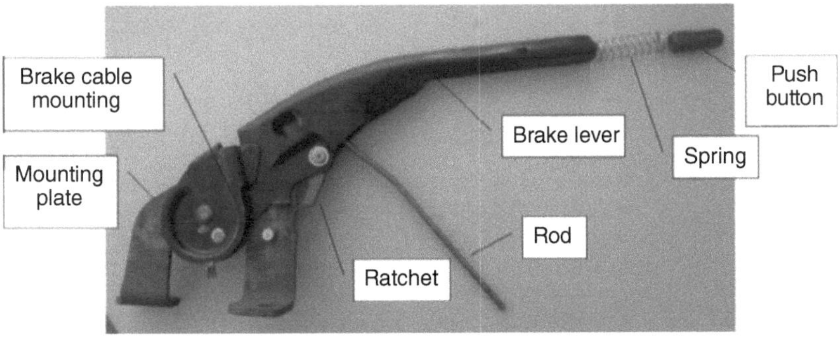

FIGURE 5.42 Parking brake lever component and its assembly (Mansor et al., 2014).

The first step in the design concept stage using TRIZ is to define the design problem and in this case, is to generate design concept of an automotive parking brake lever from composites to replace steel counterpart. Next, the design engineer needs to identify the improving and worsening parameters of the engineering system. Since replacing steel with composites results in weight reduction, it is considered as improving parameter. By doing this, it affects the component strength, reliability, and ease of manufacture and these three parameters are worsening parameters (Mansor et al., 2014). All the improving and worsening parameters (contradictions) can be found in the list of 39 engineering parameters shown in Table 5.6. The next step is to identify the appropriate solution principles using the contraction matrix between the improving and worsening parameters leading to recommendations of solution principles based on the TRIZ 40 inventive principles (Table 5.8).

Table 5.8 Contraction Matrix for the Kenaf Fiber Composites Parking Brake Lever Design (Mansor et al., 2014)

Improving Features	Worsening Features	TRIZ Solution Principles
39 Engineering parameters	39 Engineering parameters	40 Inventive principles
#2. Weight of moving object	#14. Strength	#28. Mechanics substitution #27. Cheap short-living objects #18. Mechanical vibration #40. Composite materials
	#27. Reliability	#1. Segmentation #3. Local quality #11. Beforehand cushioning #27. Cheap short-living objects
	#32. Ease of manufacture	#36. Phase transitions #1. Segmentation #28. Mechanics substitution #27. Cheap short-living objects

Table 5.9 Design Strategy Based on Identified TRIZ Solution Principles (Mansor et al., 2014)

TRIZ Solution Principles	Solution Descriptions	Design Strategy Descriptions
#1. Segmentation	1. Divide an object into independent parts 2. Make an object easy to disassemble 3. Increase the degree of fragmentation or segmentation	1. Product the component in different sections and the sections should be asymmetric as possible to simplify and ease the design and manufacturing process 2. Joining the sections using pin-and-boss method for easy assembly and disassembly process
#3. Local quality	1. Change an object's structure from uniform to nonuniform, change an external environment (or external influence) from uniform to nonuniform 2. Make each part of an object function in conditions most suitable for its operation	1. Vary the thickness of the component according to the stress concentration value. Thicker component at higher stress location points 2. Brake lever body casing designed with ribs to reinforce and strengthened the structure as well as with pin-and-boss features to provide quick and easy assembly method. Joint different functioning feature together to the same component
#40. Composite materials	1. Change from uniform to composite (multiple) materials	1. Use hybrid composition where kenaf fiber is combined with stronger and stiffer glass fiber to increase the composite strength and stiffness

Contradiction matrix recommends solution principles, and design engineer needs to look closely at the recommendations and he should decide whether or not to accept the recommendations made by the matrix. If he decides to accept the recommendations, he should choose only the most practical principles, or if he decides not to accept the recommendations, he can go through all the 40 principles again and choose the suitable principles. Once he has selected the appropriate principles, he is now in the position to generate design concepts. The aim should be to produce lighter weight part, and at the same time maintaining the strength, reliability, and ease of manufacture and he finally decides that the most relevant inventive principles are segmentation (#1), local quality (#3), and composite materials (#40) (Table 5.9).

Other Concept Generation Methods

There are many other concept generation methods developed by various experts that can be used during the design of components from composites. These methods include the 6-3-5/C-Sketch method (brainwriting),

SCAMPER checklist, extremes and inverses, problem decomposition, attribute listing, axiomatic design, requirement tree, failure mode and effects analysis, daydreaming, value engineering/value analysis, forced relationship, function analysis, inversion, collaboration, socializing, wishing, and accidental genius (Dieter and Schmidt, 2009, Martin, 2015; Pugh, 1991; Suh, 2001; Wright, 1998).

Concept Evaluation and Selection Methods for Composites

In engineering design concept evaluation is the process of comparing several design concepts and making decision which design concept is the most suitable for a product. According to French (1999) there is no specific step for design evaluation as evaluation is an ongoing and continuous activity throughout the design process. However, many experts (Cross, 1994; Dieter and Schmidt, 2009; Pugh, 1991; Ulrich and Eppinger, 2004; Wright, 1998) suggested that concept evaluation or selection is a dedicated activity to be performed by design engineers after all design concepts are developed.

Weighted Objectives Method for Design Evaluation for Composites

A weighted objectives method can be used to evaluate the best design concept. The assignment here is to evaluate the best concept of a composite boat. Five different concepts of composite boat are developed in this design. The evaluation of alternatives for the composite boat was carried out using the weighted objectives method. This method provides a means of assessing and comparing alternative designs, using different weighted objectives. This method assigns numerical weights to objectives, and numerical scores to the performances of alternative designs measured against these objectives (Cross, 1994).

The calculation of utility values for five different concepts of composite boat is shown in Fig. 5.43. Given that all five concepts are workable and fulfil the PDS, the following objectives are set:

- Low cost
- Light weight
- Manufacturability
- Material suitability
- Appearance

These are considered to have the respective relative weights of 0.2, 0.2, 0.25, 0.25, and 0.1. These figures were decided during brainstorming session carried out in the laboratory. All the concepts are assigned with values using nine-point scale. For each alternative, the utility score for each objective is multiplied by the objectives weight giving a relative utility value. If these values are added together an overall utility value for each alternative is obtained. As the values in Fig. 5.43 show, concept 3 was the best overall.

Objective	Weight	Concept-1		Concept-2		Concept-3		Concept-4		Concept-5	
		Score	Value	Score	Value	Score	Value	Score	Value	Score	Value
Low cost	0.2	5	1.00	8	1.60	7	1.40	5	1.00	7	1.40
Low weight	0.2	4	0.80	8	1.60	7	1.40	7	1.40	7	1.40
Manufacturability	0.25	6	1.50	7	1.75	8	2.00	5	1.25	8	2.00
Material realization	0.25	7	1.75	6	1.50	8	2.00	6	1.50	8	2.00
Appearance	0.1	8	0.80	6	0.60	8	0.80	8	0.80	6	0.60
Overall utility value	1.00	5.85		7.05		7.60		5.95		5.40	

FIGURE 5.43 Weighted objectives method for concept evaluation of hybrid composite boat (Misri, 2011).

Pugh Selection Method for Composites

Pugh selection method, coined by Pugh (1991) and found in Cross (1994), was a variation of weighted objectives method. It is internationally recognized as one of the important tools in CE. The main aim of the Pugh selection method is to evaluate the best design concepts after extensive concept generation process is carried out. This method uses an existing design concept as a reference called datum.

The Pugh selection method was successfully adopted in numerous design problems and it is equally suitable for design evaluation of composite products. This method was also used to select the best materials during materials selection process for composite materials as discussed in Chapter 6 on materials selection for composites.

Fig. 5.44 shows concept selection process employing Pugh method in the development of a garbage bin from composites. Six concepts were prepared, and concept 1 acts as a datum. The design of currently used product is chosen as "datum" and it is listed among other concepts. The design engineer compared each concept against the datum in fulfiling the requirements of the design. The legend "+" is used to indicate a concept is better than the datum in fulfiling a particular design requirement, "−" indicates worse than datum, and "S" is the same with datum. The concept, which scores the highest positive difference after adding "+" and "−" is selected as final design concept; in this case concept 6.

Concept						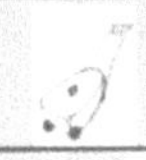
Criteria	DATUM					
Size		-	-	+	S	-
Weight		-	S	+	S	-
Safety		S	-	S	-	S
Material		S	S	S	S	S
Testing		S	S	S	S	S
Target cost		-	-	+	S	+
Manufacture and process		S	S	S	S	S
Competition		+	-	-	-	+
Life in service		S	S	S	S	S
Target market		-	-	+	S	+
Aesthetics, appearance and finish		S	-	-	S	S
Ergonomic		S	S	-	S	+
Customer		S	-	-	-	+
Results		4- 1+	7- 0+	4- 4+	3- 0+	3- 5+

FIGURE 5.44 Pugh selection method to select the best design concept for composite garbage bin.

Direct Comparison Method for Concept Design Selection of Product From Composites

After carrying out the concept evaluation based on weighted objectives method, Pugh selection method, or any other reliable method of design concept selection, there are occasions when the difference in the final score is considered very small (e.g., a total within 5%), it is recommended to perform a second evaluation. One way of doing this is by comparing two design concepts against each other in terms of all relevant criteria (Sapuan and Willmot, 1996). The design concept, which met the criteria better than the other concept, is given a value of 1 and the other is given a zero. All values are then summed up. The design concept with the highest final result is the selected concept. The evaluation criteria chosen here are mainly taken from the PDS. However, it might be necessary to reconsider dismissed concepts should the chosen one fails at a later stage. At UPM in the design of an automotive antiroll bar from hybrid composites, there was occasion after the evaluation was made, two concepts had very high scores compared to other concepts and the difference of the two scores was quite small and the direct replacement comparison method was considered as an option to solve the design problem.

Concept Design Selection Using AHP for Products From Composites

The analytical hierarchy process (AHP) is a system developed in the late 1970s (Saaty, 1980) to structure the experience, intuition, and heuristics-based decision making into a well-defined methodology based on mathematical principles (Bhusan and Rai, 2004). Hambali et al. (2009, 2011) performed design concept selection of composite automotive bumper beam using AHP.

A Case Study of Design Concept Selection Using AHP for Natural Fiber Composites Ironing Board

This section demonstrates the application of AHP, a multicriteria decision making method for performing concept design selection of daily consumer product. The product chosen was an ironing board, which is a very classical and indispensable tool to assist user during ironing of clothes. As shown in Fig. 5.45, the product consists of two main components, which are the top surface and the legs supporting it. The motivation is to replace the current ironing board top surface material with natural fiber composites, which could provide higher potential of cost saving and lightweight property without compromising the structural strength of the product.

After brainstorming session, three basic design concepts for the new product were produced which are solid-type ironing board (design concept 1)

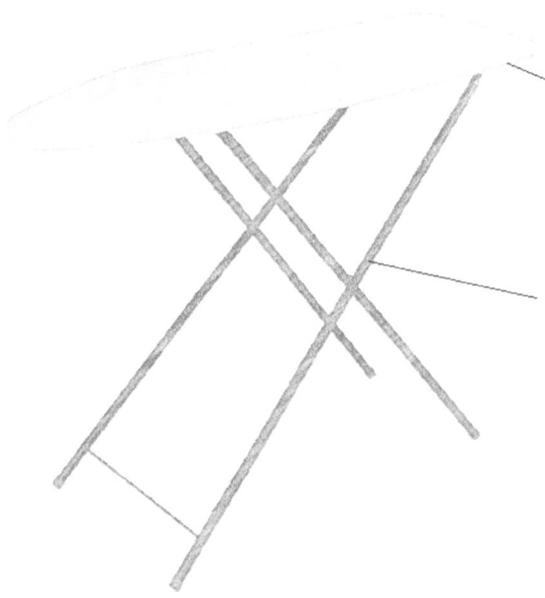

FIGURE 5.45 Example of an ironing board.
Courtesy Dr Muhd Ridzuan Mansor, UTeM, Melaka, Malaysia.

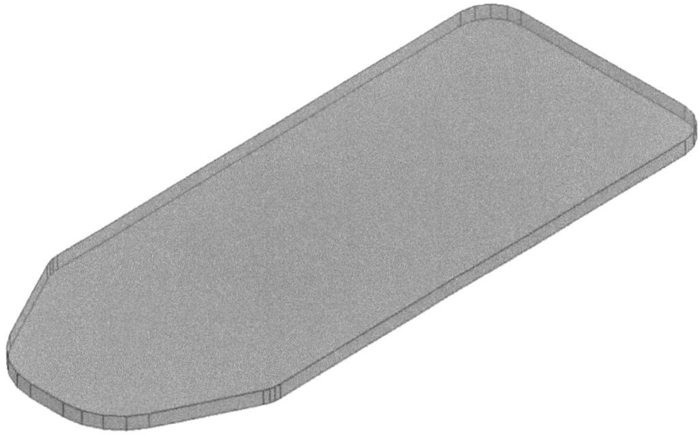

FIGURE 5.46 Design concept 1—solid ironing board.
Courtesy Dr Muhd Ridzuan Mansor, UTeM, Melaka, Malaysia.

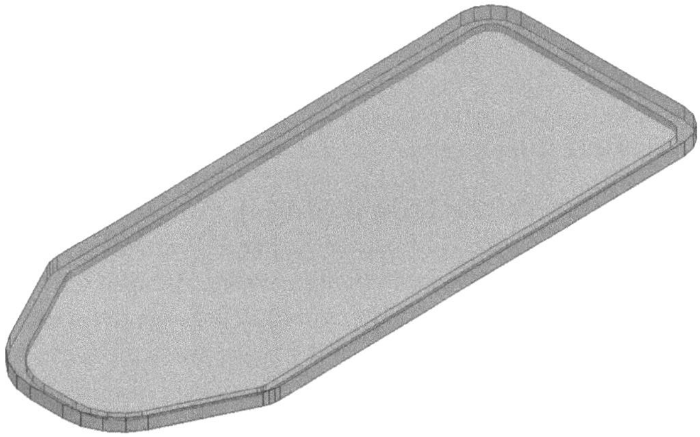

FIGURE 5.47 Design concept 2—pocketed ironing board.
Courtesy Dr Muhd Ridzuan Mansor, UTeM, Melaka, Malaysia.

(Fig. 5.46), pocketed-type ironing board (design concept 2), and finally pocketed-type ironing board with ribs (design concept 3). Design concept 2 ironing board (Fig. 5.47) was designed to reduce the volume of the material and weight of the product through the pocketed profile underneath the top surface, while design concept 3 (Fig. 5.48) was designed with added ribs to add further structural strength and stiffness to the product. All the design concepts were modeled using AutoCAD software in 3D to provide better visualization for decision makers in performing the selection process.

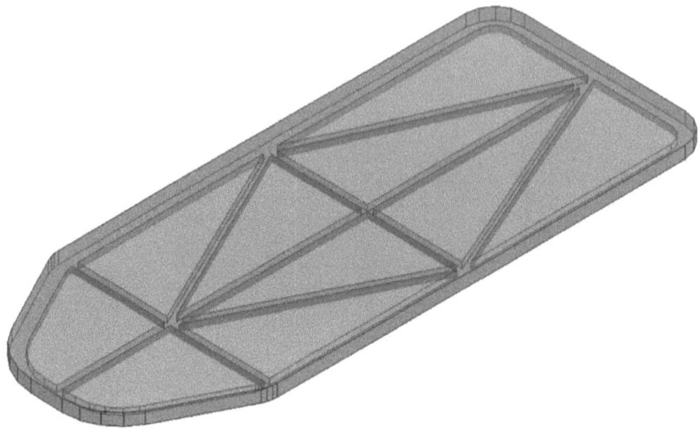

FIGURE 5.48 Design concept 3—pocketed ironing board with ribs.
Courtesy Dr Muhd Ridzuan Mansor, UTeM, Melaka, Malaysia.

In this case study, four selection criteria were chosen which are product weight, structural strength, ease of manufacture, and raw materials' cost. Each design concept was rated based on its performance using subjective judgments, such as high, medium, and low. Detail description of the selection criteria for the natural fiber composites ironing board design concept and their rating performances are shown in Table 5.10.

1. Constructing the AHP hierarchy framework
   The design concept selection process using AHP method begins with construction of the AHP hierarchy framework. As shown in Fig. 5.49, the basic hierarchy consists of three main levels, whereby the first level (level 1) corresponds to the goal of the project which in this case refers to the need to select the best natural fiber composites ironing board conceptual design. At the second level (level 2), all the selection criteria used for the selection process are listed which are the weight, strength, ease of manufacture, and raw materials cost. Finally, the hierarchy is ended with listing of all the alternatives (the concept designs which need to be selected) which is at level 3. The hierarchy can be expanded into more levels, if the selection process needs to be refined to included subcriteria. The hierarchy also showed how the judgment process flow is made, which started from level 2 (criteria) with respect to level 1 (selection goal), until ended at level 3 (alternatives) which is with respect to all of the selection criteria (level 2).
2. Performing judgments and determining the ranking of alternatives
   The AHP judgment method is initially performed using pair-wise comparison technique, whereby the decision is made by comparing between alternatives with respect to the same criteria. Similarly, the

Table 5.10 Ironing Board Selection Criteria and Design Concept Ratings

Selection Criteria	Selection Criteria Descriptions	Design Concept Rating		
		Design Concept 1	Design Concept 2	Design Concept 3
Weight	The weight of the board should be as light as possible to facilitate ease of carrying and handling by the user	High	Medium	Low
Strength	The strength of the board must be high to ensure it will not break and bend while load is applied during ironing process	Medium	Low	High
Ease of manufacture	The ease of manufacture is linked with the shape complexity, where simple shape will result in easier manufacturing process	High	Medium	Low
Raw materials cost	Low raw materials' cost is reflected by less volume of material needed to form the product	High	Low	Medium

Courtesy of Dr Muhd Ridzuan Mansor, UTeM, Melaka, Malaysia.

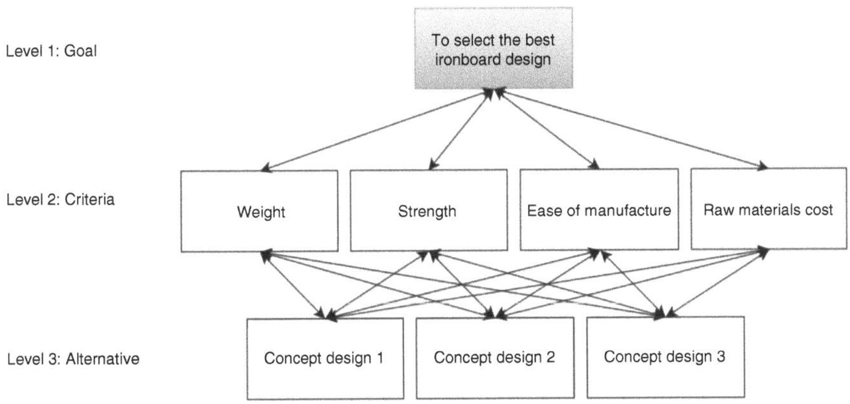

FIGURE 5.49 AHP hierarchy framework for natural fiber composites ironing board design concept selection.
Courtesy Dr Muhd Ridzuan Mansor, UTeM, Melaka, Malaysia.

pair-wise comparison method was also performed at the first stage of the selection process, to determine the weight of importance between each selection criterion with respect to the selection goal. In this case study, a computer-based AHP program called Expert Choice was used to record the judgments and later analyze the results. The use of such

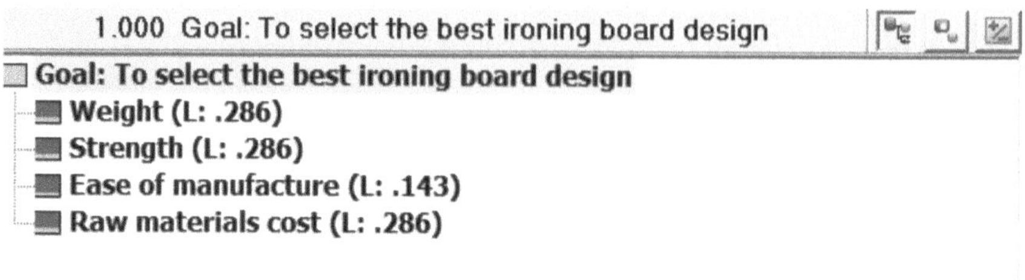

FIGURE 5.50 The weight importance of the selection criteria obtained using Expert Choice.
Courtesy Dr Muhd Ridzuan Mansor, UTeM, Melaka, Malaysia.

software enabled faster computational time and helped in analyzing performance with higher accuracy, which is very valuable when more complex decision hierarchy is involved during the decision making process. Based on the criteria weighing judgments performed by the designer as shown in Fig. 5.50, the criteria for product weight, strength, and raw materials cost resulted in equal scores between each other, while ease of manufacture scored lower importance to the final decision on the best concept design to be selected.

The next stage involved synthesizing the judgments to determine the rank of the alternatives with respect to the overall selection goal. Using the AHP method, the rank of each conceptual design is calculated based on the total eigenvector score for each alternative. As shown in Fig. 5.51, design concept 3 (ironing board with pocketed profile and ribs) scored the highest value of 0.417 (or 41.7%), followed by design concept 2 (score of 0.328 or 32.8%) and finally design concept 1 (score of 0.255 or 25.5%). This shows that design concept 3 is the best design concept suitable for the intended natural fiber composite ironing board application compared to other design alternatives. Alternatively, the results obtained also showed that design concept 3 is 1.27 times (0.417/0.328) more preferred to design concept 2, and 1.64 times (0.417/0.255) more preferred to design concept 1 based on similar selection criteria.

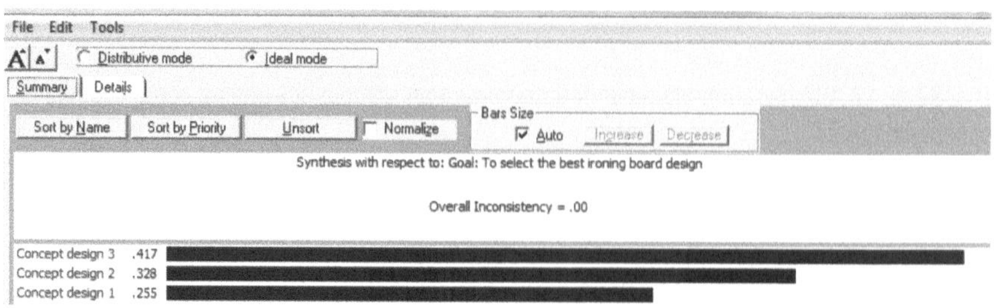

FIGURE 5.51 Final eigenvector score for the concept designs.
Courtesy Dr Muhd Ridzuan Mansor, UTeM, Melaka, Malaysia.

In addition to the overall ranking (based on eigenvector), one of the added advantages of using the AHP method for performing subjective judgments is its ability to determine the consistency of the judgments made. The capability of determining the judgment consistency value is performed by calculating consistency index, based on the eigenvalue score. Lower the consistency index obtained indicated that the judgments made are very consistent, thus ensuring high selection result accuracy and reliability, which is crucial for the successful decision making process. In AHP method, judgments which resulted in consistency index score lower the 0.1 (or 10%) are acceptable for consistent judgment results. Based on Fig. 5.51, the overall judgment consistency index score obtained in this case study was 0.0, which proved very high consistency judgments were made by the designers despite the subjective nature of the assessment rating used. The final outcome of the case study revealed that design concept 3 is the best design concept, which meets all the required selection criteria compared to other alternatives.

TOPSIS Method for Design Concept Selection of Products From Composites

TOPSIS is one of the methods used for multicriteria decision-making (MCDM). The algorithm used in TOPSIS method considers ideal and nonideal solution and helps design engineer to rank the option and to choose the best one. The detail description about this method can be found in specialized publication (Chen, 2000; Shanian and Savadogo, 2006) and it involves extensive mathematical formulations. Among the steps in TOPSIS method include determination of the normalized decision matrix, calculation of the weighted normalized decision matrix, calculation of the positive ideal and negative ideal solution, determination of the separation measures, using the n-dimensional Euclidean distance, determination of the relative clones to the ideal solution, and ranking of the preference order (Chen, 2000). Davoodi et al. (2011) used the TOPSIS method in selecting the best design concept for an automotive composite bumper beam.

CONCLUSIONS

In this chapter, a study of conceptual design of composite materials in CE is presented. A review of previous work on conceptual design in composite product development is reported. Methods used in conceptual design including preconcept generation stages, concept generation stage, and concept evaluation stage are discussed. Ten idea generation methods for composite product development are presented. Although in conceptual design for composites, many different approaches and emphases had been implemented by different researchers, this chapter only mainly focuses on design methods in conceptual design.

References

AISI, 2004. Technical Information. (American Iron and Steel Institute). Steeler Compatibility Cal Poly ME428—Senior Design. November 30.

Altshuller, G., 1998. 40 Principles: TRIZ Keys to Innovation, TRIZTools. vol. 1, Worchester, MA.

Amirruddin, A.K., 2007. Design, analysis and fabrication of a Hovercraft Hull Base prototype. Master of Science Thesis, Universiti Putra Malaysia.

Andreasen, M.M., Hansen, C.T., Cash, P., 2015. Conceptual Design Interpretations, Mindset and Models. Springer International Publishing, Cham, Switzerland.

Anon, 1962. Build the top floor first and jack it up. Engineering 193, 170–171.

Anthony, L.J., 1986. Information Sources in Engineering. Butterworth, London.

Ashby, M., Johnson, K., 2002. Materials & Design: The Art and Science of Materials Selection in Product Design. Elsevier Butterworth Heinemann, Amsterdam.

Barbero, E.J., 1999. Introduction to Composite Materials Design. Taylor & Francis, Inc., Philadelphia.

Barton, J.A., Love, D.M., Taylor, G.D., 2001. Design determines 70% of cost? a review of implications for design evaluation. J. Eng. Des. 12, 47–58.

Benyus, J.M., 1997. Biomimicry: Innovation Inspired by Nature. HarperCollins Publisher Inc., New York.

Bergers, R., 2016. Conceptual design of a composite main fitting of a cantilever type nose landing gear, Master of Science Thesis, Delft University of Technology, The Netherlands.

Bhusan, N., Rai, K., 2004. Strategic Decision Making Applying the Analytic Hierarchy Process. Springer, New York.

BioComposites Centre (BC), Bangor University, 2014. BioComposites Centre (BC), Bangor University, Bangor, UK. Available from: http://www.bc.bangor.ac.uk/products-and-services/the-bc-innovation-chain/index.php?menu=1&catid=11616&subid=11629

Boothroyd, G., 1988. Estimate costs at an early stage. Am. Mach. 132, 54–57.

Burns, L.A., Mouritz, A.P., Pook, D., Fieh, S., 2012a. Strength improvement to composite T-joints under bending through bio-inspired design. Compos. A 43, 1971–1980.

Burns, L.A., Mouritz, A.P., Pook, D., Fieh, S., 2012b. Bio-inspired design of aerospace composite joints for improved damage tolerance. Compos. Struct. 94, 995–1004.

Buzan, T., 2002. How to Mind Map®: The Ultimate Thinking Tool That Will Change Your Life. Thorsons, London.

Buzan, T., 2005. Mind Map® Handbook: The Ultimate Thinking Tool. Thorsons, London.

Chen, C., 2000. Extensions of the TOPSIS for group decision-making under fuzzy environment. Fuzzy Sets Syst. 114, 1–9.

Corbett, J., 1986. Design for economic manufacture. Ann. CIRP 35, 93–97.

Cross, N., 1994. Engineering Design: Strategies for Product Design, second ed. John Wiley & Sons, Chichester.

Davoodi, M.M., 2012. Development of thermoplastic toughened hybrid kenaf/glass fibre-reinforced epoxy composite for automotive bumper beam. PhD Thesis, Universiti Putra Malaysia.

Davoodi, M.M., Sapuan, S.M., Ahmad, D., Aidy, A., Khalina, A., Jonoobi, M., 2011. Concept selection of car bumper beam with developed hybrid bio-composite material. Mater. Des. 32, 4857–4865.

Davoodi, M.M., Sapuan, S.M., Robiah, Y., 2008. Conceptual design of a polymer composite automotive bumper energy absorber. Mater. Des. 29, 1447–1452.

Dieter, G.E., Schmidt, L.C., 2009. Engineering Design, fourth ed. McGraw-Hill, New York.

Dominy, J., 2000. Design concepts for continuous fibre composites. In: Owen, M.J., Middleton, V., Jones, I.A. (Eds.), Integrated Design and Manufacture using Fibre-Reinforced Polymer Composites. Woodhead Publishing Limited, Cambridge, pp. 233–259.

Dorworth, L.C., Gardiner, G.L., Mellema, G.M., 2009. Essentials of Advanced Composite Fabrication & Repair. Aviation Supplies & Academics, Inc, Newcastle, WA.

Easterling, K., 1990. Tomorrow's Materials, second ed. The Institute of Materials, London.

Enkel, E., Gassmann, O., 2010. Creative imitation: exploring the case of cross-industry innovation. R&D Manage. 40, 256–270.

French, M.J., 1999. Conceptual Design for Engineers, third ed. Springer-Verlag, Berlin, Heidelberg.

Hambali, A., 2009. Selection of conceptual design using analytical hierarchy process for automotive bumper beam under concurrent engineering environment. PhD Thesis, Universiti Putra Malaysia.

Hambali, A., Sapuan, S.M., Ismail, N., Nukman, Y., 2009. Application of analytical hierarchy process in the design concept selection of automotive composite bumper beam during the conceptual design stage. Sci. Res. Essays 4, 198–211.

Hambali, A., Sapuan, S.M., Tajul, A.A., Taufik, 2011. Development of conceptual design of automotive bumper beam at the conceptual design stage. J. Adv. Manuf. Technol. 5, 11–26.

Helms, M., Vattam, S.S., Goel, A.K., 2009. Biologically inspired design: process and products. Des. Stud. 30, 606–622.

Hsu, W., Woon, I.M.Y., 1998. Current research in the conceptual design of mechanical products. Comput.-Aid. Des. 30, 377–389.

Hyman, B., 1998. Fundamentals of Engineering Design, second ed. Pearson Education, Inc., Upper Saddle River, NJ.

Imihezri, S.S.S., Sapuan, S.M., Sulaiman, S., Hamdan, M.M., Zainudin, E.S., Osman, M.R., Rahman, M.Z.A., 2006. Mould flow and component design analysis of polymeric based composite automotive clutch pedals. J. Mater. Process. Technol. 171, 358–365.

Kapidžić, Z., Nilsson, L., Ansell, H., 2014. Conceptual studies of a composite–aluminum hybrid wing box demonstrator. Aerospace Sci. Technol. 32, 42–50.

Kim, W.C., Mauborgne, R., 2005. Blue Ocean Strategy: How to Create Uncontested Market Space and Make the Competition Irrelevant. Harvard Business School Press, Boston, MA.

Lee, H.B., 2004. Comparison of cost, surface roughness and time using stereolithogaphy and 3D printer for the design of composite pedal. Master of Science Thesis, Universiti Putra Malaysia.

Li, S., Zeng, Q.Y., Xiao, Y.L., Fu, S.Y., Zhou, B.L., 1995. Biomimicry of bamboo bast fiber with engineering composite materials. Mater. Sci. Eng. C3, 125–130.

Loiter, B., 1986. Manufacturing Assembly Handbook. Butterworths, Boston.

Maier, C., 1997. Material Databases 1. Materials Information Service, The Institute of Materials, London.

Mallick, P.K., 2008. Fiber-Reinforced Composites: Materials, Manufacturing, and Design, third ed. CRC Press, Boca Raton.

Mansor, M.R., 2014. Concurrent conceptual design of hybrid natural/glass fiber reinforced thermoplastic composites for automotive parking brake lever. PhD Thesis, Universiti Putra Malaysia.

Mansor, M.R., Sapuan, S.M., Zainudin, E.S., Nuraini, A.A., Hambali, A., 2014. Conceptual design of kenaf fibre polymer composite automotive parking brake lever using integrated TRIZ-morphological chart-analytical hierarchy process method. Mater. Des. 54, 473–482.

Markov, L., Cheng, R.M.H., 1996. Conceptual design of robotic filament winding complexes. Mechatronics 6, 881–896.

Marsden, W., Warde, S., 2016. Data management for composite materials data management for composite materials. Granta Technical Paper, GRANTA Material Intelligence, Granta Design Ltd, Cambridge, pp. 1–17, Available from: www.grantadesign.com/download/pdf/composites.pdf

Martin, 2015. 18 Best Idea Generation Techniques. Available from: https://www.cleverism.com/18-best-idea-generation-techniques/

Masselter, Speck, 2011. Biomimetic Fiber-Reinforced Compound Materials. In: Cavrak, M. (Ed.), Advances in Biomimetics. Intech, Rijeka, Croatia.

Mayer, R.M., 1993. Design with Reinforced Plastics: A Guide for Engineers and Designers. The Design Council, London.

Misri, S. 2011. Design and fabrication of a boat using sugar palm–glass fibre reinforced unsaturated polyester composites. Master of Science Thesis, Universiti Putra Malaysia.

Misri, S., Leman, Z., Sapuan, S.M., 2014. A total design of glass-sugar palm fibre reinforced unsaturated polyester composite of a small boat. In: Sapuan, S.M. (Ed.), Engineering Composites: Properties and Application. UPM Press, Serdang, Malaysia.

Munroe, A.S., 1995. Is your design a life sentence? Mach. Des., 156.

NBOS, 2006. What is NBOS? Available from: http://www.nbos.gov.my/posts/what-is-nbos

Neopolen® P., 2006. plastics plus. Automotive packaging innovation. Available from: http://www2.basf.de/basf2/html/plastics/images/pdfs/engl/schaum/Neopolen_e.pdf

Netten, B.D., Vingerhoeds, R.A., 1997. EADOCS: conceptual design in three phases—an application to fibre reinforced composite panels. Eng. Appl. Artif. Intell. 10, 129–138.

Norianti, M., 2016. Design and fabrication of kenaf fibre reinforced unsaturated polyester composite shoe shelf. Master of Science Thesis, Universiti Putra Malaysia.

Osborne, A.F., 1953. Applied Imagination. Charles Scribners & Sons, New York.

Pahl, G., Beitz, W., Feldhusen, J., Grote, K.H., 2007. Engineering Design: A Systematic Approach, third ed. Springer, London.

Pawlyn, M., 2011. Biomimicry in Architecture. RIBA Publishing, London.

Pugh, S., 1991. Total Design: Integrated Methods for Successful Product Engineering. Addison Wesley Publishers Ltd, Wokingham, England.

Rumble, J., 1986. National needs for materials property data—a long-range view. Proceedings of the 13th International SAMPE Symposium, 7–10 April, Covina, CA, 1166–1168.

Saaty, T.L., 1980. The Analytic Hierarchy Process. McGraw-Hill, New York.

Sanyang, M.L., Sapuan, S.M., Jawaid, M., Ishak, M.R., Sahari, J., 2016. Effect of sugar palm-derived cellulose reinforcement on the mechanical and water barrier properties of sugar palm starch biocomposite films. BioResources 11, 4134–4145.

Sapuan, S.M. 1994, Design of staking system for Straw Bales. Master of Science Report. Loughborough University, Leicestershire.

Sapuan, S.M., 1998a. Computerized information retrieval system for the postgraduate research degree levels in mechanical engineering design. AEESEAP J. Eng. Educ. 28, 27–36.

Sapuan S.M., 1998b. A concurrent engineering design system for polymeric-based composite automotive components. PhD Thesis, De Montfort University, Leicester, UK.

Sapuan, S.M., 2005. A Conceptual design of the concurrent engineering design System for polymeric-based composite automotive pedals. Am. J. Appl. Sci. 2, 514–525.

Sapuan, S.M., 2006. Using morphological chart technique for the design of polymeric-based composite automotive pedals. Discov. Innov. 18, 311–317.

Sapuan, S.M., Abdalla, H.S., and Nash, R.J. 1995. Proposed design and manufacturing techniques of polymeric based composite pedal box system. Proceedings of the 11th National

Conference on Manufacturing Research, Advances in Manufacturing Technology IX, Stockton, D., Wainwright, C. (Eds.), Leicester, 12–14 September, Taylor & Francis, London, pp. 77–81.

Sapuan, S.M., Khalina, A., Mastura, M.T., 2016. Development of biocomposites from forest products: sugar palm (Arenga pinnata merr.) based composites. Presented at the First International Conference on Bioscience and Biotechnology—2016, 12–13 January, Colombo, Sri Lanka (Book of Abstract, p. 19).

Sapuan, S.M., Maleque, M.A., Hameedullah, M., Suddin, M.N., Ismail, N., 2005. A note on conceptual design of polymeric composite automotive bumper system. J. Mater. Process. Technol. 159, 141–151.

Sapuan, S.M., Mansor, M.R., 2014. Concurrent engineering approach in the development of products: a review. Mater. Des. 2014, 161–167.

Sapuan, S.M., Mansor, M.R., 2016. Design of natural fiber reinforced-composite structures. In: Campilho, R.D.S.G. (Ed.), Natural Fiber Composites. CRC Press, Boca Raton, pp. 255–278.

Sapuan, S.M., Mollah, M.A.H., Nukman, Y., 2001. Concurrent Engineering, Product Design and Development. Prentice Hall, Petaling Jaya, Malaysia.

Sapuan, S.M., Osman, M.R., Nukman, Y., 2006. State of the art of the concurrent engineering technique in automotive industry. J. Eng. Des. 172, 143–157.

Sapuan, S.M., Willmot, P., 1996. The application of engineering design methodologies in the design of a weather protection system. ASEAN J. Sci. Technol. Dev. 13, 67–85.

Sapuan, S.M., Yusoff, N.B., 2015. The relationship between manufacturing and design for manufacturing in product development of natural fibre composites. In: Salit, M.S., Jawaid, M., Yusoff, N.B., Hoque, M.E. (Eds.), Manufacturing of Natural Fibre Reinforced Polymer Composites. Springer, Cham, Switzerland, pp. 1–15.

Seo, K.K., Park, J.-H., Park, Jang, D.-S., Wallace, D., 2002. Approximate estimation of the product life cycle cost using artificial neural networks in conceptual design. Int. J. Adv. Manuf. Technol. 19, 461–471.

Shanian, A., Savadogo, O., 2006. TOPSIS multiple-criteria decision support analysis for material selection of metallic biopolar plates for polymer electrolyte fuel cell. J. Power Sources 159, 1095–1104.

Sobey, A.J., 2010. Concurrent engineering in the context of the composite leisure boatbuilding industry. PhD Thesis, University of Southampton.

Suh, N.P., 1990. Principles of Design. Oxford University Press, Oxford.

Suh, N.P., 2001. Axiomatic Design: Advances and Applications. Oxford University Press, Oxford.

Turner, M.J., Grande, D.L., 1978. Study of advanced composite structural design concepts for an arrow wing supersonic cruise configuration. NASA Contractor Report 2825, Langley Research Center, National Aeronautic and Space Administration (NASA), Hampton, Virginia.

Ulrich, K.T., Eppinger, S.D., 2004. Product Design and Development. McGraw Hill Irwin, Boston.

Uusitalo, K., 2013. Designing in carbon fibre composites. Master of Science Thesis, Chalmers University of Technology, Gothenburg, Sweden.

Vincent, J.F.V., Bogatyreva, O.A., Bogatyrev, N.R., Bowyer, A., Pahl, A.K., 2006. Biomimetics: its practice and theory. J. R. Soc. Interface 3, 471–482.

Vullings, R., Heleven, M., 2015. Not Invented Here: Cross-Industry Innovation. Bis Publishers, Amsterdam.

Wright, I., 1998. Design Methods in Engineering and Product Design. McGraw-Hill, London.

Zwicky, F., 1948. The Morphological Method of Analysis and Construction. Interscience Publishers, New York, Courant Anniversary Volume.

CHAPTER 6

Materials Selection for Composites: Concurrent Engineering Perspective

INTRODUCTION

Materials selection is an important step in engineering design process. It is crucial for a design engineer to select the correct materials for a design that they developed so that the product can perform in the most effective manner.

Materials selection is a process normally carried out by a design engineer, materials scientist, or materials engineer in an engineering organization. Materials selection is a process of selecting the most suitable materials from a group of possible candidate materials by setting selection criteria after the intended design has been finalized. As a product becomes very complex, the selection of materials becomes more difficult to perform and the design engineer should study many functional and economic requirements before arriving at the decision to choose a particular material.

The range of materials available to design engineers is wide and constantly increasing. This poses a huge challenge to design engineers. For instance, during the selection process, a material or materials must satisfy a complex set of design requirements and must be at the same time, cost-effective to be a successful candidate. The fundamental questions are which material is the most suitable and can the selected material satisfy the imposed requirements. In materials selection, a design engineer must thoroughly understand and be familiar with the properties of various materials to satisfy the requirements of the design.

The use of material handbooks is a traditional method of performing materials selection. The use of computerized materials selection system has the added advantage of encouraging design engineers to consider and analyze their requirements for a material. It could be in the form of materials databases, Ashby charts, or any computerized materials selection tools. In recent years, materials selection has become too complicated to manage without the help of any tools and selecting the incorrect materials may incur financial and other losses.

CONTENTS

Introduction 209
Composite Materials Selection 211
Concurrent Engineering and Materials Selection ... 213
Attributes of a Good Materials Selection System 214
Materials Selection of Composite Materials Using Materials Data Book 215
Procedure-Based Versus Computer-Based Composite Materials Selection ... 215
Pugh Concept Selection Method 216
Case Study of Pugh Selection Method 216
Materials Selection Using Digital Logic Method 218
Background 218
Materials Selection Using Integrated Modified Digital Logic Method and Weighted Property Method 219

(Continued)

Composite Materials. http://dx.doi.org/10.1016/B978-0-12-802507-9.00006-4

Determination of the Weightage of the Selection Criteria Using Modified DLM............220
Ranking the Alternatives With Respect to the Selection Criteria Using WPM............221
Case Study on Thermoset Resin Materials Selection for Student Competition-based Race Car Composites Bodywork............................221
Materials Selection Using Quality Function Deployment (eQFD)...226
Materials Selection of Composite Materials Using Materials Databases.................226
Materials Selection of Composite Materials Using Ashby's Chart..228
Materials Selection of Composite Materials Using Knowledge-Based System...........229
Knowledge-Based System..229
Expert Systems...................230
Rule-Based System for Material Selection Process.233
Constraint Satisfaction and Constraint Violation............234
Case Study of Materials Selection of Composite Using KBS..........................234
Use of Exsys Corvid (Material Selection Expert System Shell) in Research..238
Java-Based Materials Selection...239
Materials Selection of Composite Materials Using Multicriteria Decision Making........241

Thus, it is hard to see how design engineers can optimize their choices without some form of computerized selection tool.

Materials selection should be conducted from the conceptual design stage of the design process. The selection must continue during the embodiment design stage and finally during detail design stage, materials selection still has to be performed. The only difference between these three stages of design is that the candidate materials are different. During the conceptual design stage, the number of candidate materials is enormous, and the criteria that define the properties of candidate materials are normally quite vague. During the embodiment design stage, the number of candidates is reduced and finally during detail design, only a few candidate materials are available for selection. During the latter two stages of the design process, the materials selection criteria should be further developed and refined.

Materials selection is an interrelated activity that requires the knowledge of various fields as shown in Fig. 6.1 (Pahl et al., 2007). All the fields mentioned by Pahl et al. (2007) are indeed important in materials selection, but if we refer back to the "total design model" we will notice that all these fields are the main inputs in design activities rather than materials selection. For instance, in the total design model developed by Pugh (1991), it is known that product design specification (PDS) is one of the important stages in design and a majority of the fields shown in Fig. 6.1 are normally addressed in PDS. In practice, the materials selection systems developed by various experts (Ashby, 2005;

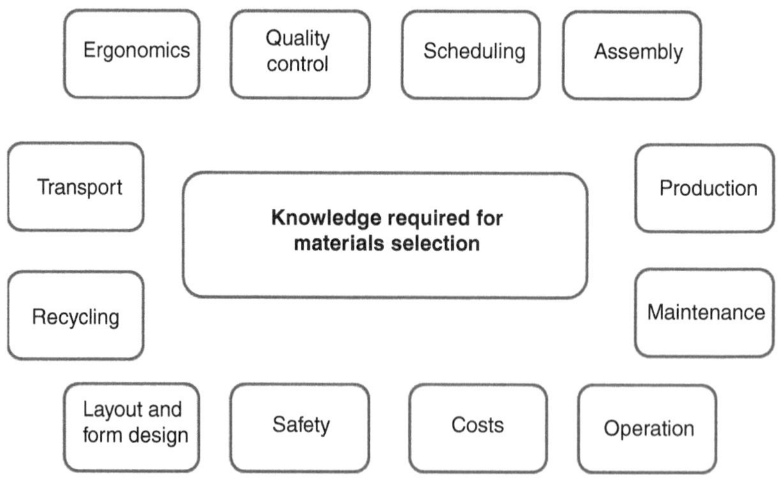

FIGURE 6.1 Knowledge required in materials selection.
Adapted from Pahl, G., Beitz, W., Feldhusen, J., Grote, K.H., 2007. In: Wallace, K., Blessing, L. (Eds.), Engineering Design: A Systematic Approach, third ed. Springer Verlag, New York.

Dieter, 2000) mainly focus on selected properties such as weight, strength, and stiffness.

In the selection of materials, inputs from engineering analysis and synthesis are very important and must go hand in hand with industrial design aspects (Ashby and Johnson, 2002). Most of the time, many computer-based materials selection systems are available to assist the designer in the selection of the most optimum materials. These computer systems can also store and process data regarding properties of materials.

The motivation of writing this chapter is to provide the readers with some information or knowledge of various materials selection tools and techniques for the selection of composite materials for engineering applications. The discussion is performed under the umbrella of concurrent engineering, that is, while designing a product, all other interrelated functions within the product development process are considered earlier.

Analytical Hierarchy Process 246
Case Study on Composite Material Selection Using AHP for Automative Brake Lever 250
Case Study on the Implementation of AHP for Environmental Assessment in Composite Material Selection for Automotive Components 254
TOPSIS Method 256
Materials Selection of Composite Materials Using Neural Network 258
Materials Selection of Biocomposites 262
Conclusions 265
References 265

COMPOSITE MATERIALS SELECTION

Composites have been used as materials in many industries such as automotive, aerospace, defense, furniture, construction, marine, sports equipment, telecommunication, railway, and agricultural industries. The primary advantages of composite materials are their high strength-to-weight and stiffness-to-weight ratios properties and their applications are suitable for non, semi and fully structural components. Selection of materials is not a straightforward task because experts have estimated huge number of different materials available at design engineer's disposal (e.g., 50,000; Ashby, 1989). However, with composites, it is more difficult as it is impossible to estimate the number of composites available nowadays because the permutation of the number and the types of fibers, matrices, additives, manufacturing processes, and composite types are almost unlimited (Fig. 6.2). Composite materials, unlike monolithic materials, comprise different constituent materials that made up the final materials and therefore materials selection process for composite materials is not a straightforward task. At the very best, the estimate can be made based on the number of commercial types of composites.

Materials selection of composite materials, compared to homogenous materials like metals and plastics is quite difficult to perform due to the anisotropic nature of the materials that required the tailoredmade consideration of products and materials. The real issues of materials selection for composites are the presence of large permutation of candidate materials based on the following:

- Different types of composite materials such as polymer composites, metal matrix composites, and ceramic matrix composites. Different types of fibers such as glass, carbon, aramid, boron, and natural fibers.

FIGURE 6.2 The need for systematic composite materials selection system.

- Different types of matrices and for polymer composites, the matrices comprise thermoplasticmaterials. Other variations in polymer matrices include polymer blends and the use of either synthetic or bio-based polymers. Please bear in mind that there are also different types of ceramic and metal matrices at our disposal.
- Different arrangements of fibers like continuous roving, woven roving, nonwoven fabrics, chopped strand mat, short fibers, and particulates.
- New advances in composite technology such as the emergence of nanocomposites, hierarchical composites, functional graded materials, smart and intelligent materials, biomimetic materials, and graphane-based composites.

- Broad spectrum in manufacturing processes for composites, for instance hand lay-up, pultrusion, filament winding, vacuum bagging, resin transfer molding, spray-up, compression molding, thermoforming, calendaring, and out of autoclave (OOA) processing. It is stated in Mayer (1993) that there are three ways of incorporating fiber into matrix according to manufacturing technology:
  - Fiber and matrix are acquired separately and the user has the choice to make composite material through fabrication process.
  - Semicured molding compound with exact mixture fiber, matrix, filler and additives are ready for fabrication. These intermediate materials like sheet molding compound (SMC), glass mat thermoplastics (GMT), and prepreg are made by the material supplier. The user has little choice on the preparation of composites.
  - Finished products ready to be made into components like pultruded profiles and filament wound shaft.

This shows the scope of materials selection for composite materials is very broad. It should be remembered that whatever has been achieved so far in the area of composite materials technology, is still very limited and there is still huge scope of research and development in the field of materials selection for composite materials.

CONCURRENT ENGINEERING AND MATERIALS SELECTION

Concurrent engineering for composite materials considers materials selection as one of the main activities because in composite design, design of product, manufacturing process, and material have to be carried out at the very early stages of the design process. Concurrent engineering has encouraged materials engineers and manufacturing engineers to take part in the product development at the early stage of design process (Dieter, 2000). Boothroyd et al. (1994) also suggested that an integral part of concurrent engineering is the systematic early selections of material and process combination for the manufacture of parts, which can then be ranked according to various criteria. In this chapter only the work on selection of materials is considered but the work on selection of manufacturing processes will be discussed in Chapter 8. Sapuan stated in his study on concurrent engineering (CE) that materials selection is considered as an important activity in CE as important as other activities like market investigation, design analysis, manufacturing consideration, component design, and assembly consideration.

In CE for composites, materials selection is considered as a very important activity. Research on materials selection for composite materials under the umbrella of concurrent engineering had been performed by researchers such as

Kim et al. (2000). In CE design environment, information and communication technology (ICT) tools were integrated where various activities of composite development are stored in a computer in different integrated modules. For instance, in the work of Kim et al. (2000), various modules of activities are stored together in a computer. Activities like materials selection are considered as a module, design analysis with finite element analysis (FEA) is stored in another module, and so on. List of input parameters can also be considered as a module called input module that provides properties specified by the design engineer. In their material selection module, some materials were selected by assigning fiber and matrix types in line with the design objectives. The polymer composite investigated was based on laminated composite approach where ply angles, stacking sequence, properties of lamina and laminate, and rule of mixture were considered.

The major issue with materials selection system for composite materials under the umbrella of CE is associated with the practical industrial application of such system.

In Chapter 4, it was reported that conceptual design only used 15% of total product cost but it determines 70% of the manufacturing cost. Therefore, conceptual design is a key activity in CE. Prasad (1996) has stated seven agents that influenced the domain of CE and they are talents, tasks, teams, techniques, technology, time, and tools (7Ts). Tools including software and hardware are very important in materials selection.

ATTRIBUTES OF A GOOD MATERIALS SELECTION SYSTEM

The following are the major attributes expected from a materials selection system:

- The materials selection system should facilitate new product development activity.
- The materials selection system should help reducing manufacturing cost and increase product quality.
- The materials selection system should be able to increase manufacturing threshold.
- The materials selection system should be able to reduce materials selection complexity.
- The materials selection system should be able to speed up human effort.
- The materials data in the materials selection system should be complete.
- The materials data in the materials selection should be universal and comparable.

- The materials selection system should be able to decrease manufacturing time.
- The materials selection system should be able to increase the efficiency of process.
- The materials selection system should be able to increase productivity.
- The materials selection system should not be using too many buzzwords, computer jargons, and acronyms so that the importance of materials, design, and engineering is not overshadowed.
- The materials selection system should be focusing more on the engineering approach, rather than too mathematical.
- The ideal materials system should have the decision-making procedures, which can be used by the industries or individuals facing with a materials selection decision.

However, these ideal attributes are not easy to achieve. We will notice that in many materials selection system reported in published literature and in this book, majority of the systems are still in a prototype stages. A lot of upgrading and improvement are still needed before they can be used commercially.

MATERIALS SELECTION OF COMPOSITE MATERIALS USING MATERIALS DATA BOOK

Selected composite material properties are documented in the composite materials handbooks and they are presented in the forms of tables and figures. Materials selection process can be performed manually using materials data books and materials handbooks. This process is a traditional method for materials selection and it does not follow any standard procedures. Many composite materials handbooks provide not only materials data but they also provide other information such as design, analysis, mechanical properties, applications, and characterizations.

PROCEDURE-BASED VERSUS COMPUTER-BASED COMPOSITE MATERIALS SELECTION

In this book, procedure-based materials selection is differentiated from computer-based materials selection. In product design and selection, there are techniques employed using computer software and others that are carried out without the help of computers; however there are some procedures that the designers need to follow to arrive at certain solutions. In this book such techniques are termed procedure-based techniques; examples include quality function deployment, Pugh selection method, and digital logic method (DLM).

PUGH CONCEPT SELECTION METHOD

Numerous concept selection techniques were developed to ease the task of designers in constructing better design concepts (Otto, 1995; Pahl and Beitz, 2013; Pugh, 1991; Thurston and Carnahan, 1992). However, Pugh concept selection approach is one of the simplest and has been widely used in the conceptual design stage. Pugh (1991) provided a simple graphical method that revolves around a matrix with rows (showing decision criteria) and columns (giving concept) as presented in Fig. 6.3. King and Sivaloganathan (1999) presented Pugh selection method in a simple four-step procedure as;

1. Select a "datum" concept so as to compare other concepts with it.
2. Score other concepts versus the datum one after the other. Each decision criteria is judged for the selected concept and the datum. If a particular concept is judged better then a "+" sign is assigned in the box, if worse then a "−" sign is added, and if of "equal value," then an "S" for same is entered.
3. Summation is made of all "+" scores, all "−" scores, and all " S" scores. No arithmetic score of all values should be given because the significance of each is not determined.
4. The process is iterative. Low scoring concepts are removed and reassessment is conducted. The concept with the highest number of "+" scores at the end is selected as the best concept.

This technique involves qualitative comparison of each alternative to a reference or datum alternative, criterion by criterion. Pugh's approach is crucial in conceptual design because it only requires the least amount of detailed information. According to Dieter (2000), Pugh selection method can also be used for materials selection.

Case Study of Pugh Selection Method

In this study, seven different natural fiber composites (NFCs) such as kenaf, sisal, flax, hemp, jute, coir, and oil palm empty fruit bunch (EFB) with fiber loading of 40% are considered as replacement materials for PP. Hence, Pugh's selection method is employed to select the best composite material for

	Concept 1	Concept 2	Concept 3	Concept 4	Concept 5
Criteria A	+		+	+	−
Criteria B	+	D	−	+	S
Criteria C	−	A	S	S	S
Criteria D	−	T	S	+	−
Σ+	2	U	1	3	0
Σ−	2	M	1	0	2
ΣS	0		0	3	−2

FIGURE 6.3 Pugh selection matrix (King and Sivaloganathan, 1999).

Table 6.1 Composite Material Selection for Automotive Components Using Pugh Selection Method

	PP	PP + Kenaf	PP + Sisal	PP + Flax	PP + Hemp	PP + Jute	PP + Coir	PP + EFB
Density (kg/m^3)		+	–	–	–	–	–	+
Tensile strength (MPa)		+	S	–	+	–	–	–
Tensile modulus (GPa)	D A	+	+	+	+	+		
Biodegradability	T	+	+	+	+	+	+	+
Toxic level	U	+	+	+	+	+	+	+
Raw material cost (US$/kg)	M	+	+	+	+	+	+	+
Manufacturing cost (US$)		–	–	–	–	–	–	–
Σ +		6	4	4	5	4	3	4
Σ –		1	2	3	2	3	4	3
Σ S		0	1	0	0	0	0	0
Ranking		First	Third	Fourth	Second	Fourth	Fifth	Fourth

automotive components. It is known from literature that polypropylene (PP) is the most used polymer in automobiles. Therefore, PP is chosen as the reference ("DATUM") material against which all the composite materials are rated. The main criteria taken into account for the selection are physical (density), mechanical (tensile strength and modulus), environmental (biodegradability and toxic level), and cost (raw material cost and manufacturing cost).

A relative score of "better than (+)," "same as (0)," or "worse than (–)" is entered in each cell of the matrix base on how the material rates as compared to the "DATUM" for each criteria. Thus, Table 6.1 manifested that most of the composites materials are highly positive, but PP + Kenaf composite has the maximum pluses. Its only negative is in manufacturing cost, as it involves extra expenses for the processing of the kenaf fibers. Therefore, PP + Kenaf is selected as the best composite material to substitute PP in automobiles.

It can be seen that Pugh selection method can be used as a simple and fast screening process against highly unfeasible materials and it enables the designer to concentrate on the best materials using other material selection methods. However, the inherent drawback of this method is that, no information is provided regarding the quality of the candidate materials. This means designers are not certain whether the material with the higher score is superior to the material with lower score regarding a particular criteria (i.e., tensile strength). According to Wang, Pugh selection method does not provide sufficient information to enable the designers to have confidence in the obtained result.

It is clear that the Pugh concept selection method is a viable method for material selection. Materials are evaluated at a high level using Pughselection method and sometimes, new combinations or changes can emerge as a result of critical discussions among team members.

MATERIALS SELECTION USING DIGITAL LOGIC METHOD

Background

Materials selection is one of the crucial stages in product design process whereby the best material suited to the intended application must satisfy multiple design and operational requirements which are often conflicting to each other. This type of decision-making situation is also referred to as multicriteria decision making (MCDM) and many methods have been developed by researchers to perform the selection process with systematic and reliable approach. One of the available methods for MCDM related to materials selection is the DLM for weighting purpose and weighted property method (WPM) to rank the materials listed for the intended application.

Farag (2006) demonstrated the application DLM and WPM in the materials selection for simple structural component for a sailing-boat mast with hollow cylinder geometry. The best material is selected among 15 candidate materials which consists of various grades of steel, aluminum alloys, titanium alloy, and polymer composites. In addition, there are five election criteria involved in decision-making process which are yield strength, elastic modulus, specific gravity, corrosion resistance, and raw material cost. The DLM method was applied to determine the weight of each criteria with respect to the selection goal whereas WPM was applied to rank the candidate materials based on their material properties. Results from the analysis showed that epoxy with 63% carbon fabric scored the highest overall WPM score compared to the other candidate materials.

Apart from that, similar application of the DLM and WPM methods was also reported by Findik and Turan (2012) on the materials selection for lightweight locomotive wagon design. In their report, the DLM method was applied to determine the weightage of all the selection criteria which comprised density, specific stiffness, corrosion resistance, wear resistance, and cost. Furthermore, the selection process also involved 13 candidate materials such as aluminum, steel, titanium alloy, zinc alloy, cuprum alloy, nickel alloy, synthetic polymer composites, and thermoplastic material. In the weighting process using DLM, they also applied a preliminary screening process using go-no go screening method to shortlist the candidate materials prior to the process of rating the selection criteria. Finally, the ranking process for all the short-listed candidate materials was performed using DLM method whereby final results from the analysis have shown that aluminum alloy is the best material to be selected for lightweight wagon construction.

The application of DLM and WPM for materials selection was also adopted by Mansor and Sapuan (2014) involving automotive parking brake lever design. In their report, both methods were applied to determine the best candidate material as material substitution for the existing steel-based component. Ashby performance index and Ashby chart were also implemented for the early screening process to obtain the most suitable candidate material which comply to the component design requirements. Six candidate materials were listed at the end of the screening exercise which all fall under the polymer composites category. Final ranking analysis using the WPM method on the candidate materials has revealed that unidirectional carbon fiber reinforced epoxy composites is the best material to be selected for the material substitution purpose.

It should be noted that all application of DLM as described previously for materials selection purpose involved conventional rating value of 1 for more important property and value of 0 for less important property when comparing the selection property hand to hand. A new rating method for the DLM was also suggested by Dehghan-Manshadi et al. (2007) to be applied for the weighting process for the selection criteria which consists of rating values of 3 for more important property, 2 for equivalent important property, and 1 for less important property. The application of the new modified DLM method with combination of nonlinear normalization was shown through a case study on materials selection for cryogenic tank design. The weighting process was performed involving seven selection criteria such as toughness, yield strength, density, and specific heat whereas the ranking process using WPM involved seven candidate materials such as Inconel 768, stainless steels, titanium alloy, and aluminum alloys. Final ranking results showed that stainless steel grade 301-FH is the best material for the cryogenic tank construction based on the design requirements. They also applied the modified DLM method for a case study on materials selection for wing structure of human powered aircraft project. The new rating system was applied for the weighting process of the criteria. Final ranking results using the WPM method have showed that carbon fiber reinforced epoxy composites scored the highest performance index value compared to the other 12 candidate materials and was chosen as the best material for the intended application.

Materials Selection Using Integrated Modified Digital Logic Method and Weighted Property Method

In general, the integration of both methods is conducted to accomplish two objectives, first is to determine the weightage or importance of each selection criteria involved in the decision-making process, and second is to determine the rank of the listed potential solutions with respect to the selection goal. The former objective is performed using modified DLM whereas the latter objective is performed using WPM.

Determination of the Weightage of the Selection Criteria Using Modified DLM

In the decision-making process involving multiple selection criteria, one of the major difficulties faced is to perform systematic judgments of the importance or weightage between each criteria with respect to the overall selection goal. This is contributed to the subjective judgments that have to be made while rating the importance of the criteria which may incur inconsistency for the rating process and reduce the reliability of the final decision obtained (Farag, 2006). The application of DLM has been proven able to determine the overall importance of each selection criteria involved in the selection process in systematic manner. In the DLM, the calculated importance of each criterion is given as the weighting factor, w. The weighting process is conducted by comparing two selection criteria at once, and the rate of importance between the compared criteria with respect to the selection goal is given the value of 1 for the more important criteria and the value of zero is given to the less important criteria. The final weighting factor is calculated by summing the total positive decision obtained by each criteria and dividing it with the total number of possible decisions, N where $N = (n-1)/2$ and n is the total number of selection criteria involved (Farag, 2006). The total weighting factor value is 1.

Despite the simple conventional DLM rating process, the rating process also appeared to impose disadvantage of the possibility to dismiss any selection criteria which scored total zero rating at the end of the evaluation process. For decision makers, eliminating one selection criteria from the required list of selection criteria for the evaluation process is not an allowed option despite its relative lower importance compared to the other selection criteria. Thus, to address this limitation, a modified rating method was proposed using 1–3 importance scale (Dehghan-Manshadi et al., 2007). During comparison between the criteria, the value of 3 is assigned for the most important criteria, value of 2 is assigned for the equal important criteria, and value of 1 is assigned for the less important criteria. The final weighting factor for each selection criteria is calculated based on Eq. (6.1).

$$\text{Weighting factor}, w = (\text{total positive decisions}/N') \tag{6.1}$$

where $N' = [2n(n-1)]$ and n = number of selection criteria.

In addition, other notable improvement of the modified DLM is the introduction of equal important rating (value of 2) for the compared criteria. The new rating value helps to cater natural ways in decision making involving intuitive judgments by decision makers. Similar to the conventional DL method, the modified DLM will also yield total weighting factor value of 1. The new rating system for the modified DLM method was also suggested to yield more accurate results as compared to the conventional DLM method (Dehghan-Manshadi et al., 2007).

Ranking the Alternatives With Respect to the Selection Criteria Using WPM

The final selection process involved the determination of the rank for each potential solution which best comply with all the selection criteria and fulfil the overall selection goal. The ranking process is performed using WPM based on the weighting factors obtained using the modified DLM. In this process, the scaled property, β for each solution is first calculated. The scaled value method allows the selection process involving selection property with varying attributes and units to be analyzed based on equivalent value simultaneously.

For the condition where high value of solution attribute is desired such as material's tensile strength and tensile modulus, the scaled property value, β is calculated using Eq. (6.2)(Mansor and Sapuan, 2014).

$$\begin{aligned}\beta &= \text{scaled property} \\ &= (\text{numerical value of property} \times 100)/\text{largest value}\end{aligned} \tag{6.2}$$

On the other hand, for condition where low value is best such as material's price and density, the scaled property value, β is calculated using Eq. (6.3).

$$\begin{aligned}\beta &= \text{scaled property} \\ &= (\text{lowest value} \times 100)/\text{numerical value of property}\end{aligned} \tag{6.3}$$

Finally, the overall importance score for each solution is given by the final weighted property index, γ value which is determined by the addition of the scale property and the weighting factor, w for each property using Eq. (6.4). The ranking of the solution is performed based on the γ score and the best solution for the selection goal is chosen based on the highest γ score.

$$\begin{aligned}\gamma &= \text{weighted property index} \\ &= \sum_{i=1}^{n} \beta_i w_i \quad (n = \text{number of selection criteria})\end{aligned} \tag{6.4}$$

Case Study on Thermoset Resin Materials Selection for Student Competition-based Race Car Composites Bodywork

The application of the modified DLM and WPM for MCDM process is demonstrated using a case study on the thermoset resin materials selection for student competition-based race car composites bodywork. The race car bodywork is fabricated using glass fiber polymer composites to attain strong and lightweight body construction as well as geometrical design flexibility and ease of fabrication for aerodynamic performance compared to the use of sheet metal. The challenge in terms of composite product design for the component is to select the best type of resin as the polymer matrix material for the composites construction. Initial screening process conducted for the type of resin (either thermoset or thermoplastic) has pointed out that the thermoset resin is the most suitable type of resin to be applied for the composites bodywork due to better mechanical performance and compatibility with the available hand lay-up manufacturing process compared to thermoplastic resin. Further literature review conducted has revealed three types of the most applied thermoset resin for automotive application, namely polyester resin, vinylester resin, and epoxy resin (Holbery and Houston, 2006). Thus, to perform the final decision in selecting which type of thermoset resin is suitable for the goal of fabricating the composites bodywork, the modified DLM and WPM are applied to rank the best solution from the listed thermoset resin based on the required performance requirements. Both methods were selected based on their capability in performing MCDM process involving multiple selection criteria and multiple solutions with varying attributes systematically and simultaneously. In general, the selection criteria selected in the case study comprised three main performance requirements, namely weight, structural performance, and cost. Fig. 6.4 shows the overall selection criteria and the corresponding material attribute for the criteria whereas Table 6.2 shows the summary of the thermoset resin material properties applied in the case study.

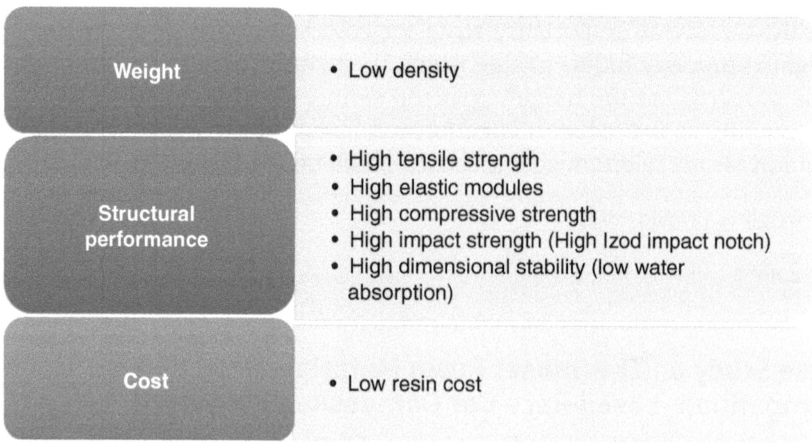

FIGURE 6.4 Overall selection criteria for composites bodywork resin material.

Table 6.2 Thermoset Resin Materials Properties (Holbery and Houston, 2006)

	Thermoset Resins		
Material Properties	**Polyester**	**Vinylester**	**Epoxy**
Density (g/cm^3)	1.2–1.5	1.2–1.4	1.1–1.4
Elastic modulus (GPa)	2–2.45	3.1–3.8	3–6
Tensile strength (MPa)	40–90	69–83	35–100
Compressive strength (MPa)	90–250	100	100–200
Water absorption (24 h @ 20°C)	0.1–0.3	0.1	0.1–0.4
Izod Impact Notch (J/cm)	0.15–3.2	2.5	0.3
Price (£/kg)	1–2	2–4	3–15

The next stage in the resin materials selection process was to determine the weightage of each selection criteria with respect to the overall selection goal. The weighting process was performed using modified DLM. Seven selection criteria were involved in the weighting process namely density, tensile strength, elastic modulus, compressive strength, impact strength, dimensional stability, and resin cost. The selection criteria cover the main elements for composites bodyworks design requirements which are lightweight for improved overall power to weight ratio of the race car and low product cost with good structural capability to ensure safe and operational performance. As mentioned in the previous section, 3 rating scales were applied in the weighting process to determine the weighting factors for each of the selection criteria. The rating scale to rate the importance of each criteria are value of 3 for the most important criteria, value of 2 for the equal important criteria, and value of 1 for the less important criteria. The overall weighting factors are calculated based on Eq. (6.1). Table 6.4 summarizes the overall weighting results for thermoset resins selection criteria using modified DLM. Based on Table 6.4, it is shown that weight, elastic modulus, tensile strength, and cost obtained the highest importance compared to the other selection criteria, followed by compressive strength, impact strength, and finally dimensional stability (represented by the water absorption property).

The final stage of the resin materials selection process for the composites bodywork product is ranking the identified potential solution (thermoset resins) using WPM. The ranking process was performed on the basis of the material properties identified based on literature for each of the candidate thermoset resins. The ranking process involved determining the scale property values for each material property using Eqs. (6.2) and (6.3) as mentioned in the previous section. Later, the overall performance index for each thermoset resin was calculated using Eq. (6.4) mentioned in the previous section. Table 6.3 summarizes the overall results of thermoset resins materials selection for race car composites bodywork performed using WPM.

Table 6.3 Weighting Results for Thermoset Resins Selection Criteria Using Modified DLM

No	Material Properties	Number of Possible Decisions [$N = n(n-1)/2$]																					Total Positive Decisions	Weighting Factor, *w*
		1	2	3	4	5	6	7	8	9	10	11	12	13	14	15	16	17	18	19	20	21		
		1-2	1-3	1-4	1-5	1-6	1-7	2-3	2-4	2-5	2-6	2-7	3-4	3-5	3-6	3-7	4-5	4-6	4-7	5-6	5-7	6-7		
1	Density (g/cm^3)	2	2	3	3	3	2																15	0.179
2	Elastic modulus (GPa)	2						2	3	3	3	2											15	0.179
3	Tensile strength (MPa)		2					2					3	3	3	2							15	0.179
4	Compressive strength (MPa)			1					1				1				3	2	1				9	0.107
5	Water absorption (24 h @ 20°C)				1					1				1			1			1	1		6	0.071
6	Izod Impact Notch (J/cm)					1					1				1			2		3		1	9	0.107
7	Price (£/kg)						2					2				2			3		3	3	15	0.179

Note: Decision ratings: 1= less important; 2= equal important; 3 = more important. Weighting factor, α = *(total positive decisions/*N'*) where* N'= [2n(n-1)] *and* n = *number of property.*

Table 6.4 Ranking of Thermoset Resins for Student Competition–Based Race Car Bodywork Using WPM

Thermoset Resin	Weighting Factors, *w*	0.179		0.179		0.179		0.107		0.071		0.107		0.179		Performance Index, γ
	Material Properties (Average)	Density		Elastic Modulus		Tensile Strength		Compressive Strength		Water Absorption		Izot Impact Notch		Price		
		B	g/cm³	*B*	GPa	*B*	MPa	*B*	MPa	*B*	24 h@20°C	β	J/cm	β	£/kg	
Polyester		92.59	1.35	49.44	2.23	85.53	65.00	100	170	50	0.20	67	1.68	100	1.5	80.05
Vinylester		96.15	1.30	76.67	3.45	100	76.00	58.82	100	100	0.10	100	2.50	50	3.0	81.88
Epoxy		100	1.25	100	4.50	88.82	67.50	88.24	150	40	0.25	12	0.30	16.67	9.0	68.25

Sample calculation: Polyester: $\gamma = 0.179 \times 92.59 + 0.179 \times 49.44 + 0.179 \times 85.53 + 0.107 \times 100 + 0.071 \times 50 + 0.107 \times 67 + 0.179 \times 100 = 80.05$

As shown in Table 6.4, results from the WPM analysis indicated that the highest performance index scored was obtained for vinylester resin (81.88), followed by polyester resin (80.05), and finally epoxy resin (68.25). This showed that vinylester is the best thermoset resin to be selected for the composites bodywork construction based on the overall selection criteria compared to the other material candidates. The results of the materials selection process also demonstrated the capability of the modified DLM and WPM in assisting product designers to perform systematic decision-making process effectively with simultaneous consideration of multiple selection criteria and multiple selection solution.

MATERIALS SELECTION USING QUALITY FUNCTION DEPLOYMENT (eQFD)

Mayyas et al. (2011) applied quality function deployment (QFD) to select material from construction of house of quality (HoQ). This method has been followed by Prasad and Chakraborty (2013) in their study for materials selection. They developed QFD-based user-friendly software to select materials. However, they only considered the top rank of criteria to evaluate the potential materials. In a case of bio-composite materials, it is important to evaluate the materials for selection on each of criteria since the properties of each natural fiber are different from another because of so many factors that may enhance or worsen its properties.

Quality function deployment for environment (QFDE) is a tool that collaborates voices of customers and environment simultaneously. QFD as originated from Japan is an approach to convert customer demands into certain quality characteristics. This method determines the quality of the products systematically at the early stage of design process which has been implemented by most of the Japan automakers including Mitsubishi Motors and Toyota Motors (Akao, 1997). Later, QFDE is developed to incorporate environmental aspects into QFD to evaluate the quality in terms of environmental aspects in product quality requirements (Masui et al., 2001). Moreover, QFDE is developed to enhance environment consciousness in the product design.

MATERIALS SELECTION OF COMPOSITE MATERIALS USING MATERIALS DATABASES

The nature of the current material data books available does not satisfy today's advanced industry requirements, as data sheets are most of the time incomplete and once published are difficult to update. Thus, designers spend long time working through tables of properties in various data books to make appropriate material selection for design. To ease the task of designers, minimize potential errors, and accommodate the increasing number of new materials, nowadays, computer databases of material properties are utilized to search for a wider range of materials (Sapuan and Abdalla, 1998). Computerized materials databases are an important part of any computer-aided system for selection (Kutz, 2002). Computer systems are used for storing and processing data regarding the properties of materials (Sapuan et al., 2011a,b). It allows designers to have easy access to several data and swift retrieval from a computer database.

For the past few decades, many selection systems were invented to enable design engineers to select suitable materials (Sapuan et al., 2002). One of the most commonly used computer system in industries for material selection is the expert system, also known as knowledge-based system (KBS) such as Plastics Computer Aided Material Selector (Plascams), PERITUS, Cambridge Materials Selector (CMS), etc.

Plascams is an expert system primarily developed for plastic material selection. This expert system helps designers to search desired plastic materials among hundreds of materials using two search routines (Maier, 1993; Sapuan et al., 2011a). PERITUS is a computer package that contains expert system for selection of polymeric, metal, and ceramic materials as well as processes for design (Sapuan, 1999). It helps to accomplish the preliminary selection of materials by manipulating the data.

Computer-aided material preselection by uniform standards (CAMPUS) is a material database that is commonly used for plastics (Sapuan, 2001). There is another database known as FUNDUS which is similar to CAMPUS, used for the selection of long fiber reinforced plastics. This system permits the distribution of material information from the producer to the designer–end user (Sapuan, 2001; Baur, 1995). Both CAMPUS and FUNDUS share common features which enable the user to access all properties for any listed product, print the data of any product(s), search for products satisfying specific property requirements within the database, select and view properties for better comparison, and arrange according to specific requirement in increasing or decreasing order (Sapuan, 2001).

A computerized materials selection system known as CMS was developed by Ashby (1993) and Cebon and Ashby (1992, 1996). The CMS displays material properties in the form of material selection charts through the use of optimization procedures. The selection process depends on implementing performance indices, a combination of material properties, which if maximized, optimizes performance. The charts are developed to present the materials, and the performance indices, so that the

most suitable selection of materials and shape can be carried out (Sapuan, 2001).

MATERIALS SELECTION OF COMPOSITE MATERIALS USING ASHBY'S CHART

Ashby chart is a powerful materials selection technique that has been internationally accepted as a tool; almost every materials selection problem that can cater to all types of materials ranging from metals to woods (Ashby, 2005). It can be applied for selection materials for composites as well. Ashby chart is a graphical tool in the form of scatter plot. The Ashby materials selection system is available in the form of commercial software called Cambridge Engineering Selector developed by Granta Design, UK. Ashby (2005) has enumerated four steps in selection of materials. From the broad spectrum of materials to a final material, the essential steps are as follows:

- translating design requirements by expressing them in terms of objectives, constraints, free variables, and function,
- screening the materials using constraints or in other words removing the materials that do not satisfy the constraints,
- ranking the materials using objective by finding the screen materials that can perform the task best and finally,
- obtaining supporting information by studying in detail the family history of the optimum materials.

Ashby chart is a well-accepted materials selection technique and had been used by many researchers and this technique has been extended in various ways. Koning (1994) had extensively used Ashby charts in solving a case-based design problem and the materials selection approach is to extend Ashby charts with fuzzy set theory. As Ashby charts are approximate, Koning (1994) through Fuzzy set theory had attempted to make the charts more precise.

Shah (2014) stated that Ashby charts are useful for four key reasons:

1. they allow quick retrieval of typical properties of a particular material
2. they allow quick comparisons of properties of different materials, revealing their comparative efficiencies
3. they facilitate the selection of the materials during the product design stage
4. they enable substitution studies exploring the potential of one material to replace another

Another advantage offered by use of the Ashby chart is that more information can be obtained when plotting the two axes. This additional information can be important when deciding on the suitable material for an application. Among the examples given by Ashby are as follows:

1. longitudinal wave velocity $(E/p)^{1/2}$, when modulus E against density p is plotted

2. fracture surface energy G_{Ic}, when fracture toughness Kl against modulus E is plotted
3. volume specific heat pC_v, when thermal conductivity h against diffusivity, a is plotted

Plotting one material properties on each axis of a materials selection chart helps to provide the best performance indices for material selection. In the example shown in Fig. 6.4, the axes are Young's modulus and density. The logarithmic scales span a range so wide that all materials are included. When data for a given material class such as metals are plotted on these axes, it is found that they occupy a field which can be enclosed in a "balloon." Ceramics also occupy a field, and so do polymers, elastomers, composites, and so on. The fields may overlap, but are nonetheless distinct. Individual materials or subclasses (like steels, or polypropylenes, PP) appear as little "bubbles" which define the ranges of their properties (Ashby and Cebon, 1993).

MATERIALS SELECTION OF COMPOSITE MATERIALS USING KNOWLEDGE-BASED SYSTEM

Knowledge-Based System

Several systems are available to aid material experts or designers in choosing suitable materials. In the orthodox technique of material selection, designers normally use different data books with tables of material properties. However, such conventional method of material selection is tremendously tedious and is unable to meet today's advanced industry requirements; the data sheets are often defective and its difficult to upgrade after been published (Sapuan and Abdalla, 1998). White (1995) highlighted the significance of computer-aided material selection over books, which normally becomes outdated prior to reaching the bookshelves. Additionally, indexing books to sort out data regarding properties of materials in the desired manner is quite difficult.

Of late, there is much emphasis on the utilization of computer systems to store and process data related to material properties. It enables the designers to easily access and retrieves information regarding suitable materials from the computer database of material properties. The availability of online access to a material database permits the concurrent engineering team which involve designers to meticulously select those materials that are stronger, lighter, and cost-effective. With the use of computer systems, optimum selection of materials can be achieved as well as wasting materials, time, and cash can also be remarkably minimized. Hence, KBS plays a crucial role in selecting the most suitable material.

KBSs are systems employed in the technique of artificial intelligence (commonly referred to as expert systems). It is described as a computer system that mimics human problem-solving using expert system and suggestion to a database of knowledge on a specific topic (İpek et al., 2013). They try to comprehend and initiate human knowledge in computer systems (Liao, 2005). Thus, KBSs are known to be human-centered computer system. Material properties can be obtained from experiment works or/and from known information. İpek et al. (2013) defined data as the results acquired from measurements that can be presented in numbers, whereas, information is the connection between objects of data and is mostly communicated in simple words.

Expert Systems

Expert systems are subdivision of applied artificial intelligence and it came into being in the mid-1960s by the artificial intelligence community (Liao, 2005). An expert system is an interactive computer-based decision-making tool that utilizes both facts and heuristics to answer difficult decision problems founded on the knowledge obtained from an expert (İpek et al., 2013). A wide range of task-specific knowledge is transferred from a human expert to a computer. Such knowledge is stored in the computer and users consult the computer through the user interface for specific advice when necessary. Upon inquiry, the computer makes inferences and arrives at a specific answer.

Expert system, also called KBS, is a computer program that simulates the reasoning of human expert in a given field of knowledge. Expert systems rely on heuristics, or the rule of thumb, to extract the information from the large knowledge base. Expert systems are finding many applications in industry including the area of design, manufacturing, trouble shooting, failure analysis material substitution, and material selection. An important advantage of expert system is its ability to capture valuable expertise and make it available to a wider circle of users. They also provide impartial recommendation and are able to search large database for optimum solutions (Farag, 1979; Weiss, 1997).

Nowadays, the use of computers in engineering design has become an acceptable route in most of the engineering projects. Expert systems are particularly suitable for the processing of unstructured scattered knowledge for the solution of complex problems. One such problem is the selection of materials for design with novel materials. In all the systems, data about the materials and their properties were processed and stored in database systems. Logical user interfaces between expert systems and database were developed. Then, the design and selection of the optimal materials were solved through experience and expert knowledge. The expert system developed could offer companies sufficient research potential to solve their practical problems (Sapuan, 1999).

An expert system is developed by creating a knowledge base that is required for the application. A knowledge base contains facts and data that are specific to a particular problem area and rules that tell how to manipulate the facts or data stored (Yingzhan, 2007). Unlike a conventional database, a knowledge base may include facts, assumptions, beliefs, expertise, and heuristic methods, which are exploratory methods for solving problems where an evaluation is made of the progress toward a goal using a series of approximate results.

How to reason result from the phenomena that existed in the system is an important question in using the expert system. According to information of the current problem, the inference engine can distinguish, select, and match the rules in knowledge base to gain results of problems (Hongwei, 2005).

At present, forward and backward reasoning are widely used; forward reasoning is the inference course from premise to conclusion, and on the contrary, backward reasoning is the inference course from conclusion to premise. Applied to the system, forward reasoning is the course from fault feature to fault name, backward reasoning is that first assume fault name, and then search whether prerequisite is satisfied or not, so it can judge whether assumption is right. We utilize forward and backward reasoning in fault diagnosis, and it has got satisfactory effects.

Expert system has been found in many applications especially in diagnosis system, system advisor; web-based advisory system, and fuzzy decision making. Guo et al. (2008) have developed diagnosis system for real time fault in pneumatic system using expert system approach. The system is very effective in improving detection and isolation capabilities of quantitative knowledge–based system.

Wang et al. (2010) used expert system approach for heavy machinery fault diagnostic system based on supporting maintenance. Machine design, assembly, processing environment, and operation as a factor influence the fault of the system. Chunming and Dawei (2007) investigated a fault diagnosis expert system that includes knowledge base, reasoning machine, explanatory, and interfaced based on transient response testing on the mixed-signal circuit using artificial neural network approach. Yang et al. (2005) proposed an expert system using a decision table based on the cause–symptom matrix as probabilistic method to aid plant operators in diagnosing the cause of abnormal vibration for rotating machinery.

KBS has three main components (Fig. 6.5) which are typically classified as

1. Knowledge base
   The knowledge base is the place which contains the rules and techniques that represent human expert knowledge in the system. The domain of a knowledge-based system is represented as frames and objects (Sapuan et al., 2002). The connection between the objects is linked together in the form of hierarchy. Information regarding biocomposites and the type of reinforcement are denoted as objects. The manufacturing techniques for biocomposites and their mechanical and physical properties are stored in the knowledge base. The data for

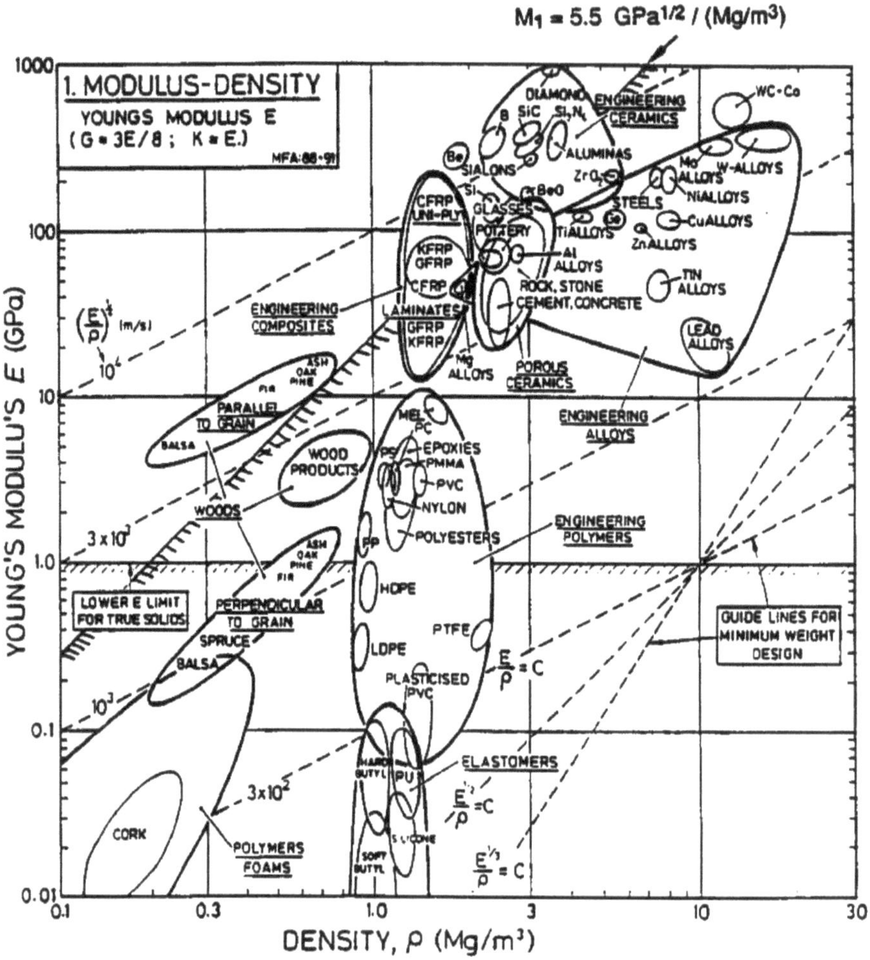

FIGURE 6.5 Sketch of a typical Ashby materials selection chart (Ashby and Cebon, 1993).

the object are collected from published journals, textbooks, online, and other sources. For the frame-based structure, the reinforcement is defined as subclass to the class biocomposite in the object hierarchical tree (Sapuan et al., 2002). In addition, the biocomposite classification is further clarified in depth as the reinforcement–matrix combination and formed as an instance.

2. Inference engine
The inference engine is an essential component in a KBS. It is also known as the "supervisor" that directs the operation upon the knowledge stored in the expert system. It stands between the user and the knowledge base. The internal operation of the inference engine is

transparent to the user. Its fundamental function is inferencing and control. The inference engine consists of program instructions that are required for any expert system to operate. It works depending on rules and arrive at the solution using reasoning method. Forward chaining is the reasoning pattern used by the inference engine. During the operation, the engine will browse through the knowledge base and find the solution with respect to the rule constraints. The engine continues its act of reasoning until the most suitable results for the constraint defined in the rules are obtained.

3. User interface
   This component of the expert system is used for two-way communication between the user and the system. The user interface interconnects and organizes data that have to be evaluated for further evaluation. If the user simply puts all the required information, the interface displays all information transaction, which occurs during the consultation. The interface can be in the form of menus, questions, or graphical icons that are displayed on the computer screen. Communication through the user interface is accomplished by using a keyboard, mouse, or trackball. Sapuan et al. (2002) used a session window of Kappa–PC interfaces to create an application interface. They designed a user-friendly interface in such a manner that when an image in the interface is clicked, it invokes the necessary function and starts running the interface process.

Rule-Based System for Material Selection Process

A rule-based system consists of information acquired from a human expert. The obtained information is represented in the form of rules, such as "If–Then." The "If–Then" rule-based technique can be employed to conduct material selection process by performing operations on data to inference to reach the most suitable conclusion. These inferences are essentially a computer program that provides a methodology for reasoning about information in the rule base or knowledge base, and for formulating conclusions (Liao, 2005). According to Sapuan and Abdalla (1998), when the material selection problem is solved, knowledge about dependences among the material data is used to achieve the optimum material. These dependences in the rule-based system have the following logic

If

The tensile strength of this material is	*high*
The tensile modulus of this material is	*high*
The density of this material is	*low*
The material cost of this material is	*low*

Then this material is selected

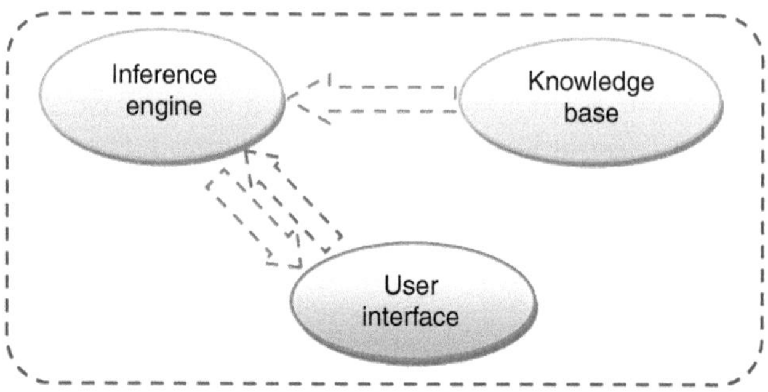

FIGURE 6.6 Expert system components.

Many rules are implemented in succession using chaining process (Sapuan and Abdalla, 1998). The chaining process takes place when a conclusion of one rule concurs with the *premise* of another rule. This is significant when different types of properties or attributes (mechanical, physical, cost, etc.) are to be evaluated (Fig. 6.6).

Constraint Satisfaction and Constraint Violation

It is important to note that each of the rules defined has a set of constraints concerning the properties of the desired material. Therefore, during the material selection process the system communicates all these constraints to examine its validity. A material is only chosen when its properties satisfy all the rules and their constraints. This implies that the chaining process of the material is terminated as soon as that particular material violates a constraint. Thereafter, the KBS searches for other materials until it identifies materials that satisfy all the constraints. At this stage, the selected materials are considered as the most appropriate materials.

Case Study of Materials Selection of Composite Using KBS

This case study extracted from the work of Sapuan and Abdalla (1998) involved the use of KBS for composite material selection. The selection of suitable materials for an automotive accelerator pedal was the main emphasis in this work. Fig. 6.7 shows the general architecture of the research work. The reasoning was carried out via decision rules whereby the rule-based system is the major tool in the material selection process. The system was used to select suitable materials for the polymeric composite automotive pedal box system (Fig. 6.8).

The KBS contained a frame-based system which acts as a database for material properties and manufacturing process capability. A number of constraints in

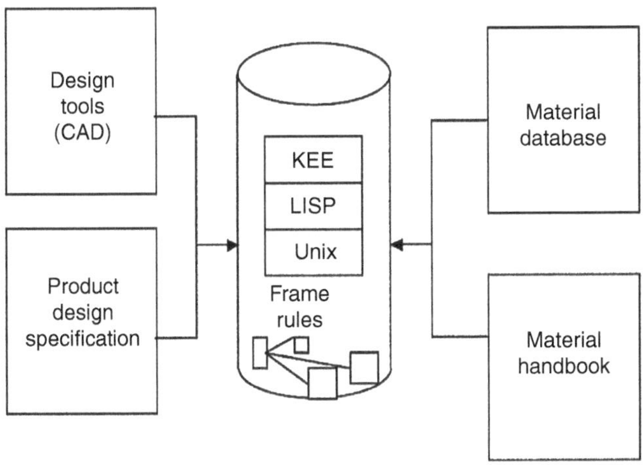

FIGURE 6.7 Structure of prototype system for material selection.

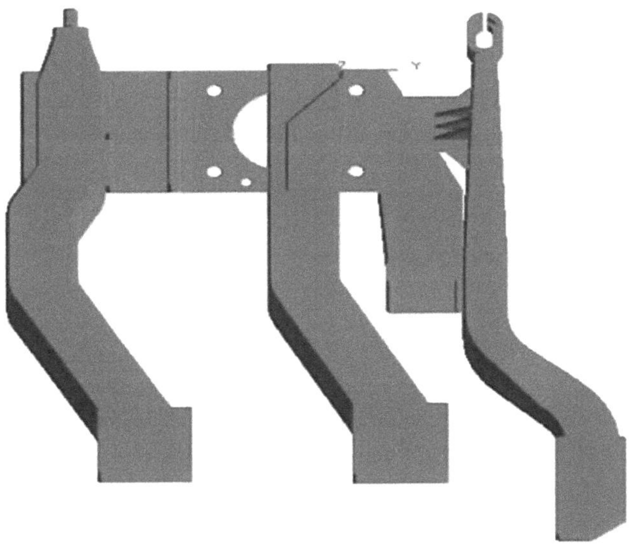

FIGURE 6.8 Automotive pedal box system.

the material selection are represented inside the expert system using the rules in KEE. In this context material property criteria have been utilized as rules to approve constraints. Using the material properties' rules, the designer can examine whether or not the proposed materials are suitable candidates using a particular component. For instance, if the designer specifies the material with specific modulus of elasticity (E_m), in a frame-based system hierarchical graph, the system will compare this modulus of elasticity with the predefined modulus of elasticity limit:

$$E_m < E_{\max}$$

Warning is given in the case of inconsistency or an invalid parameter using the constraint violation facility in KEE. Consequently, the material with this modulus of elasticity is singled out.

The materials selection is performed using rule-based system and the data contained in the knowledge-base are utilized in the selection process. The data from the material databases and other sources are input manually to the knowledge base. The selection of the material and processing parameters for the accelerator pedal depends on a number of factors including mechanical, physical, and chemical properties together with material processing and economic considerations.

The KBS tool-kit (KEE) was implemented to represent all data related to material properties in a hierarchical structure as shown in Fig. 6.9. Other types of data that are used in the study include constraint data for material selection in the form of product design specifications as shown in Table 6.5 and design data obtained from CAD and FEA systems. The selection of a material for the accelerator pedal is performed by using the following rules

If

(the corrosion resistance of this material is high) and
(the water absorption of this material is low) and
(the material cost of this material is low) and
(the tooling cost of produce part using this material is low) and
(the labor cost to produce part using this material is low) and
(the dimensional stability of this material is good) and
(the production volume to produce part using this material is high) and
(the manufacturing method to produce part using this material can make ribbing structure) and
(the modulus of elasticity of this material is high) and
(the yield stress of this material is high) and
(the toughness of this material is high) and
(the creep resistance of this material is good) and
(the fatigue strength of this material is high) and
(the density of this material is low)
Then

(This material is a candidate for accelerator pedal)

These rules are chained together using the forward chaining. Therefore, a material that has a yield stress more than the constraints yield stress for the

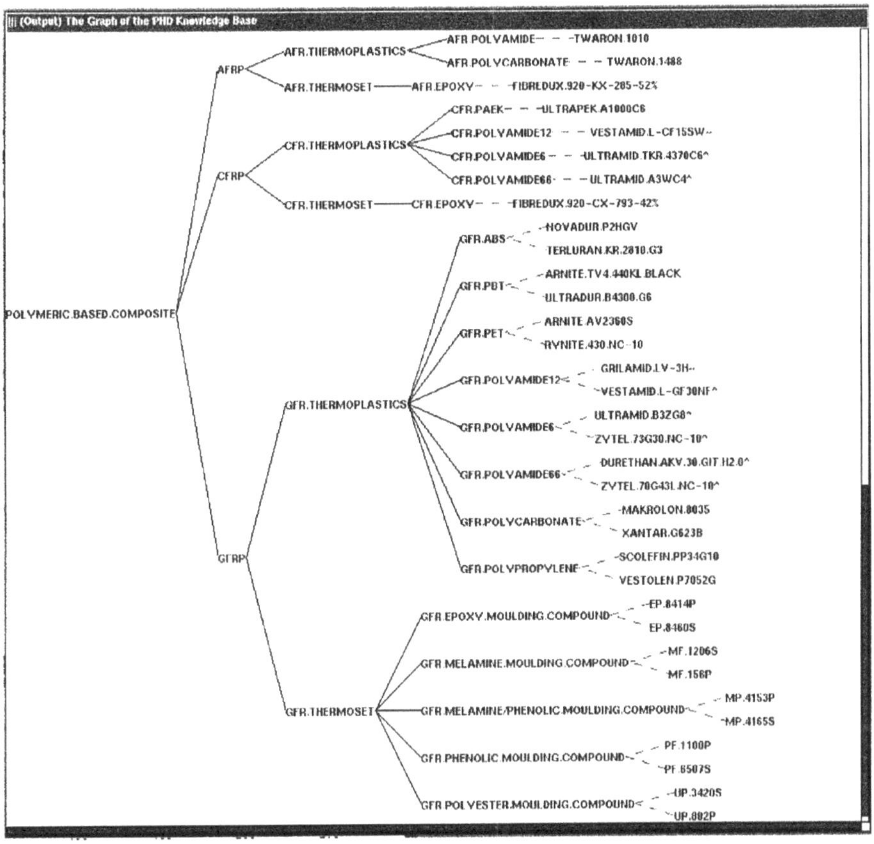

FIGURE 6.9 The hierarchical graph of polymeric-based composite materials.

Table 6.5 Product Design Specification of Accelerator Pedal
Modulus of elasticity > 5000 MPa
Yield stress > 59.7 MPa
Charpy impact strength > 30 kJ/m^2
Creep resistance = High
Fatigue strength = High
Corrosion resistance = High
Density 7 < 2.7 mg/m^3
Water absorption < 8%
The material must be recyclable
Ribbing structure can be made using the material
Manufacturing method associated with material must be high volume production
The component made from this material must be dimensionally stable
Tooling cost < £62,000
Material cost < £4.(5 per kg)
Direct labor cost < £2 per component

accelerator pedal (59.7 MPa) (Table 6.5) is considered as suitable material for this pedal. If this material satisfies the rest of the constraints, then this becomes the chosen material for the accelerator pedal. On the other hand, if any of the constraints is violated, the KBS stops checking through the rest of the rules for that material. It indicates that the chaining for these particular rules is terminated. The system then searches for the next closest material to the previous one in the hierarchical graph, and the material is checked against all the rules. This continues until a set of solutions is found.

The final phase of the selection is performed by listing four main properties of materials, such as modulus of elasticity, material cost, and yield stress. Arrow's impossibility theorem is used to give the overall ranking of the materials. This method uses the principle based on the ranked orders over a number of different attributes. It produces an overall ranking for the accelerator pedal which works by giving each material the scores for each attribute and then sums up all the scores. The material with the highest score is ranked as "High" and the material with the lowest score is ranked as "Low." The material with the highest score in this phase is selected as the material for the accelerator pedal.

Use of Exsys Corvid (Material Selection Expert System Shell) in Research

Exsys Corvid software was developed by Exsys Inc. since 1983. It has decades of experience in the artificial intelligence field and designing practical expert system development tools. Exsys corvid software is widely used in decision support, regulatory compliance, product selection and recommendation, configuration, and problem solving diagnostics, etc. Exsys Corvid concept provides heuristic and rules, chaining, confidence, Corvid Variable, Logic Blocks, and Metablock. Heuristics and rules are also known as rules of thumb in expert system environment which provide specific fact that tells how to make a part of decision, normally used IF/THEN rule.

Confidence is a special feature in Exsys Corvid, in which the rule can include confidence factor for a particular result. This enables the expert systems to make multiple recommendations with differing degree of confidence to reach a best fit in its conclusion (Exsys Corvid, 2008).

A number of researchers have reported using Exsys Corvid expert system shell in their research. Alibaba and Özdeniz (2004) developed a building element evaluation and selection system using Exsys Corvid software. They developed the building element selection by considering building elements such as retaining wall, foundation, external, and internal wall and floor. The developed system assists experienced and inexperienced architects and for all kinds of building types. Lu et al. used Exsys Corvid software in the development of portable decision support system for emergent high mountain syndrome

(HMS) management. They used questionnaire dialog related to the sickness symptoms such as anorexia, insomnia, dizziness, and lassitude or fatigue during mountain climbing to solve the climbers' problem. The developed system helps the climbers while climbing the mountain. In both research works, the goals and criteria of the decision are defined.

Ahmad and bin Yaacob (2002) used Exsys Corvid software in the decision making for diagnosis of a suspected transformer faults and the appropriate maintenance actions to be taken based on dissolved gas analysis techniques. Alibaba and Özdeniz (2004) employed Exsys Corvid to construct an expert system for the evaluation and selection of suitable building elements for construction. Kowalski et al. (2006) came up with an online academic advising website using Exsys Corvid software to provide decision-making knowledge and customized recommendations to students through an interactive interface. Pavleković et al. (2007) developed an expert system for detecting a child's gift in mathematics in the fourth grade of elementary school using Exsys Corvid software. Flior et al. (2010) studied the feasibility of using an Exsys Corvid based expert system to detect and respond to network threats and appropriately administrate a Linux-based iptables firewall in real-time.

Of late, Exsys Corvid expert systems are increasingly been utilized for material selection. This software has the ability to select suitable materials and automatically rank them in a single step (Fairuz et al., 2012). Fairuz et al. (2012) developed an expert system using Exsys Corvid software for material selection of polymeric-based composites for fishing boat components (more information to be added).

Sanyang and Sapuan (2015) also developed an expert system using Exsys Corvid software to select suitable biobased polymer materials for packaging fruits, dry food, and dairy products. *If–Then* rule based system was utilized to accomplish the material selection process whereas a score system was formulated to facilitate the ranking of selected materials. The expert system selected materials that satisfied all constraints and selection results were presented in suitability sequence depending on their scores. The expert system selected polylactic acid (PLA) as the most suitable material.

JAVA-BASED MATERIALS SELECTION

Java programming software is preferred as it is an open source high performance software application. Its platform is independent and portable, and has the ability to move easily from one platform to another. Java works in a distributed environment and it is "write once and use anywhere, anytime" (Java technology, 2014). Java is widely used for web-based application development as it is easy to deploy as hypertext transport protocol (http) for World Wide Web (www) usage.

The Java technology was seldom used in engineering design. Ginsberg et al. (2000) emphasizes the importance of implementing Java in industrial applications. They highlighted the strength of Java and its potential usage in industry. The mechanism of Java language is based on object oriented programming that allows encapsulation, inheritance, polymorphism, and operator overloading. Java Net BeansIDE has an explorer as the main interface, the project can be viewed in the explorer view, which has become a common visual metaphor in most graphical user interfaces (GUI).

Recently, Ahmed et al. developed an expert system for the selection of optimum NFC material for automotive interior components using JavaTM NetBean. Data for the physical and mechanical properties of NFC materials were accumulated from multiple sources and consolidated into a single knowledge base of NFC materials. The interior components in automobiles such as dashboard, door panel, and rear panel were considered as a case study for testing the efficiency of the system. The developed expert system further integrated with the technology of artificial neural network for classification of selected materials. Finally, the expert system was converted as web-based application for online usage with the Java applet programming and can be accessed through the web browsers worldwide.

The interface of the system was designed in a manner that it renders a user-friendly access. When the system starts, it shows the login screen as presented in Fig. 6.10. This security feature provides the authentication for the system by validating the username and password. The user types the username and password and it is compared with that stored in the database to validate the data to provide authentication to use the system. The illegal access to the system was denied with a warning message. Hence, the security of the expert system was retained.

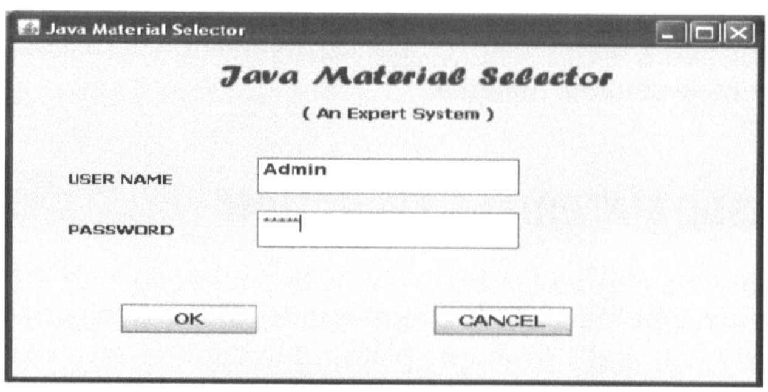

FIGURE 6.10 Login screen for user authentication.

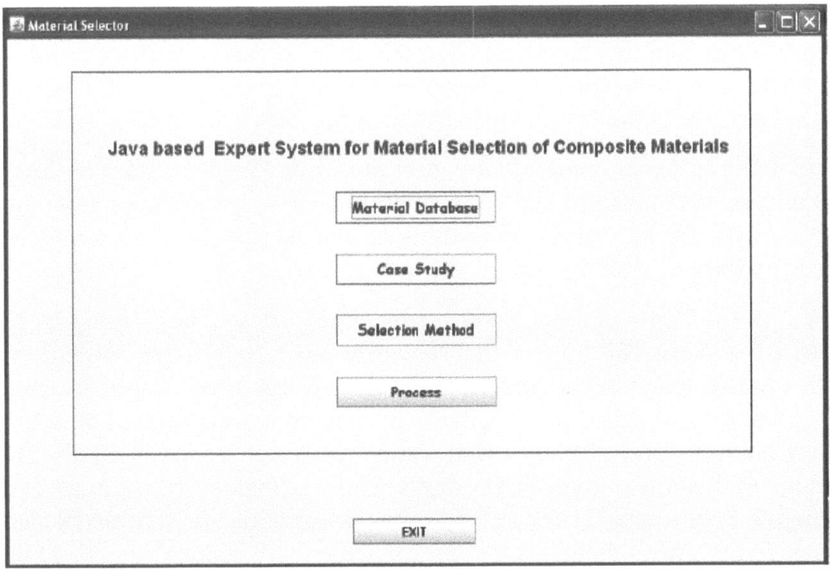

FIGURE 6.11 User interface of module screen.

The component of the expert system developed using the Java NetBean has four modules as shown in Fig. 6.11. It was designed for user-friendly access with four different interface screens.

1. Material knowledge-base module
2. Case study selection module
3. Selection method module
4. Process module

MATERIALS SELECTION OF COMPOSITE MATERIALS USING MULTICRITERIA DECISION MAKING

Materials selection is a crucial process for achieving customer satisfaction attributes. Implementing new materials in a specific industrial sector is limited by several constraints and factors. Therefore, selecting the most suitable material type for a particular application is a complex matter where keen decisions have to be taken based on pairwise comparisons as the base of decision making in various engineering applications.

NFCs became recently highly valuable alternative materials (Alves et al., 2010). Such materials have emerged in different applications because of their desired properties (Al-Oqla et al., 2015b, 2014). The advantages of such composites including light weights, high specific properties, low cost as well as recyclability and degradability characteristics made them suitable alternatives for traditional composites in modern green products (Sapuan et al., 2013; Al-Oqla et al., 2015a, 2014; Mir et al., 2013; Sahari et al., 2014).

Better evaluations of natural fibers regarding wide range criteria have several advantages. It can lead to better informative decisions regarding selecting the suitable NFC for industrial applications on one hand and enhance achieving better performance on the other.

Based on the literature, little studies considered the selections of the proper NFCs for a particular application. Sapuan et al. (2011a) have implemented AHP method to select the proper material for automotive dashboard panel. The properties of materials considered in that study were limited to density, young modulus, and tensile strength. Moreover, Cheung et al. (2009) discussed and mentioned the potential of both plant and animal for the biomedical engineering applications. They also introduced some factors to be considered in screening general materials for that branch of applications including being bio-inert, bi-functional, and bio-active.

The complete NFCs attributes and capabilities depend on the physical and chemical compositions of the inherent material. Thus, NFCs must be investigated regarding many different features and properties before being considered to be used in any particular application. Such factors of selecting the NFC products can be according to natural fiber properties (NFP), polymer base properties (PBP), composite characteristics (CC), and composite performance (CP) (Al-Oqla and Sapuan, 2014a), where different attributes and characteristics have to be investigated and taken into account to reach better evaluations of the NFC materials like physical, mechanical, chemical, biological, and environmental properties as well as toxicity, bio-stability, life cycle, durability, bio-degradability, weather resistance, occupational health and safety among others. Criteria that affect selection of natural fibers are shown in Fig. 6.12. Such criteria can be the key driver for the industry to enhance selection of the most appropriate NFC materials for a given application through better selection of the composite constituents.

Comparing various natural fiber types considering wide desired factors required precise data (Al-Oqla and Sapuan, 2014; Al-Oqla et al., 2014b, 2014; Pilla, 2011). Various properties of the available natural fibers are shown in Table 6.6 where the average values of various properties are utilized.

Comparisons between different natural fiber types regarding their specific modulus of elasticity and specific strength are shown in Figs. 6.13 and 6.14, respectively. It can be clearly shown that natural fibers vary in their physical and mechanical properties. These comparisons actually combined the fibers' modulus of elasticity and specific strength relative to their densities to give

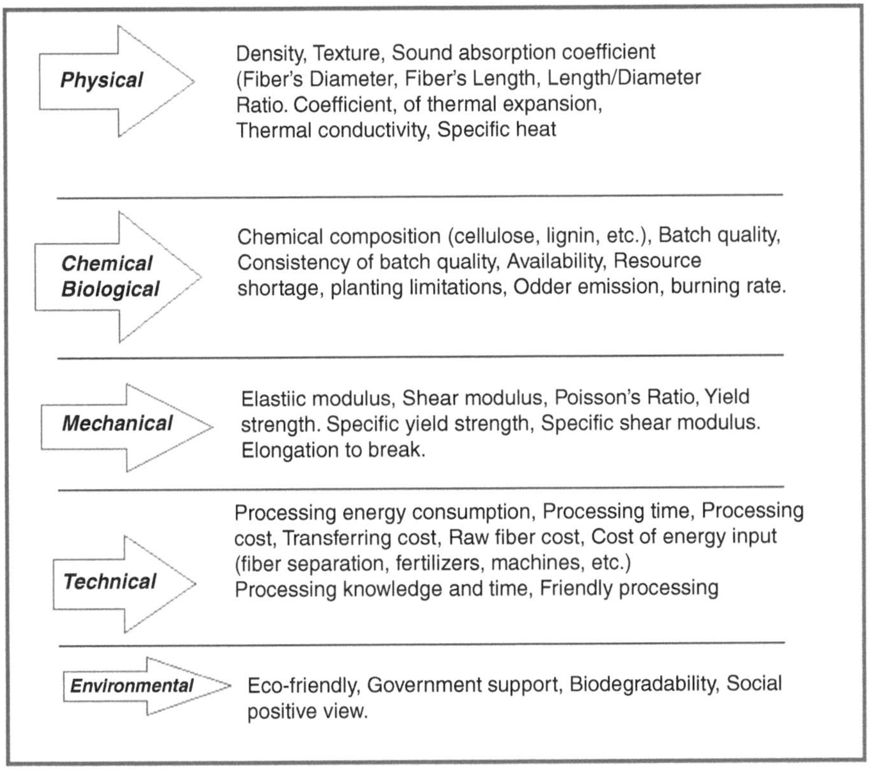

FIGURE 6.12 Selection criteria of the natural fibers for NFC materials.
Reproduced from Addin, O. Sapuan, S.M., Othman, M. Ahmed Ali, B.A., 2011. Comparison of naive bayes classifier with back propagation neural network classifier based on f-folds feature extraction algorithm for ball bearing fault diagnostic system. Int. J. Phy. Sci. 6 (13) 3181–3188.

Table 6.6 Various Properties of the Commonly Used Fibers (Al-Oqla et al., 2015b)

Fiber Type	Coir	Date Palm	Flax	Hemp	Jute	Sisal
Density (g/cm^3)	1.15–1.46	0.9–1.2	1.4–1.5	1.4–1.5	1.3–1.49	1.33–1.5
Tensile strength (Mpa)	95–230	97–275	343–2000	270–900	320–800	363–700
Tensile modulus (Gpa)	2.8–6	2.5–12	27.6–103	23.5–90	8–78	9–38
Specific modulus	4	7	45	40	30	17
Elongation to break (%)	15–51.4	2–19	1.2–3.3	1–3.5	1–1.8	2–7
Cellulose (wt. %)	32–43.8	46	62–72	68–74.4	59–71.5	60–78
Hemicellulose (wt.%)	0.15–20	18	18.6–20.6	15–22.4	13.6–20.4	10–14.2
Lignin (wt.%)	40–45	20	2.3	3.7–10	11.8–13	8–14
Moisture content (wt.%)	8	5–12.1	8–12	6.2–12	12.5–13.7	10–22
Cost per weight (USD/kg)	0.3	0.02	9.0	1.2	0.3	1.0

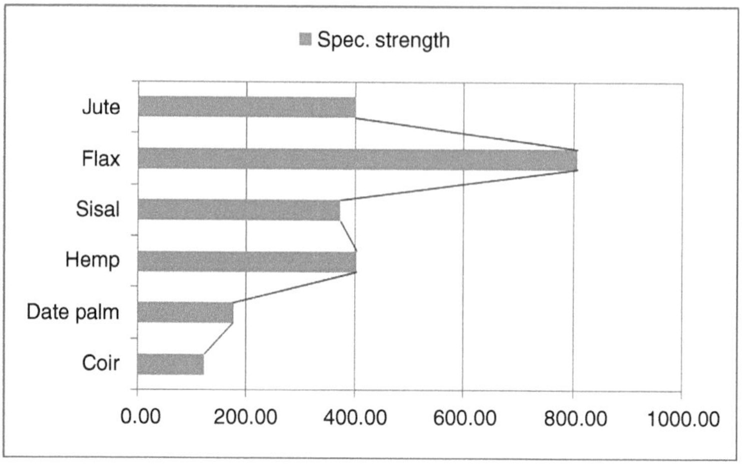

FIGURE 6.13 Comparison of different natural fiber types regarding specific strength (Al-Oqla et al., 2015b).

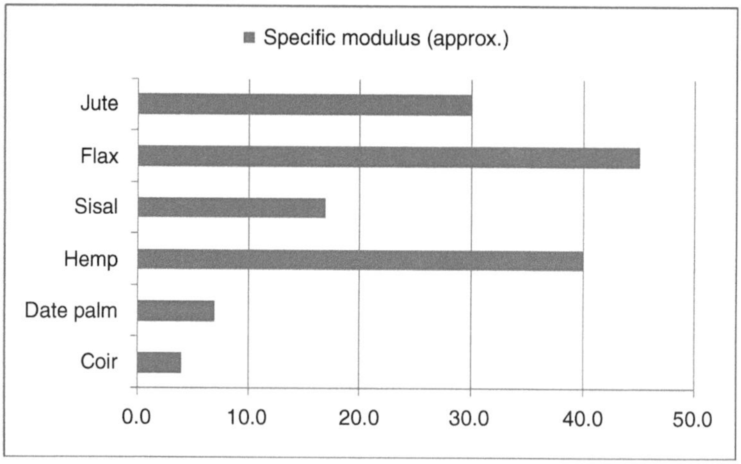

FIGURE 6.14 Comparison of different natural fiber types regarding specific modulus (Al-Oqla et al., 2015b).

more relative information about different fiber types to enhance NFCs' selection process. It is clear that flax, hemp, and jute are the best regarding these criteria (specific modulus and specific strength) in both cases of specific strength and specific modulus. However, further investigations regarding wider criteria rather than physical and mechanical properties are needed to make more realistic informative decisions regarding the best fiber type. That is, further criteria like cost might change the priority of the previous three fiber types as seen in

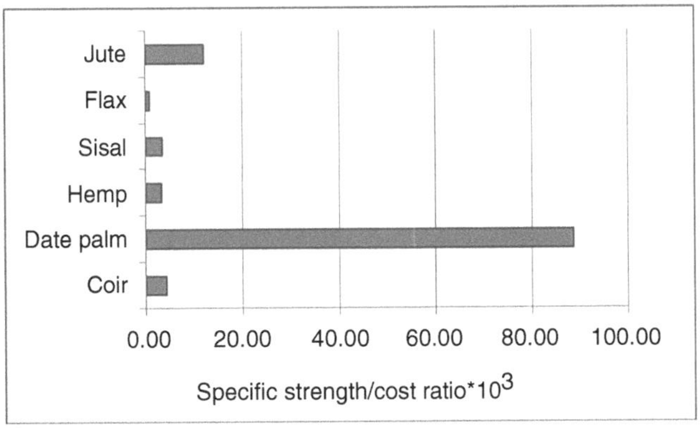

FIGURE 6.15 Comparison of different natural fiber types regarding combined specific strength to cost ratio criterion (Al-Oqla et al., 2015b).

Fig. 6.15 where the specific strength to the cost ratio comparisons were conducted. It is obvious that date palm fibers are the best type among all other types if further economic criteria are considered. It can be seen that date palm fiber is about four times better than jute.

It is realized that selecting the most appropriate natural fiber type to form NFCs is affected by several criteria and considered as a multicriteria decision-making problem.

MCDM can be applied in conditions where several contrasting criteria are present in the process. They can be further categorized into two main groups which are multiple objective decision making (MODM) and multiple attribute decision making (MADM) methods. The proceeding method is applicable to composites material selection which is normally more complex.

Screening methods are applied for direct comparison between a list of possible materials when the list has been narrowed down into two or less materials with the best characteristics. However, it is less meaningful when more than two materials are considered for selection. In this circumstance, MCDM method is more suitable to handle this by providing the most optimum selection. Examples of MCDM methods available are elimination and et choice translating reality (ELECTRE) method, technique for order preference by similarity to ideal solution (TOPSIS) method, Vlse Kriterijumska Optimizacija Kompromisno Resenje (VIKOR) method, preference selection index (PSI) method, and grey relational analysis method. In composite material selection, analytic hierarchy process (AHP) method which is an MCDM method is widely used. The AHP method is very agile and able to be successfully used together with other design tools such as QFD and other MCDM methods such as TOPSIS. It offers more accurate results to be more decisive in determining the best material for the intended product design.

ANALYTICAL HIERARCHY PROCESS

The nature of multicriteria problems is established in various engineering fields (Al-Oqla et al., 2014; Surawski et al., 2013). AHP is a carefully designed method to solve MCDM problems (Saaty, 2003). It helps solving sophisticated problems and reduce bias in the decision-making process throughout a mechanism that enables it to check the consistency of the evaluated attributes (Al-Widyan and Al-Oqla, 2014; Wong and Li, 2008; Rao and Patel, 2010).

In the AHP approach, complex decisions are to be decomposed into simple pairwise comparisons considering multiple factors or criteria. The main goal is decomposed into its constituent objectives. More precisely, the top of the hierarchy structure is occupied by the main goal but decision alternatives at the bottom, whereas the selection criteria and their subs are to be organized between the top and bottom levels. Then, a relative weight for each item in any certain level has to be assigned. Consequently, the relative importance of each factor in the model can be illustrated. Then, pairwise comparisons have to be conducted between all items within each level with respect to their parent to get a relative proportional score. The consistency of this judgment is mandatory. Thus, a consistency indicator or measurement is essential for the expert judgment validation. More precisely, to consider a matrix $a(i, j)$ as a consistent matrix, it should obey both the transitivity and reciprocity rules as in Eqs. (6.5) and (6.6).

$$a_{i,j} = a_{i,k} \cdot a_{k,j} \tag{6.5}$$

$$a_{i,j} = 1/a_{j,i} \tag{6.6}$$

where i, j, and k are any matrix item.

As a result, the pairwise comparison matrices can be expressed as

$$A = \begin{bmatrix} a_{11} & \cdots & a_{1n} \\ \vdots & \vdots & \vdots \\ a_{n1} & \cdots & a_{nn} \end{bmatrix} = \begin{bmatrix} w_1/w_1 & \cdots & w_1/w_n \\ \vdots & \vdots & \vdots \\ w_n/w_1 & \cdots & w_n/w_n \end{bmatrix}$$

The formulation of a consistent matrix can be

$$\begin{bmatrix} w_1/w_1 & \cdots & w_1/w_n \\ \vdots & \vdots & \vdots \\ w_n/w_1 & \cdots & w_n/w_n \end{bmatrix} \times \begin{bmatrix} w_1 \\ \vdots \\ w_n \end{bmatrix} = n \begin{bmatrix} w_1 \\ \vdots \\ w_n \end{bmatrix}$$

Using matrix form:

$$\mathbf{A}.\mathbf{w} = n\mathbf{w} \tag{6.7}$$

where **A**,**w** and n are the pairwise comparison matrix, the eigenvector and the matrix dimension, respectively. Moreover, Saaty (1980) introduced a consistency measurement known as consistency index (CI) which is actually a deviation of consistency as in Eq. (6.8).

$$\mathrm{CI} = \frac{\lambda_{\max} - n}{n - 1} \tag{6.8}$$

The consistency index has to be compared with a proposed random index (RI) as in Table 6.7 (Saaty, 1980) to get a consistency ratio (CR) as indicated in Eq. (6.9).

$$\mathrm{CR} = \frac{\mathrm{CI}}{\mathrm{RI}} \tag{6.9}$$

This consistency ratio is able to determine the acceptance of the subjective judgment. That is, if (CR) is less than or equal to 10%, the inconsistency of the judgment is acceptable. Otherwise, the subjective judgment has to be revised.

The AHP method had been developed by Saaty (1980) and has been considered as one of the best MCA approaches available. It uses intuition in making the selection by assessing the relative weight of multiple criteria. In the event of in availability of quantitative ratings, the designer can still recognize whether there is an outstanding criteria. Therefore, AHP is widely accepted and many researches have been conducted on improving it.

Analytic hierarchy process (AHP) method is part of the MCDM method that has been applied in concurrent engineering involving composites applications. Ariff et al. (2007) reported the use of AHP in automotive bumper beam concept design selection involving synthetic polymer composite materials at the conceptual design stage of the product. Elsewhere, various applications of AHP for materials selection for both polymer matrix composites and metal matrix composites have also be been reported involving a range of products from automotive to household items (Dweiri and Al-Oqla, 2006; Hambali et al., 2011; Mansor et al., 2013; Sapuan et al., 2011b). The AHP method is found to be applied mostly in the materials selection at early product development stage especially for the automotive related application. Apart from that, it was

Table 6.7 The Random Index (RI)

n	1	2	3	4	5	6	7	8	9	10	11	12
RI	0	0	0.58	0.9	1.12	1.24	1.32	1.41	1.45	1.49	1.51	1.58

found that the AHP method is very flexible and able to be successfully integrated with design tool such as QFD and other MCDM methods such as TOPSIS to provide more accurate results in determining the optimal material for the intended application. One of the profound advantages of the AHP method toward CE application such as for material selection purpose compared to other MCDM methods is that AHP allows for flexibility in performing quick and decision-making process where judgment on the listed alternatives can easily be made using pairwise comparison technique either by accepting inputs from either individual or from group of people (Al-Harbi, 2001; Macharis et al., 2004). Thus, the group decision–making capability in AHP where collective decisions from various experts from different stakeholders can be obtained is able to very well assist the successful application of CE considering the nature of CE itself that encompasses the integration of multiple fields in the product development activities. The group judgment is later translated into single consensus either through geometric mean or arithmetic mean methods after the pairwise comparison stage in AHP with reportedly no deviation between the methods up to 200 experts (Wu, 1998). Apart from that, the AHP method also allows for decision involving both tangible and nontangible aspects which makes it a powerful and flexible decision-making process to obtain the desired priorities and optimum decision (Rathod and Kanzaria, 2011; Rao and Davim, 2008).

It was reported that AHP was utilized for selecting the most appropriate NFCs and their constituents for various industrial applications (Al-Oqla et al., 2015a, 2014b). Several models utilizing the AHP method were developed for such purposes; Fig. 6.16 demonstrates such models for selecting the most appropriate polymer type for date palm fiber for the industrial applications.

As an output of such model the final priorities of several polymer types regarding specific mechanical properties like elastic modulus (EM), fracture toughness (FT), elongation to break (EB), yield strength (YS), and impact strength (IS) are shown in Fig. 6.16.

Such models using AHP can demonstrate the real status and priorities of several alternatives regarding various evaluation criteria in details. It can be noticed here for example that Epoxy has the highest priorities in both of elastic modulus (EM) and yield strength (YS) subcriteria and the least in the others. Moreover, polypropylene has the highest priority with respect to the fracture toughness (FT) criterion, but not the least in any. Furthermore, high density polyethylene has the highest priority regarding the impact strength (IS) criterion but the least in the yield strength (YS) one. In addition to that, low density polyethylene has the highest priority considering the elongation to break (EB) criterion, but the least in both elastic modulus (EM) and yield strength

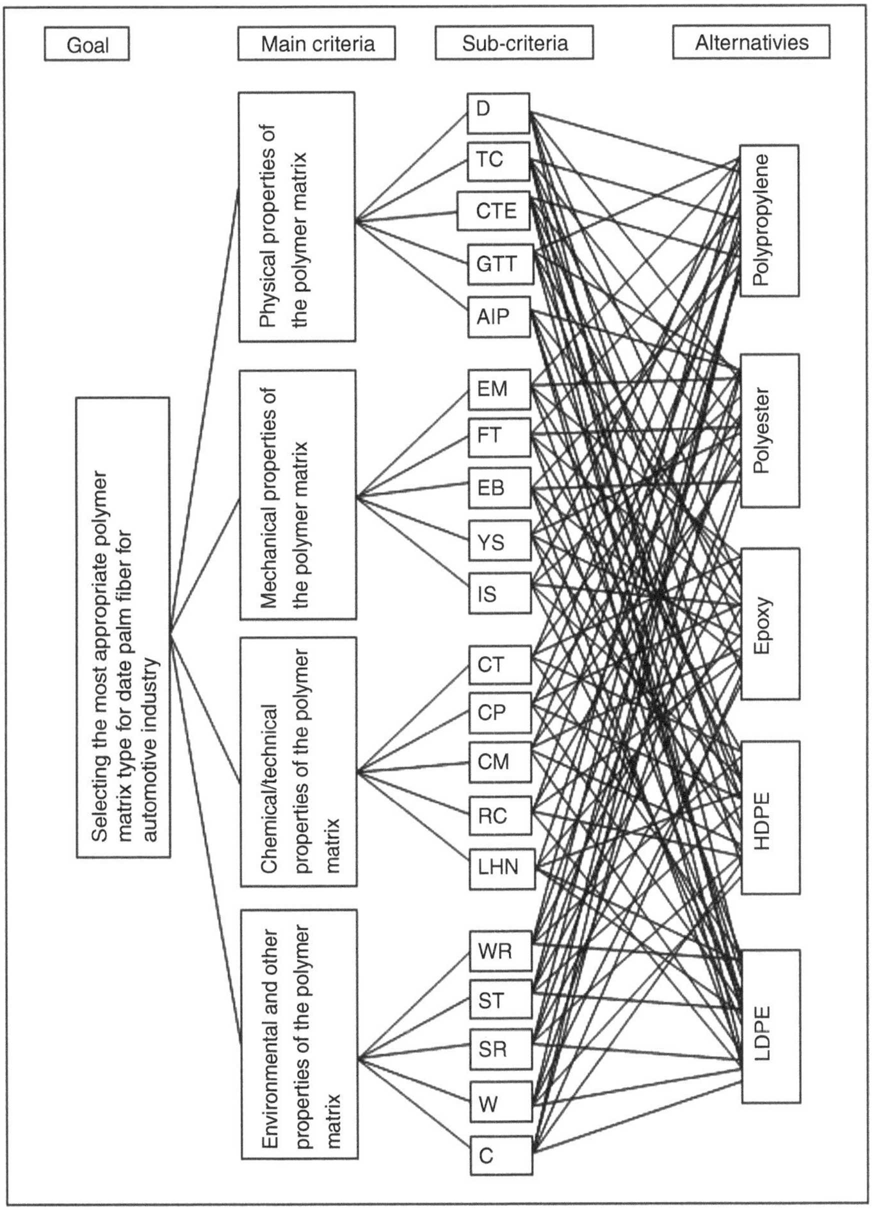

FIGURE 6.16 The AHP model for selecting the most appropriate polymer base type for the DPF (Al-Oqla et al., 2015c).

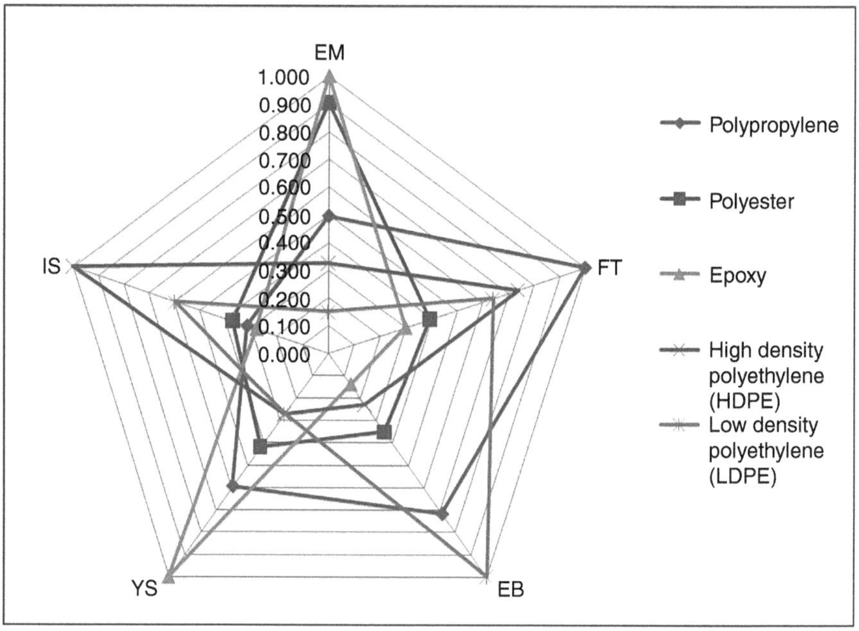

FIGURE 6.16 *(Cont.)* **Priorities of the alternatives regarding all subcriteria in the mechanical properties main criterion (Al-Oqla et al., 2015c).**

(YS) subcriteria. On the other hand, polyester does have neither the highest potential nor least one in any of the considered subcriteria of the mechanical properties category.

Such developed models using AHP utilized consistent systematic experts' judgments which eliminate the bias or prejudice in decisions and reduce human errors regarding polymer selection process. This also helps establishing a road map for the selection of the best polymer characteristics and parameters for polymer matrix composites in different industrial and engineering applications based on scientific decision-making technique.

Case Study on Composite Material Selection Using AHP for Automative Brake Lever

AHP has been widely used in various studies involving MCDM, most especially in automotive design application. In this study, the AHP method is employed in the selection of the most appropriate natural fiber to be hybridized with glass fiber reinforced polymer composites for the design of a passenger vehicle center lever parking brake component (Fig. 6.17). The new candidate material for the parking brake lever component is expected to be light in weight while at the same time able to maintain the required structural strength for

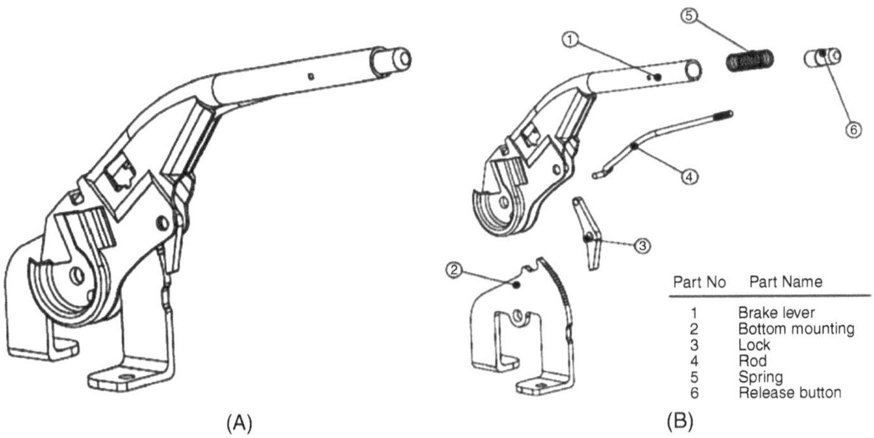

FIGURE 6.17
A passenger vehicle center lever parking brake component (A) assembly view, and (B) exploded view.

safe and effective operations compared to the current material. Furthermore, the new selected material should be able to adhere to the latest environmental policies governing automotive product design. Fig. 6.18 shows the detail AHP methodology used for the hybrid natural fiber/glass fiber reinforced polymer composites selection.

The first and foremost stage in developing the AHP method used in this study is to initiate an AHP hierarchical framework which illustrates a systematic overview of the interelationship between the study objective and the set of criteria as well as alternatives connected with it. Hence, a four level AHP framework is applied based on the product design specification elements. The study objective is defined at the AHP hierarchy level-1. The AHP structure was then extended to level-2, where all the main criteria such as functional performance, weight, and cost are defined. At level-3, the subcriteria such as strength, stiffness, density, and raw material cost for the material substitution are stated. Finally, at level-4, 13 types of natural fibers are selected as candidate material for the component design as shown in Fig. 6.19. Connecting lines are constructed to link all the information within the framework indicating their relationship with each other.

Performing pairwise comparison between the defined objectives, criteria, and alternatives within the AHP framework developed in the previous stage. This is the second stage in the selection of the most suitable natural fiber to be hybridized with glass fiber reinforced polymer composites using the AHP method. Basically, the pairwise comparison is a method to decide on the relative importance between criteria with regard to the objective and between alternatives with respect to the criteria. Therefore, at this stage, the relative importance

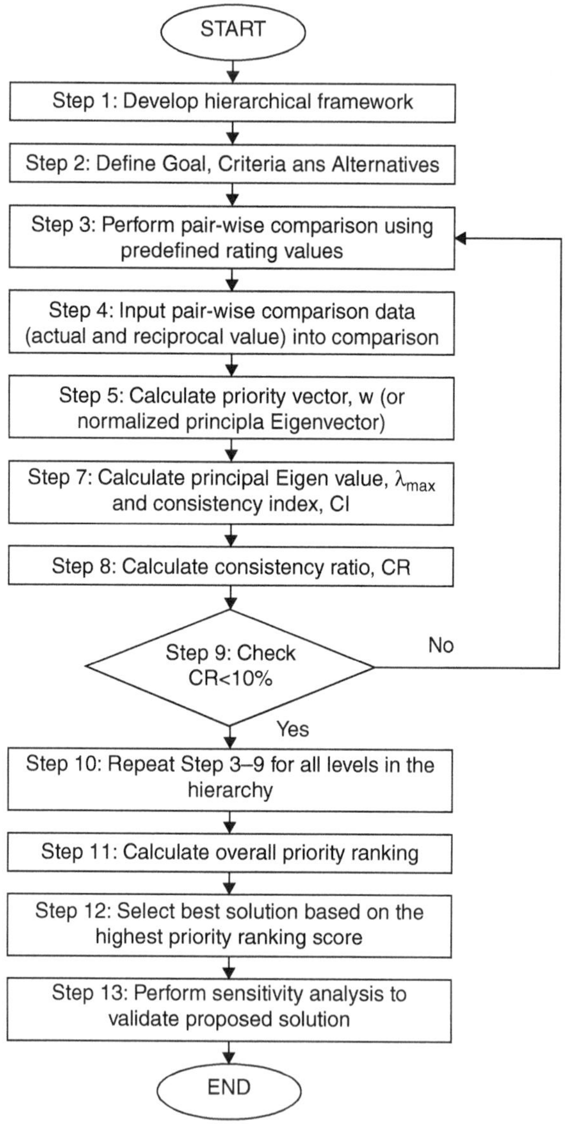

FIGURE 6.18 AHP methodology for material selection.

between two information is given a numerical value based on the predefined rating values for pairwise comparison. To help in the pairwise evaluation, material properties for natural fiber–based polymer composites available from the literature were used as reference.

The third stage in the AHP analysis is to synthesize all the pairwise comparison data into matrix form, normalizing the values within the matrix and

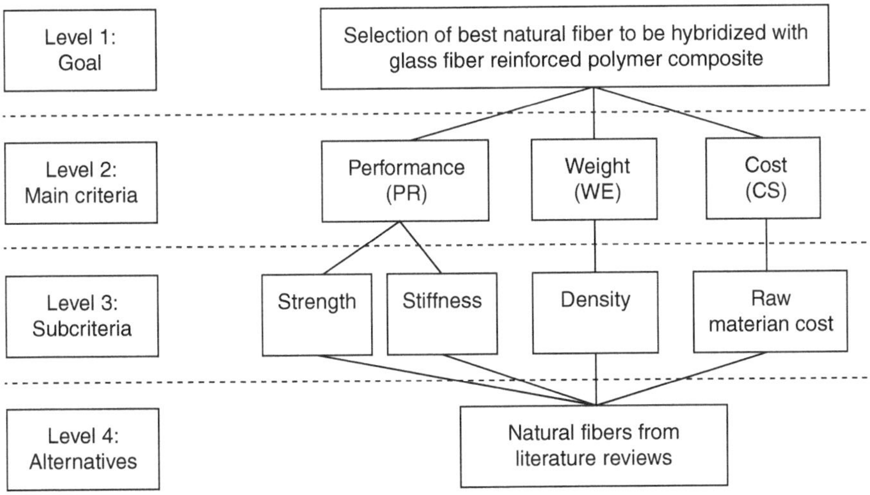

FIGURE 6.19 Developed AHP hierarchical framework.

calculating priority vector, w. Thereafter, analysis is conducted to determine whether the judgments made during the pairwise comparison are consistent or not by calculating consistency ratio (CR). The consistency verification process is regarded as one of the most advantages of the AHP method especially when subjective judgments are involved during the decision-making process (Hambali et al., 2010).

After all the aforementioned stages, the AHP analysis finally concluded that kenaf bast fiber is the most suitable candidate material (Fig. 6.20) to be hybridized with glass fiber reinforced polymer composites for the design of a passenger vehicle center lever parking brake component.

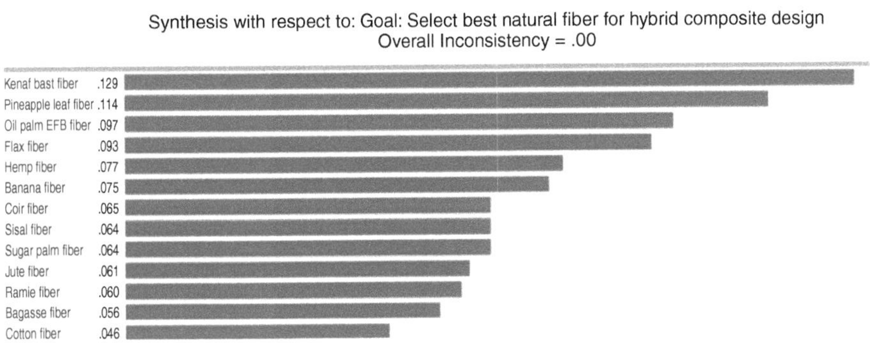

FIGURE 6.20 Overall results of the material selection.

Case Study on the Implementation of AHP for Environmental Assessment in Composite Material Selection for Automotive Components

In this brief case study, a detailed eight steps of procedure were adopted to apply the concept of AHP. The flow chart of AHP methodology is shown in Fig. 6.21. Firstly, the problem should be clearly identified and the study goal has to be defined as well. Furthermore, the decision makers have to identify the factors or criteria that will affect the selection process.

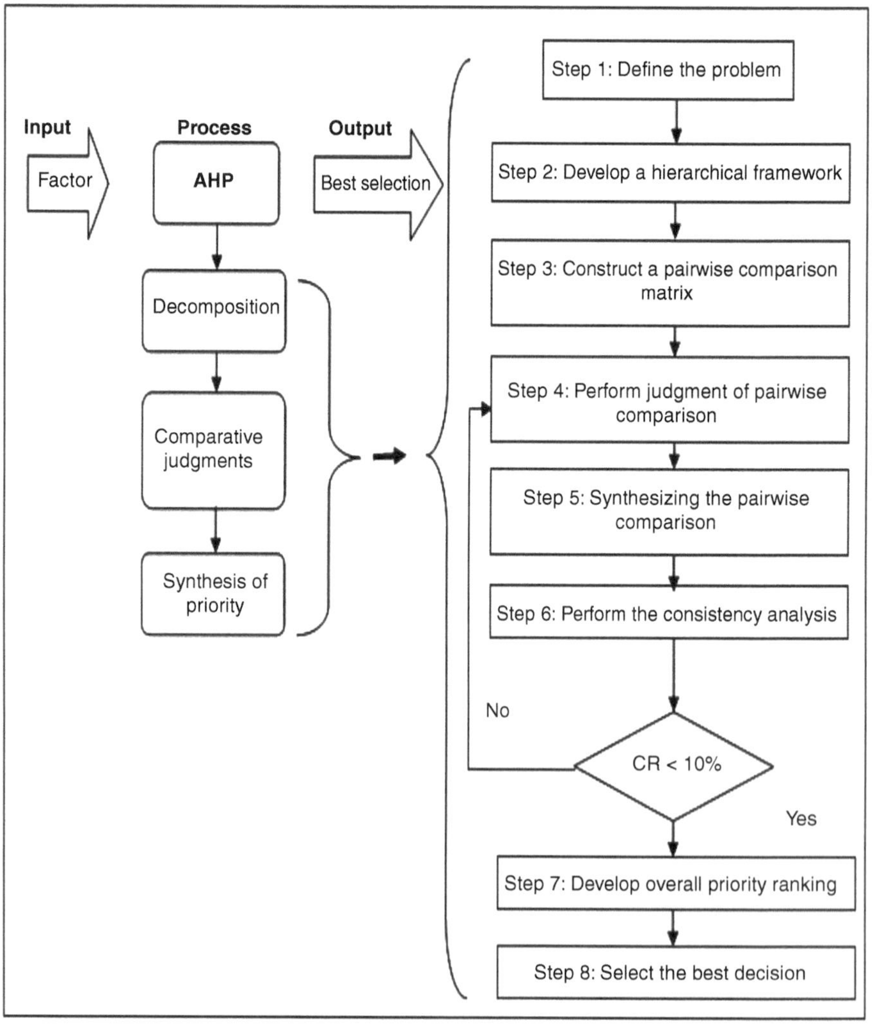

FIGURE 6.21 Process flow chart of the AHP study (Saaty and Vargas, 2012).

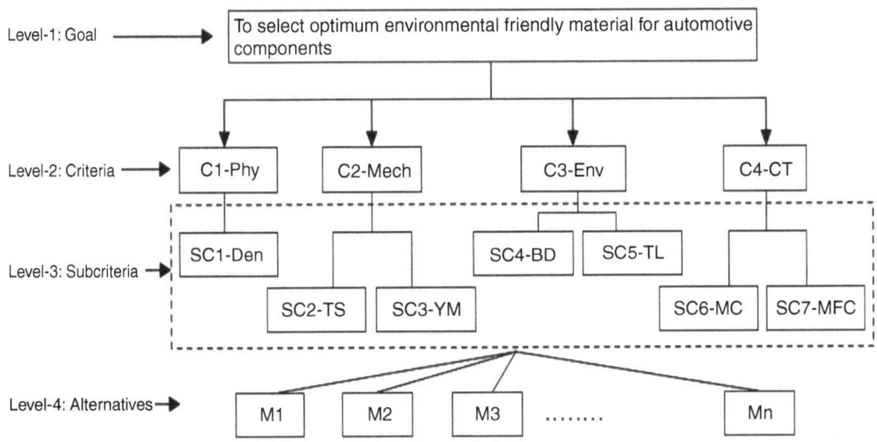

FIGURE 6.22 The hierarchical framework of AHP method.

In the second step, the hierarchical framework of the method is developed, which shows a systematic overview of the relationship between the study objective and the set of criteria, subcriteria, and alternatives or candidate materials that are involved in the selection process. In this study, the model is presented with four levels, where the top level or level-1 is the objective or goal of the study. In level-2, the criteria or the factors influencing the goal are defined (physical, mechanical, environmental, and cost). Then, level-3 comprises the subcriteria such as density (kg/m^3), tensile strength (MPa), young's modulus (GPa), biodegradability, toxic level, raw material cost(US$/kg, US$ 1 = RM 3.23), and the manufacturing cost. Finally, the level-4 or bottom level displays the set of alternatives or the candidate materials in the hierarchy. The hierarchical framework of our study is shown in Fig. 6.22.

The third step was to construct a pairwise comparison matrix by considering all the factors. In AHP, a nine-point scale was used to conduct the pairwise comparisons for each level. In the pairwise comparison the element in the corresponding level was compared with each other to calibrate the relative importance with one to another. This process compares the relative importance, performance or likelihood of two elements in the level. After accomplishing the judgment on pairwise comparison, the priority vector or eigenvector have to be analyzed. The hierarchical synthesis used to calculate the priority vector by the weight of the criteria and the sum was to be taken over all weighted eigenvector entries corresponding to those in the next lower level of the hierarchy. The next step was to perform consistency analysis which is reported as consistency ratio. The aim of this step is to guarantee that the judgment made through pairwise comparison is consistent.

The final step was to develop an overall priority ranking to determine the best alternatives. Thus, hemp + PP composite material was selected as the optimum

Rank	Materials	Initial percentage	Physical increase 20%		Mechanical increase 20%		Cost increase 20%	
1	Hemp + PP	15.0%	Hemp + PP	14.5%	Hemp + PP	16.0%	Hemp + PP	16.5%
2	Kenaf + PP	13.9%	Kenaf + PP	13.7%	Kenaf + PP	15.0%	Flax + PP	13.2%
3	Jute + PP	13.5%	Jute + PP	13.3%	Jute + PP	14.5%	Kenaf + PP	13.2%
4	Sisal + PP	13.2%	Flax + PP	13.1%	Sisal + PP	13.5%	PP	13.2%
5	Flax + PP	13.1%	Sisal + PP	13.1%	Flax + PP	13.1%	Sisal + PP	13.1%
6	PP	11.6%	PP	11.4%	PP	11.2%	Jute + PP	12.4%
7	Coir + PP	10.2%	Coir + PP	11.0%	Coir + PP	8.4%	Coir + PP	9.7%
8	EFB + PP	9.5%	EFB + PP	99%	EFB + PP	8.2%	EFB + PP	8.8%

FIGURE 6.23 Sensitivity analysis test.

material for automotive components with environmental assessment. After obtaining the final results, sensitivity analysis was conducted to fine tune the judgment and validate the results as shown in Fig. 6.23.

TOPSIS METHOD

One of the most used multicriteria decision-making methods is the technique for order preference by similarity to ideal solution (TOPSIS). Here, the best alternative would be the closest one to the positive ideal solution and the farthest simultaneously from the negative ideal one. The main principle of this method is to suggest a solution which has the shortest distance from the hypothetical one in the Euclidean space (Tavana and Hatami-Marbini, 2011). The suggested solution can sometimes have the minimum Euclidean distance from both the ideal and negative ideal (worst) solutions compared to other alternatives. TOPSIS technique tries to find solutions that are close to the ideal solution but far from the worst one. If a multicriteria decision-making problem is supposed to have n alternatives ($A_1, A_2, A_3, \ldots, A_n$) and m criteria ($C_1, C_2, C_3, \ldots, Cm$), the evaluation values of each alternative regarding each criterion can be arranged in a matrix as $A(aij)_{nxm}$. Let the vector of criteria be $W = (w_1, w_2, w_3, \ldots, w_m)$ and satisfy $\sum_{j-1}^{m} w_j = 1$. Thus, the judgment matrix for the ranking can be illustrated as Eq. (6.10)

$$A = \begin{bmatrix} a_{11} & a_{12} & \cdots & a_{1n} \\ a_{21} & a_{22} & \cdots & a_{2n} \\ \cdot & \cdot & \cdot & \cdot \\ \cdot & \cdot & \cdot & \cdot \\ \cdot & \cdot & \cdot & \cdot \\ a_{m1} & a_{m2} & \cdots & a_{mn} \end{bmatrix} \quad i = 1, \ldots, m, j = 1, \ldots, n \tag{6.10}$$

Then the main steps of TOPSIS can be summarized as

Step 1: Building the normalized judgment matrix. The normalized score r_{ij} is calculated as Eq. (6.11):

$$r_{ij} = \frac{X_{ij}}{\sum_i (X_{ij}^2)^{\frac{1}{2}}} \quad i = 1,\ldots,m,\ j = 1,\ldots,n \tag{6.11}$$

Step 2: Finding the weighted normalized judgment matrix. This can be obtained utilizing the AHP technique (i.e., AHP can be integrated with TOPSIS in this step). The weighted normalized score V_{ij} is expressed as Eq. (6.12)

$$V_{ij} = W_j . r_{ij} \tag{6.12}$$

Step 3: Determining both positive ideal (V^+) and negative ideal (V^-) solutions. The ideal and negative ideal solutions can be calculated as Eqs. (6.13) and (6.14):

$$V^+ = \{V_1^+, \ldots, V_n^+\} \quad \text{where } V_j^+ = \{\max_i(v_{ij}) \text{ if } j \in J; \min_i(v_{ij}) \text{ if } j \in J'\} \tag{6.13}$$

$$V^- = \{V_1^-, \ldots, V_n^-\} \quad \text{where } V_j^- = \{\min_i(v_{ij}) \text{ if } j \in J; \max_i(v_{ij}) \text{ if } j \in J'\} \tag{6.14}$$

where $J = (j = 1,2,\ldots,n)/j$ is set of beneficial factors and $J = (j = 1,2,\ldots,n)/j$ is set of nonbeneficial factors.

Step 4: Finding the separation measures for each alternative utilizing the n-dimensional Euclidean distance. The separation from the ideal alternative is shown as Eqs. (6.15) and (6.16):

$$S_i^+ = \left[\sum_j (V_j^+ - V_{ij})^2\right]^{\frac{1}{2}} \quad ; i = 1, \ldots, m \tag{6.15}$$

$$S_i^- = \left[\sum_j (V_j^- - V_{ij})^2\right]^{\frac{1}{2}} \quad ; i = 1, \ldots, m \tag{6.16}$$

Step 5: Finding the relative closeness to ideal solution (C_i^*) and the corresponding rank of the candidate. That is, selecting the alternative with maximum C_i^*. Such relative closeness of the alternative A*ij* can be defined as Eq. (6.17):

$$C_i^* = \frac{S_i^-}{S_i^+ + S_i^-}; 0 < C_i^* < 1, S_i^- \geq 0, S_i^+ \geq 0 \tag{6.17}$$

Comparable to other methods, TOPSIS is considered one of the most reliable techniques for selecting the best solution among set of alternatives. It usually gives similar ranking of alternatives to the AHP in NFC materials

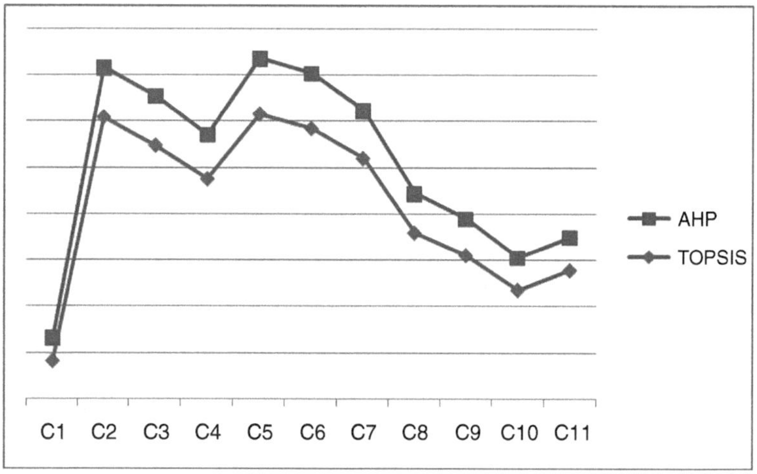

FIGURE 6.24 The trend in alternative ranking in both AHP and TOPSIS methods (Al-Oqla et al., 2015d).

selection. For instance, it was reported that for selecting the most appropriate reinforcement conditions in NFC (date palm/epoxy composite) both AHP and TOPSIS ranked the first four potential candidates (C) composites in the same way.The trend in ranking alternatives was similar in both techniques as seen in Fig. 6.24.

Applying MCDM models in the selection process of NFCs can enhance the usage of such composites in different industrial design and applications, and assist implementing such techniques to different NFCs to maximize their desired characteristics to participate in achieving better sustainable design possibilities. Such techniques are also able to rank the potential alternatives of the reinforcement conditions. Utilizing TOPSIS method is considered reliable and as valid as other MCDM techniques. The closeness of the alternatives in the final priorities demonstrates that determining the optimal alternative without using a decision-making tool is a very difficult task.

MATERIALS SELECTION OF COMPOSITE MATERIALS USING NEURAL NETWORK

Looking at the human cerebral cortex, there are approximately 100 billion neurons (El Kadi, 2006). The neuron (Fig. 6.24) is the primary unit of the nervous system. It serves as the processing unit of the nervous system for receiving and combining signals from various other neurons. The entire nervous system functionality depends on interconnections of neurons. Although the human information processing system is considered to be relatively slow, the

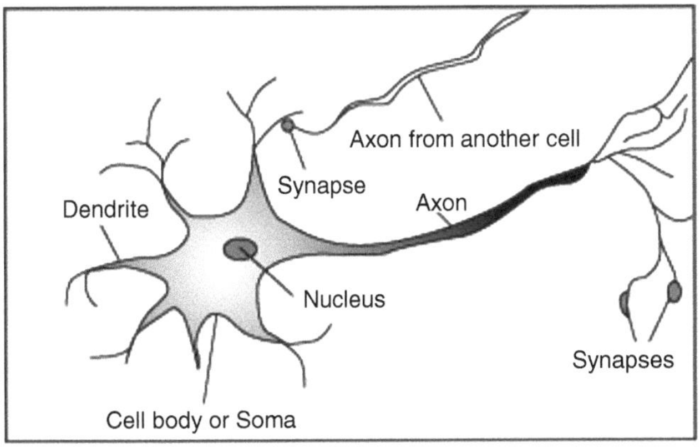

FIGURE 6.25 A biological neuron (William et al., 2008).

overall processing operation is obtained within a few hundred milliseconds. This illustrates that biological computation is accomplished with few serial steps, each taking place on a massive parallel scale. In light of this inborn parallel architecture, each of the processing units is locally connected and relatively simple (El Kadi, 2006). Hence, design experts managed to develop a computer-based neural network system that mimics the computational pattern of the nervous system.

On the basis of the aforementioned background, neural network (NN) or artificial neural network (ANN) is generally defined as a computer system that emulates the neurons of a biological nervous system or computational models of the brain (Sapuan and Mujtaba, 2010; Fig. 6.25).

ANN architecture consists of a number of layers known as input, output, and hidden layers (Fig. 6.26). The input layer is the layer where the input patterns are applied, whereas the output is obtained from the output layer. The layers between the input and output layers are referred to as the hidden layers. These particular layers assist the network to extract higher order statistics which are valuable when the size of the input layer is large (Schalkoff, 1997). There may be one or more hidden layers in the ANN and their outputs are not directly observable; hence, they bear the name hidden layers. In the architecture of ANN, the connections are between two successive layers, and the information flows in one direction, forward from the input layer to the output layer via the hidden layers. There are no cycles or loops in the network and no lateral connection between the neurons of the same layer (Sumathi and Paneerselvam, 2010).

Generally, the topological structure of all ANNs is similar. Some of the neurons interface with the real world to receive its input while the other neurons provide the world with the network's output. All the remaining neurons are

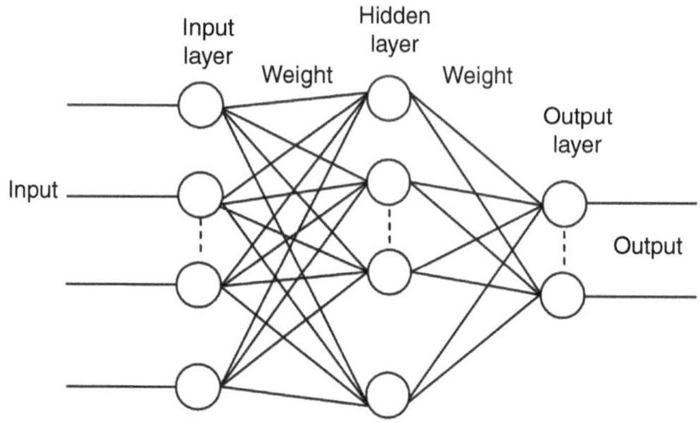

FIGURE 6.26 Architecture of neural network.

hidden. Every single input to a neuron has a weight factor of the function that determines the strength of the interconnections and thus the contribution of that interconnection to the following neurons (Hasan et al., 2009). ANNs can be trained to execute a specific function by adjusting the values of these weight factors between the neurons either from the information obtained from outside the network or by the neurons themselves in response to the input. This provides ANNs with the ability to achieve learning and memory (Hasan et al., 2009).

ANN provides practical solution for wide range of applications such as pattern recognition, classification, and optimization problems. It is also used in signal processing, speech recognition, condition monitoring, and functional approximation (Addin et al., 2011; Naji et al., 2013; Zare et al., 2013). Of late, the use of ANN for material selection of composite materials has been increasingly reported in literatures. Zhang and Friedrich (2003) carried out a review work regarding the various applications of ANN to polymeric composites, polymers, and other materials. El Kadi (2006) conducted a review work on the mechanical modeling of fiber-reinforced composite materials using ANN during the last decade. Goel and Chen (1996) reported the application of neuro-expert network for materials selection. They utilized neural network connectionist approach with back propagation for training data samples. The expert system evaluated the output of neural network on the basis of price, availability in stock, and closeness to desired material properties. In the case of Somkuwar et al. (2011), they used expert network with integration of radial basis neural network architecture for materials selection. Sensorial properties and cost of steel metal were considered for selection. They developed the expert system with visual C++ and integrated with a neural network tool box of Matlab. Smith et al. (2002) employed the inverse problem of NN in the selection of powder metallurgy materials. Related study was also reported by Cherian et al. (2000);

they developed an ANN material selection system for powder metallurgy. Furthermore, Yang et al. (2003a,b) utilized NN in conjunction with other software tools such as genetic algorithm and computer-aided design/ computer-aided manufacturing for the selection of optimum materials in composite manufacturing. Zhou et al. (2009) also proposed ANN used with generic algorithm to optimize the multiobjectives of material selection to choose suitable materials for developing sustainable products. Recently, Sapuan and Mujtaba (2010) developed a neural network–based materials selection system for selecting optimal material from natural fiber polymer composites. They implemented back propagation algorithm to correlate between the inputs and to predict the output materials. Further the predicted output was processed using multiattribute ranking method to select the most suitable material. Another work worth mentioning here is that of Ali et al. (2015); in this work, they incorporated rule-based expert system with ANN technology for NFC material classification.

In the work of Sapuan and Mujtaba (2010), they presented the basic structure of the computer-aided materials selection system using ANN as shown in Fig. 6.27. The main tools utilized in this material selection system consist of the ANN, materials data source, material database, and product design tools. MATLAB neural network toolbox was used to develop the ANN and the data were stored in an MS Excel spreadsheet.

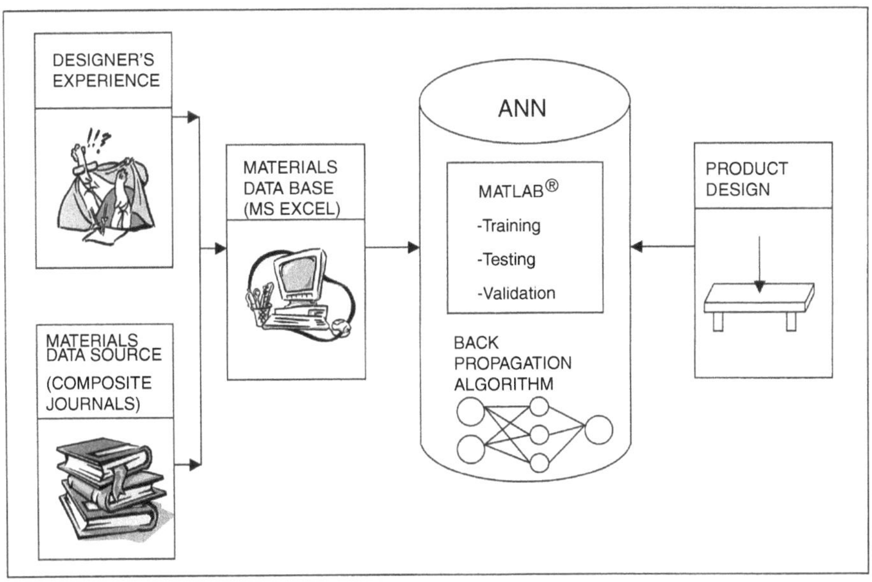

FIGURE 6.27 Structure of composite materials selection system using neural network (Sapuan and Mujtaba, 2010).

Permission pending.

MATERIALS SELECTION OF BIOCOMPOSITES

Biocomposites are broadly defined as composite materials made from natural fiber and petroleum-derived nonbiodegradable polymers (i.e., PP, PE) or biodegradable polymers (starch, PLA, PHA). Recently, biocomposites became highly valuable materials due to their environmental advantage over counterparts. The growing environmental awareness of people and the new stringent green policies enacted by governments have intensified the search and development of more environmentally friendly materials to preserve our immediate environment and public health. Natural fibers reinforced in biocomposites have tremendous advantages over traditional glass fibers that make them very competitive in modern industrial applications and particularly in automotive ones (Al-Oqla et al., 2015b). These advantages include low cost, good thermal and acoustical insulation properties, availability, enhanced CO_2 sequestration, energy recovery, reduced dermal and respiratory irritation, and reduced tool wear in machining operations (Faruk et al., 2012; Kalia et al., 2011). The large number of publications related to development, fabrication, and characterization of new materials for biocomposites reflect the growing significance of material selection for biocomposite products.

Although the concept of material selection is not a novel area in the field of concurrent engineering, there are still limited studies on implementing material selection techniques to biocomposites or even reporting significant and comprehensive data sets on biocomposites which may be utilized during materials screening and ranking (Shah, 2014). Hence, more and more material experts or researchers are now looking into this gap. Table 6.8 presents a number of recently (2010–15) performed investigation in the field of materials selection for biocomposites product, aimed for specific applications.

Sapuan et al. (2011a) reported the use of the AHP method to select the most suitable biocomposite material for automotive dashboard panel, where a database of properties for NFC materials was systematically organized. The material properties considered in their study were only the density, the Young's modulus, and the tensile strength. They evaluated 29 different types of NFC materials to rank their suitability for the automotive dashboard panel. Mansor et al. (2013, 2014a) used AHP and AHP-TOPSIS method to perform the selection of the most proper natural fiber materials to be hybridized with glass fiber reinforced polymer composites for automotive brake lever design. Furthermore, Ali et al. (2015) employed AHP approach for environmental assessment in biocomposite materials selection for automotive components. In the work of Koronis et al. (2013), they reviewed green composites as appropriate materials for automotive applications and ternary diagrams were proposed as materials screening tools.

Most recently, Al-Oqla et al. (2014, 2015a,b) have considerably contributed to the selection of natural fiber reinforced biocomposites for automotive applications

Table 6.8 Recent Publications (2010–2015) on Material Selection for Biocomposites

Authors	Work Title
Al-Oqla et al. (2015a)	Decision-making model for optimal reinforcement condition of NFCs
Al-Oqla et al. (2015b)	Selecting natural fibers for bio-based materials with conflicting criteria
Al-Oqla et al. (2014)	A novel evaluation tool for enhancing the selection of natural fibers for polymeric composites based on fiber moisture content criterion
Al-Oqla et al. (2014b)	Combined multicriteria evaluation stage technique as an agro waste evaluation indicator for polymeric composites: date palm fibers as a case study
Mansor et al. (2014a)	Application of integrated AHP-TOPSIS method in hybrid NFCs materials selection for automotive parking brake lever component
Mansor et al. (2014b)	Materials selection of hybrid bio-composites thermoset matrix for automotive bumper beam application using Topsis method
Shah (2014)	NFCs: comprehensive Ashby-type materials selection charts
Sapuan and Mansor (2014)	Concurrent engineering approach in the development of composite products: a review
Ali and Sapuan (2013)	Java-based expert system for selection of NFC materials
Mansor et al. (2013)	Hybrid natural fiber/glass fiber reinforced polymer composites material selection using analytical hierarchy process (AHP) for automotive component design
Fairuz et al. (2012)	A prototype expert system for material selection of polymeric-based composites for small fishing boat components
Sapuan et al. (2011b)	Materials selection for natural fiber reinforced polymer composites using analytical hierarchy process.
Sapuan and Mujtaba (2010)	Development of computational framework for material selection of natural fiber-reinforced polymer composite materials using neural network

using decision-making techniques. Sapuan and Mujtaba (2010) used neural network to develop prototype computational framework for the selection of natural fiber reinforced polymer composites. Elsewhere, Ali et al. (2015) presented the combined use of ANN and expert system for material classification of natural fiber reinforced composite. They reported that the implementation of expert system alone makes it difficult to scrutinize the vast selected materials. However, hybrid of expert system with neural network technology gave better results.

Hosein et al. (2011) have proposed some criteria for materials selection for ecodesign through selection of biocompatible materials and biocomposites particularly 100% biobased composites fulfil majority of these criteria. These include the use of

- Renewable materials—Natural fibers and biopolymers are renewable materials

- Residual materials from production processes—For example, wastes in the form of woodchips, from saw mill are good sources of fibers
- Retrieved components from disposed products—For example, some components from used wooden furniture made of wood composites can be retrieved and can be made into new furniture.
- Recycled materials alone or combined with primary materials—Biocomposites particular 100% biobased composites can be recycled.
- Biodegradable materials—Biopolymer and natural fibers are biodegradable materials
- Few, simple, unblended materials—These are the main feature of biocomposites
- Nonhazardous recyclable materials—Biocomposites are nonhazardous materials
- Materials with low energy consumption in extraction and transport—Extraction and transport of biocomposites consume low amount of energy

Ermolaeva et al. (2002) developed an optimization system for material selection for some engineering materials including NFCs. Because of the vast variation in matrix type, fiber type, fiber arrangement, fiber dimension, manufacturing method, selection of NFCs (Fig. 6.28) in particular and of composite materials in general are very difficult to perform.

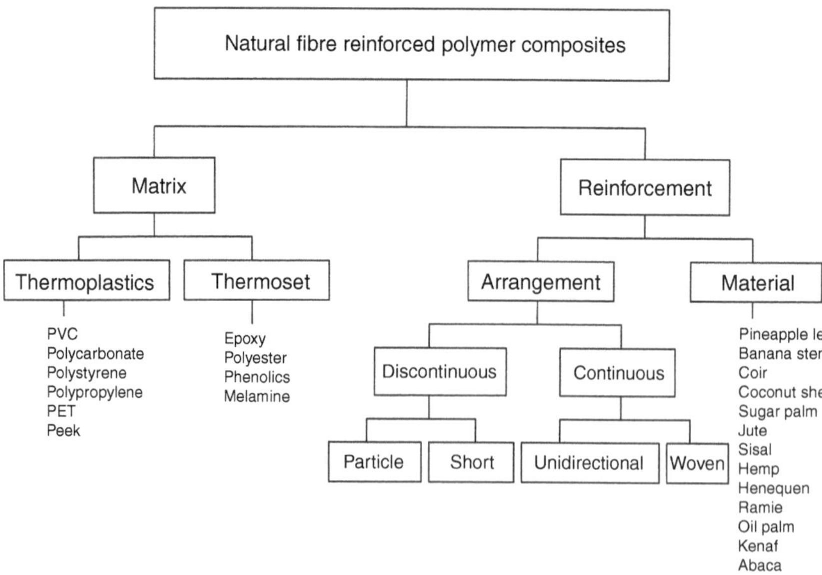

FIGURE 6.28 Materials hierarchy of NFCs (Sapuan and Mujtaba, 2010).
Permission pending.

From the literature review, material selection for biocomposite products is considered complicated. This is because the properties of natural fiber are not consistent due to the variety in the physical, mechanical, and thermal properties. Natural fiber is a hydrophilic material, and due to that, it has high sensitivity on moisture effect and it is also sensitive to heat. The heat accumulated during transportation may affect the result obtained for the composite material. This criteria need to be considered before a material is selected for biocomposites product. These possible sources of error need to be eliminated to have minimal defect on natural fiber. It is different from synthetic fiber where it can be most efficient of using material selection software where the properties are stable. It is crucial for researchers to develop a database of properties for natural fibers to assist material selection process involving composites product.

CONCLUSIONS

With regard to concurrent engineering, materials selection is an integral part of product design. It is normally carried out by design or material engineers from the early stage of the product design process. The range of materials available to design engineers is wide and constantly increasing which poses a huge challenge to design engineers. Hence, the traditional method of using material handbooks for conducting materials selection is a tedious and hectic task with many potential prejudices and mistakes. Most especially, the selection of composite materials is quite difficult to perform compared to homogenous materials like metals and plastics. However, the use of computer-aided tools for choosing composite materials helps to minimize material selection errors and accommodates the increasing number of new materials as well as prevents financial and time loss. This chapter presented a comprehensive insight of composite materials selection using computer-aided systems such as materials databases systems. Multicriteria decision-making models or tools also play significant role in the evaluation and selection of composite materials.

References

Addin, O., Sapuan, S.M., Othman, M., Ahmed Ali, B.A., 2011. Comparison of naive bayes classifier with back propagation neural network classifier based on f-folds feature extraction algorithm for ball bearing fault diagnostic system. Int. J. Phy. Sci. 6 (13), 3181–3188.

Ahmad, M.B., bin Yaacob, Z., 2002. Dissolved gas analysis using expert system. In Research and Development, 2002. SCOReD 2002. Student Conference on (pp. 313–316). IEEE.

Akao, Y., 1997. QFD: Past, present, and future. In: International Symposium on QFD, vol. 97, No. 2. Linköping University, Linköping, Sweden, pp. 1–12, 19–29.

Alibaba, H.Z., Özdeniz, M.B., 2004. A building elements selection system for architects. Build. Environ. 39, 307–316.

Ali, A.B.A., Sapuan, S.M., 2013. Java based expert system for selection of natural fibre composite materials. J. Food Agric. Environ. 11, 1871–1877.

Ali, B.A., Sapuan, S.M., Zainudin, E.S., Othman, M., 2015. Implementation of the expert decision system for environmental assessment in composite materials selection for automotive components. J. Clean. Prod 107, 557–567.

Al-Harbi, K.M.A.S., 2001. Application of the AHP in project management. Int. J. Proj. Manage. 19, 19–27.

Al-Oqla, F.M., Sapuan, S.M., 2014a. Natural fiber reinforced polymer composites in industrial applications: feasibility of date palm fibers for sustainable automotive industry. J. Clean. Prod. 66, 347–354.

Al-Oqla, F.M., Sapuan, S.M., Ishak, M.R., Nuraini, A.A., 2014. A novel evaluation tool for enhancing the selection of natural fibers for polymeric composites based on fiber moisture content criterion. BioResources 10 (1), 299–312.

Al-Oqla, F.M., Salit, M.S., Ishak, M.R., Aziz, N.A., 2015a. Selecting natural fibers for bio-based materials with conflicting criteria. Am. J. Appl. Sci. 12 (1), 64.

Al-Oqla, F.M., Sapuan, S.M., Ishak, M.R., Nuraini, A.A., 2015b. Decision making model for optimal reinforcement condition of natural fiber composites. Fibers Polym. 16 (1), 153–163.

Al-Oqla, F.M., Sapuan, S.M., Ishak, M.R., Nuraini, A.A., 2015c. Predicting the potential of agro waste fibers for sustainable automotive industry using a decision making model. Comput. Electron. Agric. 113, 116–127.

Al-Widyan, M.I., Al-Oqla, F.M. 2014. Selecting the most appropriate corrective actions for energy saving in existing buildings A/C in hot arid regions. Building Simulation, vol. 7, No. 5. Tsinghua University Press and Springer-Verlag, Berlin, Heidelberg, pp. 537–545.

AL-Oqla, F.M., Sapuan, S.M., Ishak, M.R., Nuraini, A.A., 2015d. Decision making model for optimal reinforcement condition of natural fiber composites. Fibers Polym. 16 (1), 153–163.

Alves, C., Ferr o, P.M.C., Silva, A.J., Reis, L.G., Freitas, M., Rodrigues, L.B., Alves, D.E., 2010. Ecodesign of automotive components making use of natural jute fiber composites. J. Clean. Prod. 18 (4), 313–327.

Ariff, H., Sapuan, S.M., M.S. A. Karim, M.S.A., 2007, Improvement in product development process through concurrent engineering, Proceedings of the Conference on Applications and Design in Mechanical Engineering (CADME07), 25–26 October, Perlis, Malaysia.

Ashby, M.F., 1989. Overview no. 80: on the engineering properties of materials. Acta Metall. 37 (5), 1273–1293.

Ashby, M.F., 1993. Materials selection in conceptual design. In: Hayhurst, D. (Ed.), Materials and engineering design: the next decade. The Institute of Metals, London.

Ashby, M.F., 2005. Materials selection in mechanical design. MRS Bull. 30 (12), 995.

Ashby, M.F., Cebon, D., 1993. Materials selection in mechanical design. Le Journal de Physique IV 3 (C7), C7-1.

Ashby, M.F., Johnson, K., 2002. Materials and Design: the Art and Science of Material Selection in Product Design. Butterworth-Heinemann, Oxford.

Baur, E., 1995. English version of FUNDUS material database. Reinf. Plast. 39 (4), 21.

Boothroyd, G., Dewhurst, P., Knight, W., 1994. Product Design for Manufacture and Assembly. Marcel Dekker, Inc., New York.

Cebon, D., Ashby, M.F., 1992. Computer aided materials selection for mechanical design. Metals Mater. 8 (1), 25–30.

Cebon, D., Ashby, M., 1996. Picking a winner: optimising materials selection. Mater. World 4, 646–648.

Cherian, R.P., Smith, L.N., Midha, P.S., 2000. A neural network approach for selection of powder metallurgy materials and process parameters. Artif. Intell. Eng. 14 (1), 39–44.

Cheung, H.-Y., Ho, M.-P., Lau, K.-T., Cardona, F., Hui, D., 2009. Natural fibre-reinforced composites for bioengineering and environmental engineering applications. Composites Part B: Engineering. 40, 655–663.

Chunming, L., Dawei, H., 2007. A fault diagnosis expert system base on artificial neural network for mixed-signal circuits. In Electronic Measurement and Instruments. ICEMI'07. 8th International Conference on (pp. 3-611). IEEE.

Dehghan-Manshadi, B., Mahmudi, H., Abedian, A., Mahmudi, R., 2007. A novel method for materials selection in mechanical design: combination of non-linear normalization and a modified digital logic method. Mater. Des. 28 (1), 8–15.

Dieter, G.E., 2000. Engineering Design: A Materials and Processing Approach, third ed. McGraw-Hill, Boston.

Dweiri, F., Al-Oqla, F.M., 2006. Material selection using analytical hierarchy process. Int. J. Comput. Appl. Technol. 26 (4), 182–189.

El Kadi, H., 2006. Modeling the mechanical behavior of fiber-reinforced polymeric composite materials using artificial neural networks—a review. Compos. Struct. 73 (1), 1–23.

Ermolaeva, N.S., Kaveline, K.G., Spoormaker, J.L., 2002. Materials selection combined with optimal structural design: concept and some results. Mater. Des. 23 (5), 459–470.

Exsys Corvid, 2008. Exsys Corvid Knowledge Automation Expert System Software Developer's Guide. Exsys Inc., Albuquerque, NM.

Fairuz, A.M., Sapuan, S.M., Zainudin, E.S., 2012. Prototype expert system for material selection of polymeric-based composites for fishing boat components. J. Food Agric. Environ. 10, 1543–1549.

Farag, M.M., 1979. Materials and Process Selection in Engineering. Applied Science Publication Ltd, London.

Farag, M.M., 2006. Quantitative methods of materials selection. In: Kurtz, M. (Ed.), Handbook of Material Selection. third ed. John Wiley and Sons Inc., New York, pp. 466–488.

Faruk, O., Bledzki, A.K., Fink, H.P., Sain, M., 2012. Biocomposites reinforced with natural fibers: 2000–2010. Prog. Polym. Sci. 37 (11), 1552–1596.

Findik, F., Turan, K., 2012. Materials selection for lighter wagon design with a weighted property index method. Mater. Des. 37, 470–477.

Flior, E., Anaya, T., Moody, C., Beheshti, M., Han, J., Kowalski, K. (2010). A knowledge-based system implementation of intrusion detection rules. In Seventh International Conference on Information Technology: New Generations (ITNG), 2010 pp. 738–742. IEEE.

Ginsberg, M., Hauser, J., Moreira, J.E., Morgan, R., Parsons, J.C., Wielenga, T.J., 2000. Panel session: future directions and challenges for Java implementations of numeric-intensive industrial applications. Adv. Eng. Softw. 31 (8), 743–751.

Goel, V., Chen, J., 1996. Application of expert network for material selection in engineering design. Comput. Ind. 30 (2), 87–101.

Guo, B., Qi, F., Fu, G., 2008, A knowledge-based diagnostic system for pneumatic system. Paper presented at 2008 International Symposium on Knowledge Acquisition and Modeling, Dec. 21-22, Wuhan, China.

Hambali, A., Sapuan, S.M., Ismail, N., Nukman, Y., 2010. Material selection of polymeric composite automotive bumper beam using analytical hierarchy process. J. Central South Univ. Technol. 17 (2), 244–256.

Hambali, A., Sapuan, S.M., Rahim, A.S., Ismail, N., Nukman, Y., 2011. Concurrent decisions on design concept and material using analytical hierarchy process at the conceptual design stage. Concurrent Eng-Res A 19 (2), 111–121.

Hasan, M., Hoque, M.E., Sapuan, S.M., 2009. Application of artificial neural network in composites materials. Composite Materials Technology: Neural Network Applications, 1.

Hongwei, Z., 2005. The Design and Realization of Expert System Based on Function Detection, Paper presented at the Fifth International Conference on Power Transmission & Distribution Technology, Oct 12–24, Beijing, China.

Holbery, J., Houston, D., 2006. Natural-fiber-reinforced polymer composites in automotive applications. J. Min. Metals Mater. Soc. 58 (11), 80–86.

Hosein, M., Zarandi, F., Mansour, S., Hosseinijou, S.A., Avazbeigi, M., 2011. A material selection methodology and expert system for sustainable product design. Int. J. Adv. Manuf. Technol. 57, 885–903.

İpek, M., Selvi, İ.H., Findik, F., Torkul, O., Cedimog˘lu, I.H., 2013. An expert system based material selection approach to manufacturing. Mater. Des. 47, 331–340.

Java technology, 2014. Available from: http://www.java.com/en/about/

Kalia, S., Kaith, B.S., Kaur, I. (Eds.), 2011. Cellulose Fibers: Bio-and Nano-Polymer Composites: Green Chemistry and Technology. Springer Science & Business Media.

Kim, J., Kim, C., Hong, C., Hahn, H.T., 2000. Development of concurrent engineering system for design of composite structures. Compos. Struct. 50, 29–309.

King, A.M., Sivaloganathan, S., 1999. Development of a methodology for concept selection in flexible design strategies. J. Eng. Des. 10 (4), 329–349.

Koning, J., 1994, A Fuzzy Material Selection System for Mechanical Design, Robotic Institute, Carnegie Mellon University, Pittsburgh, USA, Report number CMU-RI-TR-94-22.

Koronis, G., Silva, A., Fontul, M., 2013. Green composites: a review of adequate materials for automotive applications. Compos. B 44 (1), 120–127.

Kowalski, K., Goetz, J., Alam, M., 2006. Intelligent on-line advising with expert system shell. In World Conference on E-Learning in Corporate, Government, Healthcare, and Higher Education, vol. 2006, pp. 687–694.

Kutz, M. (Ed.), 2002. Handbook of materials selection. John Wiley & Son, New York, pp. 365–368.

Liao, S.H., 2005. Expert system methodologies and applications—a decade review from 1995 to 2004. Expert Syst. Appl. 28 (1), 93–103.

Mansor, M.R., Sapuan, S.M., 2014. Automotive composite hand operated parking brake lever design with a weighted property index method. In: Sapuan, S.M. (Ed.), Engineering Composites: Properties and Applications. UPM Press, Serdang, pp. 248–264.

Macharis, C., Springael, J., De Brucker, K., Verbeke, A., 2004. PROMETHEE and AHP: The design of operational synergies in multicriteria analysis.: Strengthening PROMETHEE with ideas of AHP. Eur. J. Oper. Res. 3, 307–317.

Maier, C., 1993. Material Witness. British Plastics and Rubber.

Mansor, M.R., Sapuan, S.M., Edi Syams, Z., Abd Aziz, N., Hambali, A., 2014a. Application of integrated AHP-TOPSIS method in hybrid natural fiber composites materials selection for automotive parking brake lever component. Aust. J. Basic Appl. Sci. 8 (5), 431–439.

Mansor, M.R., Sapuan, S.M., Hambali, A., Zainudin, E.S., Nuraini, A.A., 2014b. Materials selection of hybrid bio-composites thermoset matrix for automotive bumper beam application using Topsis method. Adv. Environ. Biol. 8, 3138–3442.

Mansor, M.R., Sapuan, S.M., Zainudin, E.S., Nuraini, A.A., Hambali, A., 2013. Hybrid natural and glass fibers reinforced polymer composites material selection using analytical hierarchy process for automotive brake lever design. Mater. Des. 51, 484–492.

Masui, K., Sakao, T., Inaba, A. 2001. Quality function deployment for environment: QFDE (1st report)-a methodology in early stage of DfE. In: Environmentally Conscious Design and Inverse Manufacturing, 2001. Proceedings EcoDesign 2001 Second International Symposium on. IEEE, pp. 852–857.

Mayyas, A., Shen, Q., Abdelhamid, M., Shan, D., Qattawi, A., Omar, M., 2011. Using quality function deployment and analytical hierarchy process for material selection of body-in-white. Mater. Des. 32, 2771–2782.

Mir, S.S., Nafsin, N., Hasan, M., Hasan, N., Hassan, A., 2013. Improvement of physico-mechanical properties of coir-polypropylene biocomposites by fiber chemical treatment. Mater. Des. 52, 251–257.

Mayer, R.M., 1993. Design with Reinforced Plastics: A Guide for Engineers and Designers. The Design Council, London.

Naji, S., Zainuddin, R., Kareem, S.A., Jalab, H.A., 2013. Detecting faces in colored images using multi-skin color models and neural network with texture analysis. Malays. J. Comput. Sci. 26 (2), 101–123.

Otto, K.N., 1995. Measurement methods for product evaluation. Res. Eng. Des. 7 (2), 86–101.

Pahl, G., Beitz, W., 2013. Engineering design: a systematic approach. Springer Science & Business Media, London.

Pahl, G., Beitz, W., Feldhusen, J., Grote, K.H., 2007, In: Wallace, K., Blessing, L. (Eds.), Engineering Design: A Systematic Approach, third ed., Springer Verlag, New York.

Pavleković, M., Zekić-Sušac, M., Đurđević, I., 2007. Expert system for detecting a child's gift in mathematics. Osijek, 98.

Pilla, S., 2011. Handbook of bioplastics and biocomposites engineering applications. Scrivener Publishing, Salem.

Prasad, B., 1996. Concurrent Engineering Fundamental: Integrated Product and Process OrganizationPrentice Hall Upper Saddle River, NJ.

Prasad, K., Chakraborty, S., 2013. A quality function deployment-based model for materials selection. Mater. Des. 49, 525–535.

Pugh, S., 1991. Total design: integrated methods for successful product engineering. Addison-Wesley Publishing Company, Wokingham, England.

Rao, R., Davim, J., 2008. A decision-making framework model for material selection using a combined multiple attribute decision-making method. Int. J. Adv. Manuf. Tech. 35, 751–760.

Rao, R., Patel, B., 2010. A subjective and objective integrated multiple attribute decision making method for material selection. Mater. Des. 31, 4738–4747.

Rathod, M.K., Kanzaria, H.V., 2011. A methodological concept for phase change material selection based on multiple criteria decision analysis with and without fuzzy environment. Mater. Des. 32, 3578–3585.

Saaty, T.L., 1980. The Analytic Hierarchy Process: Planning, Priority Setting, Resources Allocation. McGraw-Hill, New York.

Saaty, T.L., Vargas, L.G., 2012. Models, methods, concepts & applications of the analytic hierarchy process. Springer, New York.

Sahari, J., Sapuan, S.M., Zainudin, E.S., Maleque, M.A., 2014. Physico-chemical and thermal properties of starch derived from sugar palm tree (Arenga pinnata). Asian J. Spectro. 26, 955.

Saaty, T.L., 2003. Decision-making with the AHP: why is the principal eigenvector necessary. Eur. J. Oper. Res. 145 (1), 85–91.

Sanyang, M.L., Sapuan, S.M., 2015. Development of expert system for biobased polymer material selection: food packaging application. J. Food Sci. Technol. 52, 6445–6454.

Sapuan, S.M., 1999. A computer-aided material selection for design of automotive safety critical components with novel materials. Malays. J. Comput. Sci. 12, 37–46.

Sapuan, S.M., 2001. A knowledge-based system for materials selection in mechanical engineering design. Mater. Des. 22, 687–695.

Sapuan, S.M., Abdalla, H.S., 1998. A prototype knowledge-based system for the material selection of polymeric-based composites for automotive components. Compos A 29A, 731–742.

Sapuan, S.M., Jacob, M.S.D., Mustapha, F., Ismail, N., 2002. A prototype knowledge-based system for material selection of ceramic matrix composites of automotive engine components. Mater. Des. 23, 701–708.

Sapuan, S.M., Mun, N.K., Hambali, A., Lok, H.Y., Fairuz, A.M., Ishak, M.R., 2011a. Prototype expert system for material selection of polymeric composite automotive dashboard. Int. J. Phys. Sci. 6, 5988–5995.

Sapuan, S.M., Kho, J.Y., Zainudin, E.S., Leman, Z., Ahmed Ali, B.A., Hambali, A., 2011b. Materials selection for natural fibre reinforced polymer composites using analytical hierarchy process. Ind. J. Eng. Mater. Sci. 18, 255–267.

Sapuan, S.M., Mansor, M.R., 2014. Concurrent engineering approach in the development of composite products: a review. Mater. Des. 58, 161–167.

Sapuan, S.M., Mujtaba, I.M., 2010. Development of computational framework for material selection of natural fiber-reinforced polymer composite materials using neural network. In: Sapuan, S.M., Mujtaba, I.M. (Eds.), Composite Materials Technology: Neural Network Application. CRC Press, Boca Raton, pp. 317–340.

Sapuan, S.M., Pua, F.L., El-Shekeil, Y.A., Al-Oqla, F.M., 2013. Mechanical properties of soil buried kenaf fibre reinforced thermoplastic polyurethane composites. Mater. Des. 50, 467–470.

Schalkoff, R.J., 1997. Artificial Neural Networks. McGraw-Hill Higher Education, ISBN 0-07-057118-X.

Shah, D.U., 2014. Natural fibre composites: comprehensive Ashby-type materials selection charts. Mater. Des. 62, 21–31.

Smith, L.N., German, R.M., Smith, M.L., 2002. A neural network approach for solution of the inverse problem for selection of powder metallurgy materials. J. Mater. Process. Technol. 120 (1), 419–425.

Somkuwar, V., Bhagoriya, J.L., Khaira, H.K., 2011. An expert system for aid in material selection process using artificial neural network. Int. J. Adv. Eng. Appl. 3, 169–171.

Sumathi, S., Paneerselvam, S., 2010. Computational intelligence paradigms: theory & applications using MATLAB. CRC Press.

Surawski, N.C., Miljevic, B., Bodisco, T.A., Brown, R.J., Ristovski, Z.D., Ayoko, G.A., 2013. Application of multicriteria decision making methods to compression ignition engine efficiency and gaseous, particulate, and greenhouse gas emissions. Environ. Sci. Technol. 47 (4), 1904–1912.

Tavana, M., Hatami-Marbini, A., 2011. A group AHP-TOPSIS framework for human spaceflight mission planning at NASA. Expert Syst. Appl. 38, 13588–13603.

Thurston, D.L., Carnahan, J.V., 1992. Fuzzy ratings and utility analysis in preliminary design evaluation of multiple attributes. J. Mech. Des. 114, 648–658.

Wang, Y.H., Deng, C., Xiong, Y., Wu, J., 2010. A mixed expert system for fault diagnosis. Paper presented at 2010 IEEE 17th International Conference on Industrial Engineering and Engineering Management (IE&EM), 29–31 Oct, Xiamen, China.

Weiss, C.H., 1997. Theory based evaluation: past, present, and future. New directions for evaluation 1997 (76), 41–55.

White, P.J., 1995. Materials selection: is the machine really smart enough?. In: Proceedings of Society of Plastics Engineers Conference, Hartfort, Connecticut.

William, A.Y., William, S.H., Gary, R.W., 2008. Determining Hall of fame status for major league baseball using an artificial neural network. J. Quant. Anal. Sports 8, 4.

Wong, J.K., Li, H., 2008. Application of the analytic hierarchy process (AHP) in multi-criteria analysis of the selection of intelligent building systems. Build. Environ. 43 (1), 108–125.

Wu, F., 1998. Simland: a prototype to simulate land conversion through the integrated GIS and CA with AHP-derived transition rules. Int. J. Geogr. Inf. Sci. 12, 63–82.

Yang, S.Y., Girivasan, V., Singh, N.R., Tansel, I.N., Kropas-Hughes, C.V., 2003a. Selection of optimal material and operating conditions in composite manufacturing. Part II: complexity, representation of characteristics and decision making. Int. J. Mach. Tools Manuf. 43 (2), 175–184.

Yang, S.Y., Tansel, I.N., Kropas-Hughes, C.V., 2003b. Selection of optimal material and operating conditions in composite manufacturing. Part I: computational tool. Int. J. Mach. Tools Manuf. 43 (2), 169–173.

Yang, Z.R., Thomson, R., Mcneil, P., Esnouf, R.M., 2005. RONN: the bio-basis function neural network technique applied to the detection of natively disordered regions in proteins. Bioinformatics 21, 3369–3376.

Yingzhan, K., 2007. Design of the Fault Diagnosis Expert System for the Fire Control System. Paper presented at the 6th International Conference on Electronic Measurement and Instruments, Aug. 16–18, Xi'an, China.

Zare, M.R., Seng, W.C., Mueen, A., 2013. Automatic classification of medical X-ray images. Malays. J. Comput. Sci. 26 (1), 9–22.

Zhang, Z., Friedrich, K., 2003. Artificial neural networks applied to polymer composites: a review. Compos. Sci. Technol. 63 (14), 2029–2044.

Zhou, C.C., Yin, G.F., Hu, X.B., 2009. Multi-objective optimization of material selection for sustainable products: artificial neural networks and genetic algorithm approach. Mater. Des. 30 (4), 1209–1215.

CHAPTER 7

Design for Sustainability in Composite Product Development

CONTENTS

Introduction................273

Design for the Environment or Ecodesign..................273

Evolution of Design-for-Environment to Design-for-Sustainability (D4S)...274

Design for Sustainability (D4S)..........................275

Three Key Elements of Sustainability.................276

Product Development and Environmental Sustainability......................277

Product Development and Social Sustainability......................277

Product Development and Economic Sustainability......................279

Life-Cycle Thinking in Sustainable Product Development......................280

Design for Sustainability of Natural Fiber Composite Materials...................281

(Continued)

INTRODUCTION

Concurrent engineering (CE) is an integrated approach in design and manufacturing of a product where all manufacturing and other related issues are considered in the conceptual design stage of the design process (Hambali et al., 2010; Sapuan and Abdalla, 1998; Sapuan and Mansor, 2014; Sapuan et al., 2006). CE is also known as the design for "X" (DFX). DFX is interpreted as either design for excellence (Bralla, 1996) or "X" here is normally referred to as all "abilities" (Babu et al., 2008), such as maintainability, serviceability, reliability, inspectability, manufacturability, recyclability, disposability, and sustainability. Design for sustainability means 'to adapt our present design to a lifestyle that meets present needs without compromising the needs of future generations' (Ashby, 2005).

'Sustainability' envelopes the integration of environmental, social, and economic concerns (Küçüksayraç, 2015; Spangenberg et al., 2010). Hence, the concept of D4S in the composite industry requires that the design process and resulting composite product/material consider not only environmental concerns as in the case of Ecodesign (Karlsson and Luttropp, 2006), but also social and economic concerns. This implies that D4S traverses beyond how to develop 'green' composite products but it encompasses how to satisfy consumer needs in a more sustainable manner. Nowadays, it is recommended for material or design experts in the composite industry to implement D4S concept in their long-term product innovation strategies to minimize the undesirable environmental, social, and economic impacts in the product's supply chain and throughout its life cycle (Crul et al., 2009).

DESIGN FOR THE ENVIRONMENT OR ECODESIGN

The term design for the environment (DfE) has several other synonyms such as ecodesign (mainly used in Europe), green design (GD), environmentally conscious design (ECD), or life-cycle design (LCD), depending on where it is used.

Composite Materials. http://dx.doi.org/10.1016/B978-0-12-802507-9.00007-6

Environmental Sustainability of Natural Fiber Reinforced Composites......282
Socio-Economic Sustainability of Natural Fiber Reinforced Composites......283
Design for Sustainability Concept in Ford Automotive Company......284
Case Study on D4S of Bioplastic for Food Packaging......285
Design for Economic Impacts......285
Design for Environmental Impact......286
Design for Societal Impacts......287
Natural Fiber Composites: Sugar Palm Fiber Sugar Palm Starch Biocomposites......287
Design for Manufacturability......291
Summary......291
References......292

The prime objective of introducing DfE in traditional design procedures is to reduce the overall environmental impact of products and services through the use of life-cycle perspective (Kurk and Eagan, 2008). Charter and Tischner (2001) defined DfE or Ecodesign in the following words:

> "sustainable solutions are products, services, hybrids or system changes that minimize negative and maximize positive sustainability impacts – economic, environmental, social and ethical – throughout and beyond the life-cycle of existing products or solutions, while fulfilling acceptable societal demands/needs".

DfE is a concept that integrates multifaceted aspects of design and environmental considerations (Karlsson and Luttropp, 2006). The concept of DfE accompanied the increased global awareness of environmental problems in the late 1960s and early 1970s. This concept grew with the concept of design for "X," where "X" can represent recyclability, manufacturability, durability, and so on. The merits for businesses promoting environmental considerations during the earliest stages of product design are emergence of new, significant opportunities for cost savings and reduced liability (Kurk and Eagan, 2008). According to the U.S. Office of Technology Assessment, 70% of the costs of development, manufacture, and use of a product are decided in the first phases of a product design (U.S. Congress, 1992). Furthermore, DfE helps to reduce manufacturing cycle times, distinguish products, and provide a competitive advantage in markets valuing environmental attributes.

EVOLUTION OF DESIGN-FOR-ENVIRONMENT TO DESIGN-FOR-SUSTAINABILITY (D4S)

Most applications of DfE or Ecodesign mainly focus on addressing environmental issues. However, as the objectives of sustainable development and concepts of sustainability become more widespread, the "*E*" is gradually replaced with "*S*" for sustainability, along with migration toward more comprehensive design improvements (Kurk and Eagan, 2008). In simple terminology, Ecodesign is an approach dealing primarily with environmental and economic effects (Eco-efficiency) based on a life-cycle analysis of cost and impacts [life-cycle analysis (LCA), life-cycle costing]. On the contrary, D4S is understood to address all facets of sustainability, looking at bigger systems and asking more fundamental questions about consumption and production (Spangenberg et al., 2010). D4S traverses beyond Ecodesign by including social and economic factors into design coupled with environmental impacts (Fig. 7.1). It provides more opportunities for stakeholders to take part in product development. D4S also assesses long-term and global impacts based on the key elements of sustainable development instead of evaluating only the short and medium term

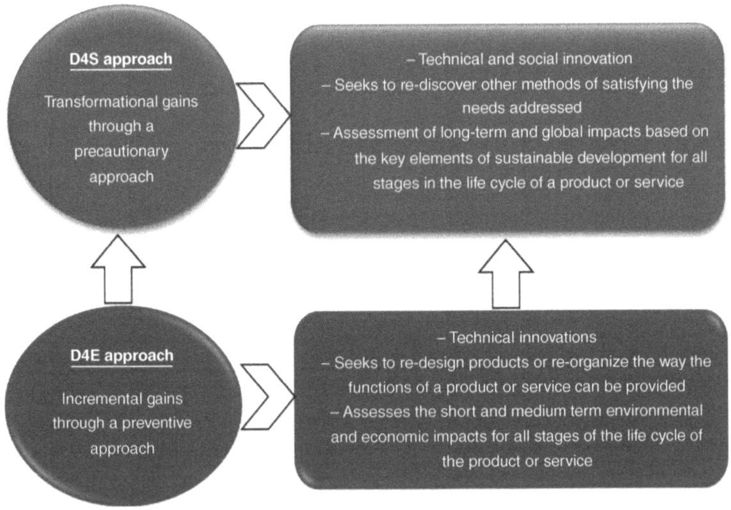

FIGURE 7.1 From DfE to D4S approach.
Adapted from Spangenberg, J.H., Fuad-Luke, A., Blincoe K., 2010. Design for Sustainability (DfS): the interface of sustainable production and consumption. J. Clean. Prod. 18, 1485–1493.

environmental and economic impacts for each and every stage of the life cycle of the product as in the case of DfE.

DESIGN FOR SUSTAINABILITY (D4S)

In today's global market, there is an increasing need for companies worldwide to innovate products and processes in order to (1) keep up with market competitive pressure; (2) increase productivity within the region or globally; and (3) secure or widen market share. Product innovation is considered as one of the cardinal strategic options at the disposal of composite firms, supply chains, and integrated industrial sectors to effectively compete in the current dynamic global market (Crul and Diehl, 2006).

Increasing global concerns regarding environmental degradation issues such as climate change, pollution, and rapid biodiversity loss as well as problems facing the society pertaining to poverty, health, employment, security, and inequity draw the attention of many companies to sustainability. The terms sustainable production and consumption have been taken up by the World Summit for Sustainable Development, Governments, Industry and Civil Society (Crul and Diehl, 2006). Adaptation of D4S is one of the most significant options available for enterprises to safely address these concerns. D4S is not limited to the narrow concept of Ecodesign or DfE; it covers broader concepts including sustainable product-service systems, systems innovations, and other life cycle–based efforts. In other words, D4S approach considers environmental, social, and economic concerns as primary element in long-term product innovation strategy (Crul and Diehl, 2006).

Three Key Elements of Sustainability

According to the World Commission for Environment and Development (WCED), sustainable development is defined as (Butlin, 1987)

> "development that meets the needs of the present without compromising the ability of future generations to meet their own needs".

Apparently, this definition does not explicitly mention about the environment; it highlighted the well-being of people as an environmental quality. From this simple description emerged a fundamental ethical principle: the responsibility of present generations to future generations (Vezzoli and Manzini, 2008). The meaning of sustainability encompasses the need to address economic development, social equity, and environmental protection as equally essential goals (Verghese et al., 2012). Hence, the key elements of sustainability are social, environmental, and economic; also referred to as people, planet, and profit as shown in Fig. 7.2.

These three key elements of sustainability are referred to as the "*triple bottom line*" (TBL) (Elkington, 1997). The concept of TBL was coined to translate the principle of sustainable development into something meaningful for businesses as well as encourage a more complete approach to sustainability. Elkington

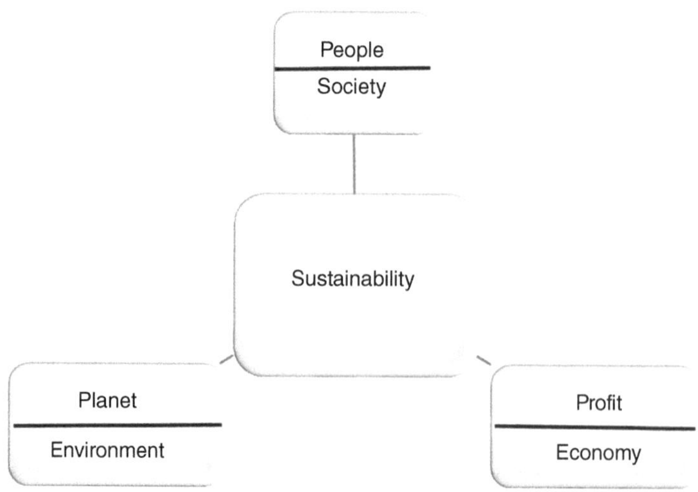

FIGURE 7.2 The three key elements of sustainability.

(1997) stated, *"society depends on the economy–and the economy depends on the global ecosystem, whose health represents the ultimate bottom line."*

According to Ljungberg (2007), there are three areas which are significantly associated with sustainable development; namely, environment, equity, and futurity. The term futurity in this context implies that the impact from development must be considered in a long perspective to minimize the impact for future generations. Therefore, when developing a new product, it is advisable to take note of the three main elements of sustainable development to obtain a suitable balance in the best possible way. The criteria for optimizing the sustainability in products and services are mainly functionality, environmental impacts, social impacts, economic impacts, market demand, quality, customer requirements, technical feasibility, compliance with legislation, and different specifications (Ljungberg, 2007).

Product Development and Environmental Sustainability

Back in the late 1980s and early 1990s, sustainability was mainly related to environmental issues, most especially pollution prevention, more toward designed to treat waste and polluting streams. Thereafter, it gradually grew to production improvements through concepts such as cleaner production, clean technology, and co-efficiency. Subsequently, it matured to product impacts, whereby the entire product life cycle is taken into consideration (Kurk and Eagan, 2008). To resolve environmental issues accompanying product development and consumption processes, concepts like Ecodesign or DfE were initiated and implemented. The latest advancement in product design is the concept of D4S, which goes further than Ecodesign or DfE by tackling social and economic concerns beside the usual environmental concerns. Environmental impacts of products and services can be classified into three major divisions: (1) ecological damages, (2) human health damages, and (3) resource depletion (Crul et al., 2009) (as shown in Fig. 7.3). All three facets of environmental impacts should be considered to adapt a holistic approach in addressing these concerns during the product design phrase.

Product Development and Social Sustainability

Social responsibility was a forgotten aspect in product design and development. Owing to numerous negative reports from the media on child labor, companies operating "sweatshops," workers' rights, and indigenous people over the last decades, social aspects of sustainability are gradually taking center stage in corporate strategies. Nowadays, corporate social responsibility is increasingly being integrated in corporate strategies besides environmental and economic issues. Attending to social concerns should be the priority of all stakeholders including but not limited to investors, local communities, and

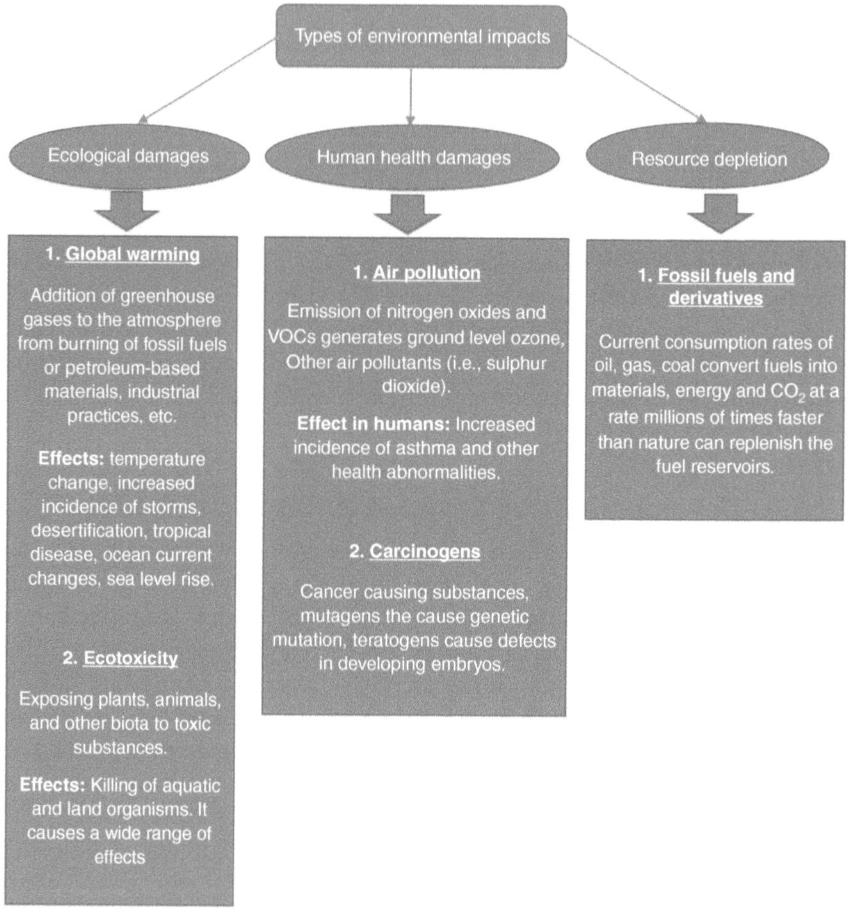

FIGURE 7.3 Types of environmental impacts (Crul et al., 2009).

participants in a supply chain (Crul et al., 2009). Fig. 7.4 shows the different aspects of social impacts.

Appropriate knowledge on social impacts pertaining to sustainable product production can assist companies to further improve their positive societal impacts and minimize the negative. Such achievement can be attained by proper societal impact assess as well as design and/or modify available product designs (Crul et al., 2009). D4S is intended to cater for societal concerns because products and production processes pose various impacts on people in diverse ways which are often ignored by stakeholders. Exploitation of labor, child labor, and resource conflict are mostly true of international operating businesses with labor laws different from that of the host country. Therefore, the introduction of social concerns in product development and processes has

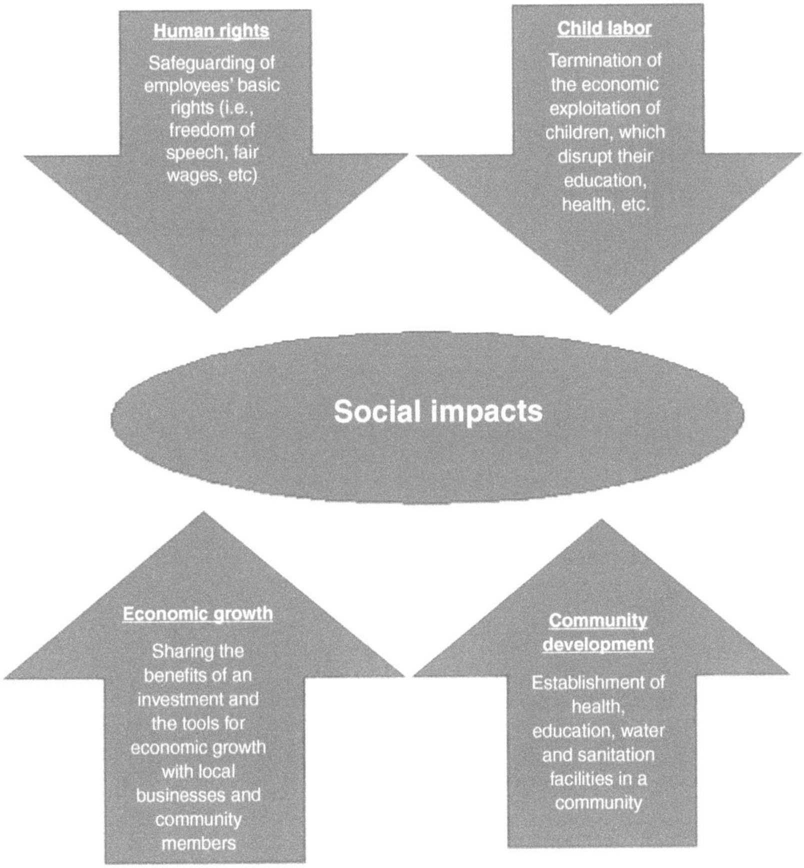

FIGURE 7.4 Different aspects of social impacts (Crul et al., 2009).

compelled companies to adapt ethical labor practices, in which employees are fairly and ethically treated.

Product Development and Economic Sustainability

Profit making is the overarching goal of every business. Hence, with the ambition of maximizing profits and cutting down expenses, many companies fail to consider environmental issues and social responsibilities in their corporate strategies. D4S can aid companies in attaining economic growth without totally compromising environmental issues and social benefits (Spangenberg et al., 2010). Sustainability concept in product development can minimize production expenses by increasing resource efficiency, enhancing the quality of the produced products, creating better marketing opportunities, strengthening customer loyalty, and opening up virgin markets (Crul et al., 2009).

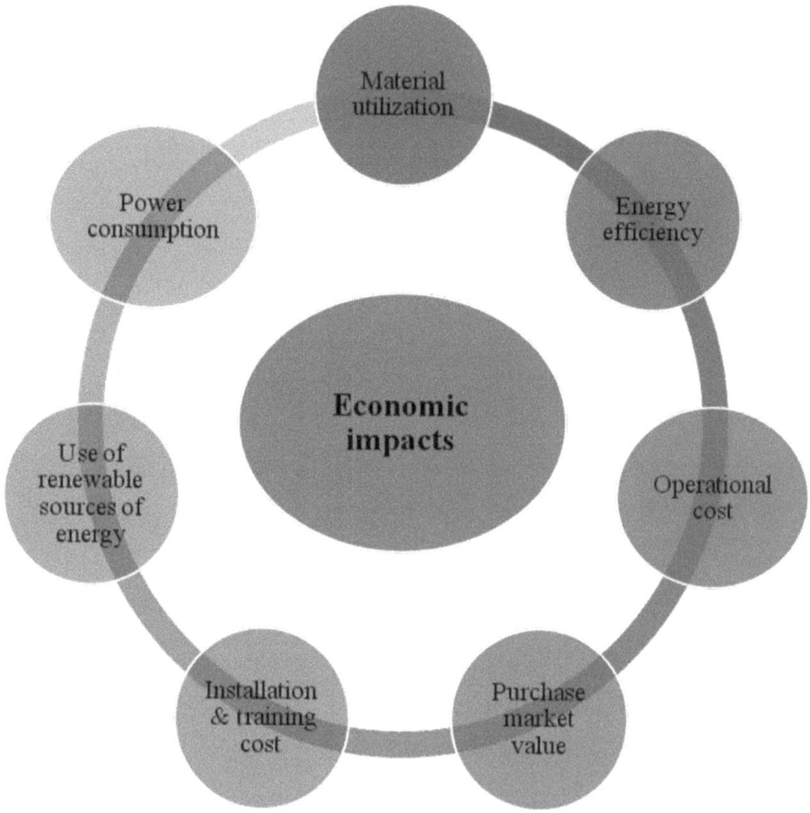

FIGURE 7.5 Different aspects of economic impacts (Mayyas et al., 2012).

In addition, D4S concept ensures the overall reduction in material and energy use to further promote more resource efficient production, which helps in securing profit margins. The various aspects of economic impacts of sustainability are presented in Fig. 7.5.

Life-Cycle Thinking in Sustainable Product Development

The D4S approach for product development rests on taking a life-cycle view of the product. The products life cycle commences with the extraction, processing, and supply of the raw materials and energy required for manufacturing the product, its distribution, use ,and its ultimate disposal (Verghese et al., 2012). Hence, LCA of a product is defined as a method that is used to evaluate the environmental impacts related to a product from the initial stage to end-of-life or cradle-to-grave (Mayyas et al., 2012). Sundin (2004) divided the life-cycle assessment into four major stages: the material extraction, manufacturing, use and disposal as shown in Fig. 7.6. On the other hand, Ashby (2009)

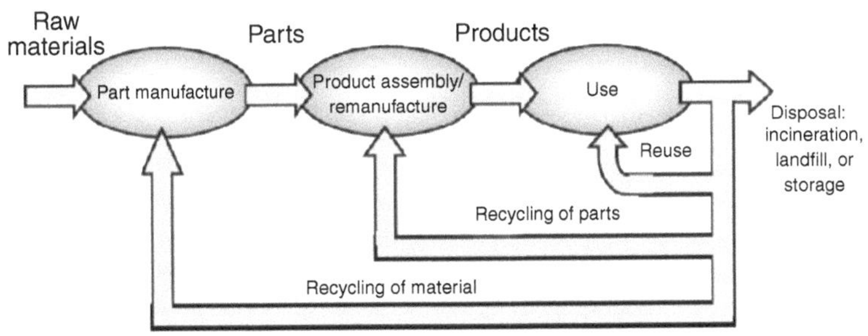

FIGURE 7.6 Life-cycle assessment into four major stages (Mayyas et al., 2012).

considered transportation as one of the main stages to be accounted during the LCA. It is worth noting that the LCA methodology has two main drawbacks: the diversity and variation in materials, processing techniques, usage durations, and disposal route (Mayyas et al., 2012; Omar, 2011). Nevertheless, LCA helps to generate product environmental life-cycle map and identify improvement options. Their goal is to optimize the "system" as a whole, which needs supply chain partnerships to achieve better and long-term environmental benefits that avoid creating new environmental impacts or "burden shifting." Burden shifting takes place when there is a shift of environmental impacts from one point in the supply chain to another (Verghese et al., 2012). Thus, the environmental challenge for D4S is to design products that minimize environmental impacts during the entire product life cycle.

DESIGN FOR SUSTAINABILITY OF NATURAL FIBER COMPOSITE MATERIALS

Up to date, glass fiber reinforced composites are widely utilized in many industries such as aircraft, automotive, construction and furniture, due to their numerous advantages. On the other hand, glass fibers are known to cause skin, eyes, and upper respiratory tract irritation (Sapuan et al., 2014). These inherent health hazards coupled with negative environmental impacts associated with the use of glass fibers fuelled the extensive search for healthier, safer, and cheaper fiber. Natural fibers are potential substitute for glass fibers, the latter is less abrasive to tooling and minimal respiratory problems for workers. Above all, they are environmentally friendly, abundantly available, cost effective, and good load bearing ability.

Natural fiber composites are composite materials with the fiber reinforcements obtained from plant, mineral, or animal sources and they reinforced the polymer composites. Plant-based fibers have gained popularity in recent years for their usage as reinforcements or fillers in polymer composites. Among the

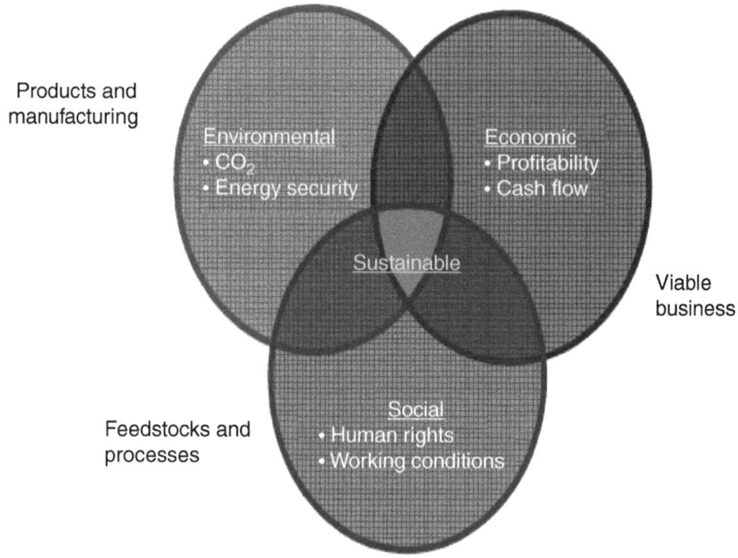

FIGURE 7.7 Environmental, economic, and social concerns related to natural fiber composites (Lee, 2013).

fibers reported to be used as reinforcements include kenaf, jute, sisal, henequen, hemp, flax, and date palm fibers. In tropical regions, fibers like oil palm, banana pseudo-stem, pineapple leaf, sugar palm, coconut (coir), sugarcane (bagasse), roselle, betel nut, cocoa pot husk, durian skin, and rice husk are used as fibers in polymer composites (Sapuan, 2014).

One of the main focuses in CE in recent years is the design for sustainability. While fulfiling the needs of future generation in mind, natural fiber composites have very important role to play. As far as composites are concerned, design for sustainability can be addressed by 100% bio-composites or natural fiber reinforced bio-polymer composites. Natural fiber reinforced polymer composite materials are discussed in the light of the three elements of sustainability (as shown in Fig. 7.7) later in this chapter.

Environmental Sustainability of Natural Fiber Reinforced Composites

Biocomposites are highly potential sustainable materials owing to their abundant availability from renewable resources, lower production cost, lower greenhouse gas emissions, less health effects, minimized waste volume on landfills and better growth in agricultural and chemical industries. They are remarkable for emitting low carbon footprint which is associated with their incorporation of CO_2-absorbing natural resources, whereas the production of conventional synthetic fibers or plastics generates large amount of carbon emissions (Dittenber and GangaRao, 2012). Moreover, manmade fibers as well as petroleum-based matrix commonly used in developing conventional composites require several hundreds of years to degrade, which leads to huge waste disposal problems. Recycling could have been a better option; however, recycled plastics are mostly contaminated by impurities and incompatible materials. The use of biopolymers as matrix in polymer composite materials can ease the ongoing waste disposal problems attached to nondegradable petroleum-based plastics. For instance, pure PLA simply degrades to CO_2, water, and methane. In addition, PLA can be retransformed to lactic acid, which can be purified and then utilized to produce PLA with the same mechanical properties (Dittenber and GangaRao, 2012; Henton et al., 2005).

It is believed that the employment of natural fibers to reinforce polymer composites enhances the environmental sustainability of composite materials. However, the concept of sustainable composite materials requires further clarification because it seems problematic to properly define. With the growth of environmentally conscious society, new products are developed with prefix terms such as "environmentally friendly" or "sustainable". Such claims (environmentally friendly and sustainability) can only be endorsed by considering the entire life cycle of the product; from the cradle (i.e., raw material extraction) to the grave (disposal). In that context, many attempts were undertaken to quantify the environmental sustainability that natural fiber composites offer. Mohanty et al. (2001) reported that the energy consumed for the processing of natural fibers was 20%–25% lower than for synthetic fibers. Similarly, Mueller and Krobjilowski (2003) found that it took 30%–40% less energy to produce a natural fiber nonwoven fabric than to prepare a glass fiber mat. Elsewhere, the life-cycle energy assessments conducted by Mutnuri et al. (2010) revealed that 40%–60% less energy was consumed by natural fiber reinforced polymers compared to glass fiber reinforced polymers, taking into account raw material extraction, raw material processing, raw material transportation, and manufacturing of the composite. More interestingly, the LCA studies of Patel and Narayan (2005) found that energy savings at the use stage of biomaterials could be even higher than at the processing stage. Finally, Patel and Narayan (2005) highlighted that green polymers and natural fiber composites can save at least 20 MJ of nonrenewable energy per kilogram of polymer and eliminate at least 1 kg of CO_2 per kilogram of polymer, as well as minimize most other environmental impacts by at least 20%.

Socio-Economic Sustainability of Natural Fiber Reinforced Composites

In fact, the current market trend of natural fiber reinforced composites is quite promising. The increased demand for natural fibers as reinforcement material for polymer composites used in various engineering applications (i.e., infrastructure, automotive, packaging, etc.) has significant socio-economic impacts. High demand for natural fibers would create an industrial crop source for the economic improvement of rural and agricultural-based communities. Such development would empower rural communities to tackle their poverty, lack of employment, and other social issues with minimal or zero assistance from foreign bodies (Dittenber and GangaRao, 2012). It also creates the opportunity for local villagers to learn advanced techniques and skills for proper natural fiber processing methods, which could lead to the development of locally manufactured natural composites. Alternatively, the prepared natural fibers could be supplied to high-technology engineering companies for manufacturing commercial composite components or products. This is a stepping stone for the local natural fiber industry to grow from simple to more technical applications.

Design for Sustainability Concept in Ford Automotive Company

Generally, metals are swiftly being replaced with plastics in the automotive manufacturing industry owing to their lightweight. Petroleum-based plastics and synthetic fibers are dominantly used in making various composites for automotive interior applications. These nonbiodegradable plastics, fibers, and composites total about 25%–40% of vehicle weight (Sanyang et al., 2016a). Despite the good properties and performance associated with these materials, automakers are extensively searching for renewable materials for better sustainability. Renewable materials such as natural fibers promote less CO_2 emission and energy saving besides reducing the weight of vehicles. Hence, the incorporation of natural fiber as reinforcement agent in polymer composites for developing automotive components can help in avoiding the negative environmental impacts of automotive production as well as reduce production cost.

Since the 1930s to the present day, natural fibers, biopolymers, and their composites have been playing a crucial role in the manufacture of automotive parts and components. Henry Ford always had the belief that agriculture and industry are natural partners. In other words, automotive components can be grown in farms. On this basis, he initiated the use of agricultural products as automobile materials. Fords develop many motor components from agricultural-based materials which greatly contribute to the splendor of ford-built cars and trucks (Lee, 2013). The company embraces the culture of using recyclable and renewable materials wherever feasible (technically and economically). Ford was the first car manufacturer to incorporate biobased plastics in paints, enamels and molded plastics parts. Ford also integrated natural fibers like hemp, wood pulp, cotton, flax, and ramie in different components of the vehicle (Brosius, 2006; Drzal et al., 2001).

In 1937, Ford produced 300,000 gallons of soy oil for use in car enamels. Afterwards, straw, flax, and soy bean meal were used in the manufacture of their car's body whereas its tires were made from golden rod latex (Phillips, 2008). The interior storage bins of the 2010 Ford Flex's third-row contained 20% wheat straw biofillers. Ford also used wheat straw in making vehicle steering wheels. Nowadays, the use of biobased materials in Ford manufactured vehicles is on the increase year by year and model by model (Lee, 2013). Ford's latest vehicles such as Ford Mustang, Expedition, F-150, Focus, Escape, Escape Hybrid, Mercury Mariner, Lincoln Navigator, Lincoln MKS, and Taurus are built with soy-based polyurethane foam seat cushions, seat backs, and seat headrests (Phillips, 2008).

The 2010 Ford Escape and Mercury Mariner were also designed with soy-based polyurethane foam headliner. The company further developed gaskets from soy foam together with 25% recycled tires in 14 vehicle platforms (Lee, 2013). The amalgamation of soy materials in Ford vehicles (smorgasbord of biobased materials, from the soy-based seating foam and body panels to the corn-based carpet mats, canvas roof, and tires) can help to reduce the consumption of petroleum by more than 5 million Ib annually and this, in turn, reduces CO_2 emission by more than 20 million Ib annually (Lee, 2013).

In 2014, Ford and Heinz (United State food processing company) have matched to investigate the use of tomato skins for automotive applications. Scientists at Heinz have been searching for innovative ways to recycle and reuse peels, stems and seeds from the more than two million tomatoes the company uses every year to produce Heinz tomato ketchup (European Bioplastics Bulletin, 2014).

Case Study on D4S of Bioplastic for Food Packaging

Many recent researches are conducted on CE/D4S of natural fiber reinforced bio-polymer composites in food packaging. Due to the difference in the nature of design, the products are designed in the form of film or flexible plastics, unlike the conventional plastics. The major elements contributing to the D4S for biobased food packaging are discussed beneath.

Design for Economic Impacts

Use of Renewable Materials

Besides minimizing the accumulation of plastics in the environment, the use of biodegradable packaging materials also helps to shrink reliance on fossil fuel for better sustainability (Rhim et al., 2013). Till now the packaging industry heavily depends on crude oil for the production of plastics. This clearly manifested how dependent the packaging and plastic industry are on crude oil. Consequently, the increasing price of crude oil has significant influence on the plastic market. Hence, to overcome the dependence on petroleum-based polymers, attempts have been made to utilize 100% renewable and biodegradable biopolymer in the production of biopackaging materials. Biodegradable plastics from renewable resources help to preserve the nonrenewable fossil fuel resources and further enhance sustainable development. The biobased plastic market is gradually exiting its infancy and taking over the petroleum-based plastic market at a growth rate of 30% annually (Arvanitoyannis, 1999; Fang and Fowler, 2003).

Design for Environmental Impact

Minimize Waste Disposal: Reusability, Recyclability, and Biodegradability

Single-use consumer packaging materials produced from nonbiodegradable plastics represents huge volume of the size of a typical landfills. It is believed that the amount of packaging waste generated in industrialized countries in a single day is sufficient to fill up a space equivalent to the Sears Tower (Chicago, USA), which was once the tallest building in the world (Imam et al., 2012). In view of the numerous environmental problems posed by petroleum-based plastics, governments of many developed countries enacted environmental policies that will help mitigate the current scenario. Most of these laws are constituted to device means of minimizing the use of nonbiodegradable plastics which continued to accumulate in landfills or disposed in water bodies, which has serious impact on marine life and community health. This provides an excellent opportunity for biobased plastics to be adopted for the replacement of petroleum-based plastics in the packaging industry.

Due to the unsustainable waste disposal in landfills, governments of several nations established laws to promote the use of recycled and/or biobased green products. Take the case of the "producer pays" principle, it was designed to encourage manufacturers to take responsibility for their products throughout their whole life cycle (Fowler et al., 2006). Such laws will not only stimulate enhancement of product recyclability, but will also offer better opportunities for the utilization of biobased materials as raw material for the manufacturing of these products. In the case of Selangor state, in Malaysia the initiation of "Selangor no plastic bag day' every Saturday," which was launched in 2010 (Fig. 7.8) has motivated the public to resort to recyclable bags and also promote the use of environmentally friendly plastics made from biobased resources. The campaign has assisted in saving about 5 million plastic bags within 2 years (2010–12). Thus, the state government was planning to expand the effort to 3 days, weekly. More and more countries and states follow in the banning of grocery plastic bags, which are responsible for the so-called "white pollution" across the globe (Nampoothiri et al., 2010). This trend has provided biobased plastics the opportunity to substitute conventional plastics.

FIGURE 7.8 Logo for 'Selangor no plastic bag day' every Saturday.

Design for Societal Impacts

Health and Wellness Effect

Several companies or institutions also have designed sustainability policies in their Health Safety and Environment (HSE) policies, which highlight the firm's environmental responsibilities to both its employees and to the society they are operating. Most of these firms are of the perception that the adoption of these policies, which may possibly include the use of biobased materials, will elevate their reputation in the eyes of the consumers, who have demonstrated their growing awareness and concern for environmental issues. For example, Universiti Putra Malaysia (UPM) is advocating for a "Green Campus" and among the established policies is "just to say no to plastic bag" as shown in Fig. 7.9. Hence, the UPM Health Centre (particularly the pharmacy) abstains from providing plastic bag to patients. These stringent policies promote the production and usage of biobased plastics.

In summary, the main driving factor for the commercialization of biodegradable plastics in many countries is government laws related to environmentally benign products, in the form of a ban or restrictions and the recommendation of certain type of products. Most of the time, these laws are meant to address specific environmental concern (Ren, 2003).

Natural Fiber Composites: Sugar Palm Fiber Sugar Palm Starch Biocomposites

The widespread usage of sugar palm fibers in rural communities for producing several local products is due to their high durability and resistance to sea

FIGURE 7.9 UPM advocates for a "Green Campus."

water. The fiber contributes significantly to human livelihood in many South East Asian countries such as Malaysia, Indonesia, Thailand, and Philippines. Martini et al. (2012) reported that different ethnic groups in Indonesia utilized sugar palm fiber for different purposes depending on their socio-economic activities, market opportunities, and availability of other natural resources. The fiber can be manually spun unidirectional to make ropes, or woven into mats. Traditionally, it is believed to be a suitable material for making ship ropes, brushes, and brooms. Fig. 7.10 shows the image of sugar palm fiber extracted from the sugar palm trunk.

FIGURE 7.10 Extracted sugar palm fiber.

Sugar palm fiber is identified by the villages as one of the best options for traditional roofing and bridge construction which can withstand tropical climate for many years (Ticoalu et al., 2013). In this way, sugar palm fiber contributes in meeting human needs by producing safe, secure, and affordable housing. The use of sugar palm fibers in roofing and other structural applications instead of wood has multidimensional advantages; (1) prevent deforestation, (2) lower construction cost, and (3) less labor. Hence, sugar palm fibers can also be considered as a green construction material that conforms to the concept of "Green Housing" while satisfying the concept of sustainability.

The UPM composites team together with villagers of Kuala Jempol, Negeri Sembilan, Malaysia are exploring the feasibility of establishing a sugar palm fiber–based composite industry, which would allow the villagers to manufacture various products from their own raw materials or they could sell to manufacturers. Either way, this initiation would generate additional financial revenue as well as create employment opportunities for the villagers. With the help of UPM research team, the possibility of using sugar palm fiber–based composites for more technical applications such as automotive components, small boat fabrication, packaging films and food containers. Fig. 7.11 shows the step by step extraction of sugar palm starch for the development of sugar palm starch–based films.

A number of reviews reported sugar palm fiber as an available natural fiber with competitive properties to other commercial natural fibers and their usage in developing green products have gradually accelerated over the years. The low density of sugar palm fibers provides them with relatively good specific mechanical and physical properties which favors their employment in manufacturing automotive components. In fact, this will help in manufacturing lightweight cars which are more fuel economical. So far, the use of natural fibers is only limited to interior automotive components. This time around, UPM composite research team developed glass/sugar palm fiber reinforced polyurethane hybrid composites for external automotive component application. The fabricated hybrid composites are to be utilized on the automotive front antiroll bar (ARB) design for typical Malaysian sedan vehicles (Proton) (Sanyang et al., 2016a). This innovation would serve as a stepping-stone for Malaysian automotive manufacturers to go green and be more sustainable.

Sugar palm fibers can serve as alternative resource for a sustainable future and have the potential to be commercially manufactured into various biocomposites for various engineering applications. In the past, the author studied sugar palm fibers reinforced petroleum-based polymer composites such as epoxy, unsaturated polyester, phenol formaldehyde, and high impact polystyrene. To overcome the dependence on petroleum-based polymers, Sanyang et al. (2016b) and Sahari et al. (2013) developed 100% renewable and biodegradable

FIGURE 7.11 Step by step extraction of sugar palm starch for the development of sugar palm starch–based films.

composite material by reinforcing sugar palm starch with sugar palm fiber. Such biocomposites are suitable for environmental sustainability. The development of sugar palm fiber reinforced sugar palm starch composites could provide 100% biodegradable and renewable composites options to sustainable packaging. Therefore, the utilization of sugar palm fiber, biopolymer, and their composites in the packaging industry would pose a positive impact on local sugar palm farmers, reduce dependency on fossil fuel, enhance environmental quality by developing a sustainable resource supply chain, and considerably decrease greenhouse gas emissions through a better CO_2 balance.

Design for Manufacturability

Manufacturing Methods

Many plastic and packaging industries ventured into biobased plastics using conventional plastic processing technologies. The only consideration is that the processing parameters of the equipment need to be adjusted to suit the characteristics of each bioplastic type. This helps the company to save vast amount of money.

The techniques for manufacturing biopolymers are all established polymer-manufacturing methods. However, the control and application of these techniques must be different to march with certain factors associated with exploiting the advantages of biopolymers. Although the manufacturing procedures show specific fundamental similarities, the major varying factor depends on whether a thermoset or thermoplastic biopolymer is to be processed (Jamshidian et al., 2010).

Injection molding is one of the most commonly used technologies for packaging plastic manufacturing. The processing conditions for biopolymers using injection molding have less damage to polymer. In contrast, the most difficult in continues processes such as extrusion, is the process in which the extrudate is stretched like in the case of film blowing. The limiting factors for the processing of biopolymers and petroleum-based polymers are all the same (i.e., degradation at higher temperature and shear). However, these limits are somehow narrower at the upper limits for biopolymers. When these upper limits are exceeded, it results in the degradation of biopolymers, which leads to mold defects such as discoloration, weld lines, or strong odor in the final product (Jamshidian et al., 2010; Johnson et al., 2003).

SUMMARY

In this chapter, the concept of design for sustainability is presented, covering the three key dimensions of sustainability. D4S is envisage as a promising problem-solving journey in which designers attempt to address environmental and economic impacts of products from the resource extraction stage to the product's end-of-life while paying full consideration on societal impacts as well. One of the main distinguishing approaches between D4E and D4S is that the latter gives due attention to social issues besides environmental and economic concerns. D4S aims to improve human well-being and profitability without compromising the utilization of the most suitable technology, appropriate materials, and manufacturing processes for the development of sustainable products. This chapter portrays natural fiber composites as a promising green material that has high positive impacts on nature, economy, and society. However, D4S has received little attention into the design profession as far as natural fiber composite development and applications are concern. Hence, there is a great need for the composite industry to adapt D4S approach to help in building a sustainable tomorrow for our future generation.

References

Arvanitoyannis, I., 1999. Totally-and-partially biodegradable polymer blends based on natural and synthetic macromolecules: preparation and physical properties and potential as food packaging materials. J. Macromol. Sci. 39, 205–271.

Ashby, M.F., 2005. Materials Selection in Mechanical Design, third ed. Elsevier Butterworth-Heinemann, Oxford.

Ashby, M.F., 2009. Materials and the Environment: Eco-informed Material Choice, first ed. Elsevier Butterworth- Heinemann, Oxford.

Babu, B.J., Prabhakaran, R.D., Agrawal, V.P., 2008. DFX analysis applied to composite products. J. Reinforced Plast. Compos. 27, 287–312.

Bralla, J.G., 1996. Design for Excellence. McGraw-Hill, New York.

Brosius, D., 2006. Natural fiber composites slowly take root. Compos. Technol. 12, 32–37.

Butlin, J., 1987. Our Common Future: The World Commission on Environment and Development. Oxford University Press, Oxford.

Charter, M., Tischner, U., 2001. Sustainable Solutions: Developing Products and Services for the Future. Greenleaf Publishing, Sheffield.

Crul, M.R.M., Diehl, J.C., 2006. Design for Sustainability: A Practical Approach for Developing Economies. UNEP/Earthprint, Paris.

Crul, M., Diehl, J.C., Ryan, C., 2009. Design for Sustainability—A Step-by-Step Approach. UNEP, Paris.

Dittenber, D.B., GangaRao, H.V., 2012. Critical review of recent publications on use of natural composites in infrastructure. Compos. Part A 43, 1419–1429.

Drzal, L.T., Mohanty, A.K., Misra, M., 2001. Bio-composite materials as alternatives to petroleum-based composites for automotive applications. Magnesium 40, 1–3.

Elkington, J., 1997. Cannibals with Forks: The Triple Bottom Line of 21st Century. Capstone, Oxford.

European bioplastics bulletin. issue 4/ 2013. Available from: http://en.european-bioplastics.org/wp-content/uploads/2013/newsletter/Issue4_13.pdf

Fang, J., Fowler, P., 2003. The use of starch and its derivatives as biopolymer sources of packaging materials. J. Food Agric. Environ. 1, 82–84.

Fowler, P.A., Hughes, J.M., Elias, R.M., 2006. Biocomposites: technology, environmental credentials and market forces. J. Sci. Food Agric. 86, 1781–1789.

Hambali, A., Sapuan, S.M., Ismail, N., Nukman, Y., 2010. Material selection of polymeric composite automotive bumper beam using analytical hierarchy process. J. Central South Univ. Technol. 17, 244–256.

Henton, D.E., Gruber, P., Lunt, J., Randall, J., 2005. Polylactic acid technology. Nat. Fibers Biopolym. Biocompos. 16, 527–577.

Imam, S.H., Glenn, G.M., Chiellini, E., 2012. Utilization of biobased polymers in food packaging: assessment of materials, production and commercialization. Emerg. Food Packag. Technol. 21, 435–468.

Jamshidian, M., Tehrany, E.A., Imran, M., Jacquot, M., Desobry, S., 2010. Poly-Lactic Acid: production, applications, nanocomposites, and release studies. Comprehens. Rev. Food Sci. Food Saf. 9, 552–571.

Johnson, R.M., Mwaikambo, L.Y., Tucker, N., 2003. Rapra review reports. Biopolymers 43, 3–26.

Karlsson, R., Luttropp, C., 2006. EcoDesign: what's happening? An overview of the subject area of EcoDesign and of the papers in this special issue. J. Clean. Prod. 14, 1291–1298.

Küçüksayraç, E., 2015. Design for sustainability in companies: strategies, drivers and needs of Turkey's best performing businesses. J. Clean. Prod. 106, 455–465.

Kurk, F., Eagan, P., 2008. The value of adding design-for-the-environment to pollution prevention assistance options. J. Clean. Prod. 16, 722–726.

Lee E.C., 2013. Bio-based materials for durable automotive applications. Available from: http://www.lawbc.com/share/bcs2013/Molecules%20to%20Market/lee-presentation.pdf

Ljungberg, L.Y., 2007. Materials selection and design for development of sustainable products. Mater. Des. 28, 466–479.

Martini, E., Roshetko, J.M., van Noordwijk, M., Rahmanulloh, A., Mulyoutami, E., Joshi, L., Budidarsono, S., 2012. Sugar palm (Arenga pinnata (Wurmb) Merr.) for livelihoods and biodiversity conservation in the orangutan habitat of Batang Toru, North Sumatra, Indonesia: mixed prospects for domestication. Agroforest. Syst. 86, 401–417.

Mayyas, A., Qattawi, A., Omar, M., Shan, D., 2012. Design for sustainability in automotive industry: a comprehensive review. Renew. Sustain. Energy Rev. 16, 1845–1862.

Mohanty, A.K., Misra, M., Drzal, L.T., 2001. Surface modifications of natural fibers and performance of the resulting biocomposites: an overview. Compos. Interfaces 8, 313–343.

Mueller, D.H., Krobjilowski, A., 2003. New discovery in the properties of composites reinforced with natural fibers. J. Ind. Text. 33, 111–130.

Mutnuri, B, Aktas, C.J., Marriott, J., Bilec, M., GangaRao, H., 2010. Natural fiber reinforced pultruded composites. Proceedings of COMPOSITES 2010, Las Vegas, Nevada, 9–11 February.

Nampoothiri, K.M., Nair, N.R., John, R.P., 2010. An overview of the recent developments in polylactide (PLA) research. Bioresource Technol. 101, 8493–8501.

Omar, M.A., 2011. The Automotive Body Manufacturing Systems and Processes. John Wiley & Sons, West Sussex.

Patel, M., Narayan, R., 2005. How sustainable are biopolymers and biobased products? The hope, the doubt, and the reality. In: Mohanty, A.K., Misra, M., Drzal, L.T. (Eds.), Natural Fibers, Biopolymers, and Biocomposites,. CRC Press, Boca Raton, pp. 833–854.

Phillips, A.L., 2008. Bioplastics boom. Am. Sci. 96, 109–110.

Ren, X., 2003. Biodegradable plastics: a solution or a challenge? J. Clean. Prod. 11, 27–40.

Rhim, J.W., Park, H.M., Ha, C.S., 2013. Bio-nanocomposites for food packaging applications. Prog. Polym. Sci. 38, 1629–1652.

Sahari, J., Sapuan, S.M., Zainudin, E.S., Maleque, M.A., 2013. Thermo-mechanical behaviors of thermoplastic starch derived from sugar palm tree (*Arengapinnata*). Carbohydr. Polym. 92, 1711–1716.

Sanyang, M.L., Sapuan, S.M., Jawaid, M., Ishak, M.R., Sahari, J., 2016a. Recent developments in sugar palm (Arenga pinnata) based biocomposites and their potential industrial applications: a review. Renew. Sustain. Energy Rev. 54, 533–549.

Sanyang, M.L., Sapuan, S.M., Jawaid, M., Ishak, M.R., Sahari, J., 2016b. Effect of sugar palm-derived cellulose reinforcement on the mechanical and water barrier properties of sugar palm starch biocomposite films. BioResources 11, 4134–4145.

Sapuan, S.M., 2014. Tropical Natural Fibre Composites: Properties, Manufacture and Applications. Springer, Singapore.

Sapuan, S.M., Abdalla, H.S., 1998. A prototype knowledge-based system for the material selection of polymeric-based composites for automotive components. Compos. Part A 29, 731–742.

Sapuan, S.M., Mansor, M.R., 2014. Concurrent engineering approach in the development of composite products: a review. Mater. Des. 58, 161–167.

Sapuan, S.M., Osman, M.R., Nukman, Y., 2006. State of the art of the concurrent engineering technique in the automotive industry. J. Eng. Des. 17, 143–157.

Sapuan, S.M., Sanyang, L., Sahari, J., 2014. Development and properties of sugar palm fiber reinforced polymer composites. In: Thakur, V.K., Kessler, M.R. (Eds.), Green Biorenewable Biocomposites: From Knowledge to Industrial Applications. Apple Academic Press, Oakville, Canada.

Spangenberg, J.H., Fuad-Luke, A., Blincoe, K., 2010. Design for Sustainability (DfS): the interface of sustainable production and consumption. J. Clean. Prod. 18, 1485–1493.

Sundin E., 2004. Product and process design for successful remanufacturing. Linköping University, Linköping, Sweden, PhD Dissertation.

Ticoalu, A., Aravinthan, T., Cardona, F., 2013. A review on the characteristics of gomuti fibre and its composites with thermoset resins. J. Reinforced Plast. Compos. 32, 124–136.

U.S. Congress, 1992. Green Products by Design: Choices for a Cleaner Environment, Office of Technology Assessment. (Report no. OTA-E-541), U.S. Government Printing Office, Washington, DC.

Verghese, K., Lewis, H., Fitzpatrick, L., 2012. Packaging for Sustainability. Springer Verlag, London.

Vezzoli, C., Manzini, E., 2008. Review: design for sustainable consumption and production systems. Syst. Innov. Sustain. 1, 138–158.

Product Design Specification (PDS) of Composite Pedal Box System[1]

In the early stages of conceptual design, the PDS of the polymer-based composite automotive pedal box system was prepared. The design of automobiles and their components is invariably evolutionary and does not change radically from one model or component to the next. In the PDS of polymer-based composite pedal box system, many similarities are found between polymer-based composites and that of metal counterpart. The detail descriptions of the elements of PDS considered in the design are listed as follows.

PERFORMANCE

- All three pedals could be assembled onto a mounting bracket, or one pedal to a bracket or two pedals in one bracket.
- The mounting bracket could in turn be secured to bulkhead of the automobile.
- The design should consider the method of pivoting pedals to the mounting bracket, where the bracket could incorporate gussets to keep pedals in place.
- The automobile has prepositioned bolts or bolt holes on the bulkhead to secure the pedal box system to an automobile.
- The mounting bracket could have holes or slots, where they could be fixed on the bulkhead of the automobile.
- Reinforcing ribs and webs could be designed at or between sections of the pedals to reduce the weight and to withstand stresses applied to the pedal.
- The pedal box system should be resistant to wear and must have sufficient bearing strength.

[1]The first PDS is concerned with the design of glass fiber reinforced polymer composite automotive pedal box system (Sapuan, 1998) and the PDS of this product is shown in Appendix A.

Composite Materials. http://dx.doi.org/10.1016/B978-0-12-802507-9.00008-8

- The design should consider possible failure through fatigue, creep, and shear.
- The force on the brake should not be less that 2700 N with the maximum deflection of 15 mm.
- The force of the clutch pedal should be less than 1000 N with the maximum deflection of 20 mm.
- The force on the accelerator pedal should be less than 375 N with the maximum deflection of 10 mm.
- The brake pedal should be depressed to apply the brakes and be capable of being depressed by the driver's foot.
- The brake pedal should be fixed totally or partly to the right of the longitudinal axis of the driver's seat.
- Actuation of one pedal should not be hindered by the depression of other pedals.
- The clutch pedal should be depressed to declutch and the final movement may also, when facility is provided, apply as the brake.
- It must be capable of being depressed by the driver's left foot.
- Accelerator pedal depression should increase speed, when the automobile is moving.
- Releasing the accelerator pedal should reduce the speed.
- The accelerator pedal should be capable of being depressed for gentle and gradual acceleration or deceleration of automobile.
- By releasing all pedals, there should be automatic return to the original positions by means of springs or similar devices.

ENVIRONMENT

The pedal box system should be capable of being used in all weather conditions and be corrosion resistant.

It should be resistant to fuel spillage and degradation by ultraviolet radiation.

The percentage of water absorption of the material should be less than eight percent (8%).

TARGET PRODUCT COST

- The cost should include material, tooling, labor, and equipment costs.
- The aforementioned total cost should be less than the cost of manufacturing a metal pedal box system.
- The tooling cost should be less than RM* 1,275,000.

*RM = Malaysian Ringgit 1 USD = RM 4.14

- The direct material cost should be as low as possible (the cost of RM 24/kg is proposed but the cost could be increased if the suitable material cannot be obtained at this price).

SIZE AND WEIGHT

- In view of the limited space available for the driver's feet, the dimensions should be kept as small as possible.
- The length of all pedals should not exceed 0.5 m.
- Accelerator pedal weight: less than 260 g.
- Clutch pedal weight: less than 500 g.
- Brake pedal weight: less than 500 g.
- Mounting bracket weight: less than 750 g.

MATERIALS

- The materials used for all components should be polymer-based composite as this is the requirement for weight reduction.
- However, the pedals, mounting bracket, and other elements may involve some metal reinforcement, molded into them or applied after molding.
- The density all materials used in all elements of pedal box system should be lower than the density of aluminum, i.e., 2.7 Mg/m^3.
- The impact strength of the material should be greater than 30 kJ/m^2.
- The material should be recyclable.

ERGONOMICS

- The distance between the steering wheel and brake pedal should be kept to approximately 600 mm.
- The pedal design should not cause fatigue to the feet of the driver.
- The design should provide comfort and enough space for installing and removal of the pedal box system.
- The pedal layout and construction should ensure a comfortable position for the driver.
- The design should take into account factors such as women wearing high heeled shoes.

SAFETY

- The components should be free from sharp edges.
- The system must comply with all relevant parts of the local and international legislations.

- Pedals should not be too short to make it difficult for a driver to depress them in an emergency.
- They should be large enough to prevent the foot from slipping.
- The pedal pads should have enough treads to prevent slippage.
- The distances between pedals should not be too close to each other.
- The clutch pedals should operate with as little force as possible to avoid the driver moving way from the seat.

Product Design Specification Document for Automotive Parking Brake Lever (Mansor, 2014)

PDS Elements	Parameters for Parking Brake Lever Design: Current Model (Proton Wira)	Parameters for Parking Brake Lever Design: New Design Intent (Hybrid Kenaf/Glass Fiber Reinforced Polypropylene Composite Parking Brake Lever)
1. Performance 1. Strength 2. Stiffness Able to achieve safe and functional operation requirement during engagement and disengagement of the parking brake where bending load is applied	Able to withstand hand control force ≤ 400 N without failure (FMVSS 571.135.7.12) *failure: free from deformation, cracks, and other defects	1. Able to withstand hand control force ≤ 400 N without failure (FMVSS 571.135.7.12) *failure: free from deformation, cracks and other defects
2. Weight 1. Density. To comply with the EU carbon emission directives, the weight of the component is minimized compared to the current component to gain lightweight advantage while still maintaining the required structural strength for safe and functional operation. Material density is directly related to the component weight. 2. Size. The parking brake lever size considers the parking brake lever ratio as well as to comply with conditions as stipulated in the ECE 74/60 (Clause 5.3) and ECE-R21 (Clause 5.3) for parking lever's hip point location with respect to the front seat hop point, and EEC 78/632 (Clause 5.3.2) for parking brake lever edge radius and cut section area.	Proton Engineering Specification for parking brake lever assembly (confidential): Parking brake lever weight = 0.36 kg using SPHC steel (Hot Rolled Mil Steel Plates, Sheets, and Strip) : (equivalent to JIS 3131 and AISI 1012 hot-rolled steel)	1. Product weight should be less than 0.36 kg 2. Parking brake lever ratio is 8.0 ± 0.5 3. Parking lever's hip point must be higher than front's seat hip point (rear most condition) (ECE 74/60 Clause 5.3 and ECE-R21 Clause 5.3) 4. Brake lever shall have minimum cross-sectional area of 6.75 cm^2 measured in a horizontal plane, and not more than 6.75 cm from the furthest projecting part measured in the vertical direction (EEC 78/632 Clause 5.3.2) 5. The brake lever radius of curvature must not be less than 3.72 mm (EEC 78/632 Clause 5.3.2)

Composite Materials. http://dx.doi.org/10.1016/B978-0-12-802507-9.00009-X

3. Standard Specifications US National Highway Traffic Safety Administration (NHTSA), Federal Motor Vehicle Safety Standards (FMVSS) No. 571.135: Light vehicle brake system or ECE R13H (European Standard) for light passenger vehicle with Gross Vehicle Weight Rating (GVWR) ≤ 3500 kg	1. US FMVSS 571.135: S.7.12.2: Test condition and procedures for parking brake Proton Engineering Specification for parking brake lever assembly testing (confidential):	1. The design must be tested according to the US FMVSS 571.135: S.7.12.2 for test condition and procedures for parking brake
4. Disposal 1. Reuse and recyclability 2. Safe for disposal The brake lever design must meet both item (i) and (ii) requirements to meet EU vehicle end-of-life requirements for percentage of components to be reuse or recycled and safe for disposal [Based on EU DIRECTIVE 2000/53/EC (Article 2)]	Proton Engineering Specification for parking brake lever assembly (confidential): Directive 2000/53/EC: ■ The component and material do not contain any lead, mercury, cadmium, or hexavalent chromium ■ 100% recyclable	1. The parking brake lever material must not contain any lead, mercury, cadmium, or hexavalent chromium (EU Directive 2000/53/EC) 2. The parking brake material is 100% reusable and recyclable (EU Directive 2000/53/EC) 3. The parking brake material is safe for disposal (EU Directive 2000/53/EC)
5. Environment 1. Corrosion resistance 2. Resilience to water absorption Able to be used in all weather conditions resistance and resilience to water absorption to avoid any corrosion happening on the component surface which will deteriorate the component structural strength and maintain dimensional stability especially for long term operation		1. The brake lever must be corrosion resistance to fuel spillage and degradation by ultraviolet radiation. 2. The brake lever must be resilience to water absorption where the percentage of water absorption must be less than 8%
6. Target Product Cost 1. Raw material cost 2. Manufacturing cost To achieve low product cost compared to metallic brake lever, low cost raw material is selected and the product is able to be manufactured in various shapes using high volume manufacturing processes to reduce the manufacturing cost		1. Total raw material cost must be less than USD 5/kg 2. The product must be able to be manufactured in various shapes using high volume manufacturing processes. The tooling cost must be less than GBP 250,000 or USD 400,000

Glossary

3D printing It is one of the rapid prototyping techniques to develop a 3D object. In this technique, the object is produced by stacking several layers of materials. The data from computer 3D model is connected to the 3D printer to create the rapid prototype.

Advanced composites Advanced composites are composites with advanced fiber reinforcements such as aramid and carbon. They are normally more expensive than "fiberglass," have very high strength-to-weight ratios and normally used in advanced applications like in aerospace industry.

Aesthetically pleasing The term is normally used by design engineering to show a product is beautiful.

Analytic hierarchy process (AHP) AHP is also known as analytical hierarchy process and it is a method to mathematically and psychologically analyze and organize decisions, which are normally complex in nature. The method uses pairwise comparisons to provide weightage to several criteria and to rank the alternative solutions. Sensitivity analysis is also used to test if the solution is implementable and robust.

Analytic network process (ANP) It is a multicriteria decision making tool similar to AHP where both AHP and ANP use pairwise comparisons for weighting and ranking the alternatives. However, it is different from AHP as it structures decision-making problem as a network while AHP structures the decision-making problem in terms of goal, decision criteria, and alternatives.

Anisotropic material A material that exhibits different properties in different directions.

Aramid It is a fiber whose name is derived from a combination of aromatic polyamides produced by Du Pont with excellent strength-to-weight properties. It is generally used in aerospace and defense industries.

Artificial neural network A computer system or computational model inspired in the natural or biological neurons like the brain.

Ashby chart Also known as material property chart. It is a materials selection tool in the form of a chart developed by M.F. Ashby. It enables the user to select materials after the required properties have been decided. It is not recommended to be used in detail design calculation.

Automated fiber placement It is also known as advanced fiber placement. This fabrication method involved heating and compacting resin preimpregnated fibers on complex tooling mandrels in an automated manner used for fabricating components with complex shapes.

Automated tape layering It is composite fabrication process that involved rapid deposition of resin preimpregnated composite tapes to a flat or curved tool in an automated manner used for fabricating complex components.

Bidirectional fiber composite It is a composite where the fibers are arranged in both 0 and 90 degree direction and they are normally made in the form of mats.

Biocomposites Biocomposites are the composites made from natural fibers and polymer matrix. The matrix is either synthetic or bio-based polymers such as poly(lactic acid) (PLA).

Composite Materials. http://dx.doi.org/10.1016/B978-0-12-802507-9.00010-6

Biodegradability It is the ability of materials to decay through the action of living organisms.

Biodegradable packaging It is a packaging that can be consumed by microorganisms after its disposal.

Biomimetics Mimicking the element and systems of nature in generating ideas or concepts in product development. It can be applied to fields of endeavor such as military technology.

Biomimicry Mimicking the elements and systems of nature in generating ideas or concepts in product development. It is specifically focused on developing sustainable solutions.

Biopolymers Biopolymers are polymers that are biodegradable and manufactured by a living organism (such as proteins and polysaccharides).

Blue ocean strategy (BOS) It is a new business strategy which focuses on uncontested market and makes the competition irrelevant. It emphasizes on the concept of value creation in the business environment.

Boron fiber Boron fiber is a high strength, lightweight material that is used as reinforcement of composites mainly used in advanced aerospace components. However, its use in the recent years has been replaced with advanced fibers like carbon and aramid.

Brainstorming It is an idea generation technique conducted in a group to generate as many design concepts as possible. This session is led by a coordinator and each member of the group normally gives ideas spontaneously and during the session, judgment is suspended.

Bulk molding compound (BMC) It is a ready to mold semicured composite made from catalyzed thermoset polyester and strands of short chopped glass fibers mixed in a mixture. It is used to produce small, intricate products using injection molding and compression molding. It is also known as dough molding compound (DMC).

Cambridge Engineering Selector (CES) It is a PC-based commercial software developed in Cambridge, UK mainly for the purpose of selection of materials. It relies heavily on the use of Ashby charts and it enables design engineers to select, explore, and apply materials property data.

Capturing design intent It is the process of capturing how the solid model should be created and how it should behave when it is changed.

Carbon fibers Carbon fibers (also known as graphite fibers) are synthetic fibers with the diameter of approximately 5–10 μm and comprised mainly carbon atoms.

Carbon fiber reinforced composites Composites in which the reinforcing phase is carbon in the form of fiber.

Carbon nanotubes (CNTs) These materials are allotropes of carbon with a cylindrical nanostructure. These materials can be combined with polymer composites to form carbon nanotubes reinforced polymer composites.

Ceramic matrix composites (CMC) Composites in which the reinforcing phase is enveloped in a ceramic material. Example: silicon carbide fibers reinforced aluminum oxide.

Chemical vapor deposition Chemical vapor deposition (CVD) is a chemical process used to produce high performance materials such as CMCs.

Chemical vapor infiltration (CVI) It is a ceramic engineering process to produce fiber reinforced CMCs. In this process ceramic matrix is infiltrated into fiber preforms using reactive gases at high temperature.

Chopped strand mat It is a mat made from randomly arranged glass fibers flattened into a sheet.

Compatibility The ability of two or more substances to be mixed with each other to obtain a high homogeneity of composition.

Compocasting Compocasting is a liquid state process for fabricating metal matrix composites in which the particulate, whisker, or discontinuous fiber reinforcements are mixed with the molten metal while being vigorously agitated. The uniform mixing during stirring is achieved by allowing the molten metal to cool down to a more viscous two-phase solid-liquid state.

Composite It is a material made from two or more constituent materials with distinctly different physical and chemical performance. However, when they are combined, a material is produced with attributes different from the individual constituent materials. The individual constituent materials remain separate and distinct within the final product.

Compression molding The method of molding a material already in a confined cavity by applying pressure and usually heat.

Concept evaluation It is also known as concept selection. It is a process of comparing, analyzing, and sequentially eliminating design concepts to make decision on the most suitable concept.

Concept generation It is a process of generating ideas during conceptual design stage. This task is normally conducted using several established idea generation techniques proposed by different design experts for the sake of addressing the customer needs.

Concept initiation Concept initiation is the same as concept generation.

Conceptual design This is an important phase in design process where 70% of the cost of the product is determined during this stage. Generally, it comprises concept generation and concept evaluation and it is done before the embodiment or detail design stage.

Concurrent engineering It refers to a philosophy used in product development, where design engineers, manufacturing engineers, and personnel in other departments work in an integrated manner from the outset of the design process with the goals of achieving higher quality product, produced at shorter time and lower cost.

Concurrent engineering team It refers to the team involved in carrying out the product development process using concurrent engineering principle and the team members must have broad range of professional skill and professional knowledge workers from different disciplines.

Constraint satisfaction It is a term commonly used in the field of artificial intelligence. When the variables satisfy all constraints and as a result the solution is found, this situation is called constraint satisfaction.

Constraint violation As opposed to constraint satisfaction, constraint violation occurs when the variables violate any one of the constraints and the process of finding solution is ceased.

Continuous fiber Continuous fiber is a finite length fiber, with the length as short as several meters to several thousand meters in length. It is called continuous fiber because the length is much greater than the width or thickness of composites.

Contradiction matrix It is a tool in TRIZ, which is able to simplify the process of selecting the most appropriate principle, out of 40 inventive principles, to resolve a specific contradiction found in the product.

Conventional mixing and pressing It is one of solid state processing techniques of metal matrix composites (MMCs).

Crashworthiness It is the ability of a structure to protect its occupants in air and space crafts, rail and road vehicles and ships and submarines during an impact in the interest of the safety of the occupants. It is also related to energy absorbing system subjected to dynamic loading.

Cross-industry innovation Cross-industry innovation is an idea generation technique based on analogy and the transfer of the idea obtained from one industry to another industry.

Curing It is a process of toughening or hardening of polymer from liquid to solid by cross-linking of polymer chains or polymerization, encouraged by electron beams, initiators, heat, or chemical additives.

Design concepts Design concepts are ideas generated from concept generation session to be proposed for further evaluation in concept evaluation stage. Generally limited number of design concepts (typically four or five concepts) is proposed further evaluation.

Design for "X" It is just a generic term to represent concurrent engineering where "X" may represent several "abilities" such as manufacturability, sustainability, recyclability, maintainability, or reliability.

Design for manufacture (DFM) It is also known as design for manufacturability or design for manufacturing. It is a design approach, where all the downstream activities are considered in the very stage of design process for the sake of facilitating ease of manufacture. It is done by addressing the manufacturing problems during the conceptual design stage, as any problems that are fixed at the outset of product development process do not incur so much cost.

Design for sustainability (D4S) Design for sustainability, also known as sustainable product design, is a method to improve product quality, market opportunities, environmental performance, and social benefits keeping in mind the well-being of future generation.

Detail design It is a stage in design process performed after embodiment design is completed where detail drawings, detail calculations, detail information on materials and manufacturing process are carried out.

Diffusion bonding Diffusion bonding is a solid-state processing technique for MMCs, works based on solid-state diffusion principle. In this method, the atoms of surfaces of two solid metals mingle with each for extended period at high temperature and pressure. The metal matrix is in the form of foils and it is in contact with fiber reinforcement and after consolidation the MMC is made.

Digital logic method (DLM) DLM is one of the tools used by design engineer for materials selection process. In this method relative importance of goals is determined. Weighting factors are assigned to materials properties. Performance index of each candidate material is determined and the materials are finally ranked.

Direct comparison method It is concept evaluation method where only two final concepts, which have scores very close to each other from the first stage of design evaluation, are compared. The concept which satisfies better any of the design requirements is given the score 1 while the other concept, zero. The concept which gets better total scores is selected as the most suitable concept.

Dough molding compound It is the same as BMC.

Embodiment design It is a stage between conceptual and detail design in the design process. Refinement of final design concept is done in the form of schematic drawings, with calculations, dimensions, tolerances, and potential materials and manufacturing processes.

Environmentally conscious design See design for sustainability.

Exfoliation–adsorption It is a manufacturing route to produce nanocomposites using solvent systems. In these composites, the fiber is layered silicate and mixed with polymer matrix. Layered silicates are dispersed in solvent. The polymer adsorbs into silicate to form nanocomposites.

Exploitation of existing system A study of existing product or analogous products, earlier version of one's own company product or competitor products with similar function that can be used to generate new design concepts.

Extending search space method It is a method to remove mental block in creative thinking and techniques that can be used for this purpose include Why?Why?Why?, random input and counter-planning.

Extrusion Extrusion is a fabrication process for metals, polymers, and polymer composites used to fabricate components of simple or very complex fixed cross-sectional profiles. In this process, a material is pushed through a die of the intended cross-sectional profile.

Failure mode and effect analysis (FMEA) Failure mode and effects analysis is the method to identify failure modes, and their causes and effects in components and subsystems and the information are recorded in a FMEA worksheet.

Fiber Single strand of material used as the main constituent in composites.

Fiber orientation Direction of fiber alignment in a nonwoven fiber composite.

Filament winding An automated process which involves winding resin treated continuous filaments under tension over a mandrel in a pattern designed to achieve maximum strength in a certain direction. This process is normally suitable to make components with circular cross section.

Filler In composites, filler is a secondary material (e.g., talc, $CaCo_3$, and kaolin) added to basic materials (fiber and matrix) to alter the physical mechanical, thermal, or electrical properties of the composites.

Filled composites Composites with reinforcement phase are in the form of fillers or particulates such as oil palm filler particles.

Finite element analysis Finite element analysis (FEA) is a computation program that uses the finite element method to perform engineering calculations and to find how applied stresses affect the design. FEA can be used to various engineering problems such as static, dynamic, and buckling problems.

Four actions framework This is a tool in blue ocean strategy, that is, eliminate, reduce, raise, and create that can be used for concept generation.

Gallery method Gallery method is a concept generation method that uses a mixture of physical and mental activity in generating ideas. The participants move past the ideas (as in an art gallery) and final design concept is decided by the team after the ideas are discussed.

Gel coat It is a material such as a quick-setting resin used for surface layer of a mold to provide a high-quality visible surface appearance of fiber reinforced polymer composite products.

Glass fiber reinforced polymer (GRP or fiberglass) A combination of polymer material and fibrous glass with better mechanical properties as compared to the base material.

Glass mat thermoplastics (GMT) GMT is a glass fiber reinforced thermoplastic composite sheet used as intermediate material in compression and press molding processes. It is a thermoplastic equivalent of sheet molding compound.

Graphene-based composites Graphene is allotrope of carbon in the form of a 2D honey-comb lattice in which one atom forms each vertex. Combining graphene with polymers forms graphene-based composites.

Green composites Green composites are environmentally friendly composites that do not pose major environmental pollution such as fully biodegradable fiber reinforced biopolymer composites.

Green design See design for sustainability.

Hand lay-up A traditional, labor intensive, open mold composite manufacturing process to produce one-off or low volume products.

Honeycomb materials Honeycomb materials are used as cores for sandwich structures. They are made by localized bonding of stacked strips of material. The most common cell shape is hexagonal.

Hybrid composites Composites with the fibers are made from two or more materials normally carried out to optimize the properties of composites.

Idea generation techniques Idea generation techniques are used to generate ideas in solving problems and in engineering design it is normally referred to as concept generation. Many methods have been invented by experts in facilitating idea generation.

Impact strength It is the resistance of a material to fracture under impact or shock loading, expressed in terms of the amount of energy absorbed before fracture. In composites, Charpy, Izod, and drop-weight impact tests are common forms of impact tests.

Inference engine It is a tool in artificial intelligence, in particular in expert system. It takes advantages of logical rules such as using ruled-based system (*if–then*) to deduce new knowledge.

Information gathering It is an important activity before concept generation is carried out. Gathering as many information as possible from different sources to help design engineer to perform design task is required.

Injection molding It is a process of injecting polymers or polymer composites (molten polymer and chopped fibers) into a closed mold under pressure and the composites are allowed to cure.

Injection molding compound (IMC) It is specially formulated molding compound in the forms of granules with appropriate optimized constituents to be used in injection molding process.

In situ intercalative polymerization It is a fabrication process for intercalated sheet polymer nanocomposites by swelling of layered silicates and polymerization.

Interfacial bonding It is the bonding in the area between fiber and matrix in composites. If the bonding is strong, good performance of composite can be obtained but if there is weakness in this bonding, some modifications are needed to improve the properties of composites.

Internal mixer It is a machine used to mix fiber and matrix to form intermediate materials. These materials are placed in a mold for subsequent molding process, for example, compression molding to produce fiber reinforced composites.

Isotropic It is the material that exhibits properties that do not change with the direction of measurement. The properties of isotropic material are independent of the axis.

Kevlar Kevlar is the registered trademark of aramid developed at Du Pont (see aramid).

Knowledge-based system It is a branch of artificial intelligence that uses reasoning and logics as well as knowledge base in solving complex problems.

Lamina It is a single ply or layer in a laminate, with all fibers having the same fiber orientation.

Laminate A composite made by bonding together two or more layers or laminae of material(s) in an ordered stacking sequence.

Laminated composites See laminate.

Lanxide process Lanxide process is one of the processes to fabricate ceramic matrix composites (MMC). Ceramic matrix is produced when molten metal and a gas reacts, for example, molten aluminum and oxygen react to form alumina. Then particles are placed in the liquid metal to form CMCs.

Life cycle analysis It is an analysis to assess environmental impacts in all stages of a product's life from extraction of raw materials to disposal/recycling.

Life cycle design See Design for sustainability

Liquid infiltration under gas pressure It is a fabrication process to produce MMCs. It uses molten metal as matrix to infiltrate the preform (fiber) by pressure of the gas.

Mandrel A tool that serves as a core around which composites are laid up or wound and processed.

Manufacture In total design model, manufacture is a stage in design process carried out as soon as detail design stage is completed.

Market investigation It is the first stage activity in a total design model before product design specification (PDS) is prepared and in this stage all information and knowledge related to design are gathered.

Materials characterization It is a process of measuring material structures and properties.

Materials selection It is the process of identifying and selecting materials performed by a design engineer, from a set of candidates after product design has been completed.

Matrix The continuous phase from homogeneous material, which envelops the reinforcing phase, that is, fibers or filaments in a composite material.

Matrix transfer molding It a liquid state processing of CMC that can be used to produce component like tube. It involves inserting fiber in the form of cloth in a mold, transferring matrix to the mold, and melting it at high temperature and pressure to form CMC.

Melt intercalation It is a method to produce nanocomposites, where molten polymer is mixed with organophilic clay in injection molding or extrusion process at high temperature and the polymer is intercalated/exfoliated to form nanocomposites.

Metal matrix composites (MMCs) Composites in which the reinforcing phase is enveloped in a metallic material.

Mind mapping It is also known as concept mapping. It is an idea generation technique that takes information from one's brain. It is a note-taking exercise out of one's mind utilizing keywords and colorful images as opposed to writing long list of notes.

Moisture contents It is the quantity of moisture contained in a material; in our case composite material.

Mold flow analysis Mold flow analysis is a computer simulation program that simulates materials like polymer and polymer composites into a mold and it also simulates the conditions during actual processing.

Molding compound It is a type of compound suitable for molding. It is a mixture of polymer with catalysts, fillers, pigments, and fibers specially formulated and made to be used in composite manufacturing processes like injection and compression molding.

Mold release agent A release agent is a lubricant or chemical applied to mold surfaces to facilitate ease of release of the molded component during molding process.

Morphological chart Morphological chart is a chart used in morphological analysis of a product. The morphological chart method provides facilities to generate ideas by listing the subfunctions of a product and providing subsolutions to each subfunction. The final concept is decided by choosing a subsolution for each subfunction and combining all subsolutions.

Multicriteria decision making It is decision-making tool used for the condition where multiple conflicting or contrasting criteria are present during the evaluation process.

Multidirectional fiber composites They are composites with the fibers arranged in 0, 90, and ± 45 degree directions and they are normally made in the form of mats.

Nanocomposites Nanocomposites are high performance composites made from a combination of polymer matrix and nanoparticles (the size of nanometers, nm). Important properties such as mechanical, electrical, and thermal properties of these materials are superior to their constituent materials acting alone.

Natural fibers See natural fiber composites.

Natural fiber composites Composites with the fibers derived from natural sources particularly plant based such as coconut fiber (coir), banana pseudo-stem fiber, sisal, ramie, hemp, jute, flax, pineapple leaf fiber, oil palm fiber, and sugar palm fiber. The major advantages of natural fibers include environmentally friendly, low density, low cost, abundance and renewable.

Nonwoven fabric composites They are polymer composites with the reinforcements are made from fabrics produced by grouping long fibers either chemically, mechanically (spunlacing, needle punching and paper-making techniques), or by heat.

Out-of-autoclave (OOA) processing Out of autoclave (OOA) pre-preg composite processing is an alternative process to the traditional high pressure autoclave curing process commonly used in the development of composite aerospace parts. OOA is a process that produces autoclave-quality parts using a process called vacuum bag only (VBO).

Part consolidation It means fabrication of one-piece composite component to replace multiple parts either previously made from composites, metals, or other materials. To achieve that, the need to have separate parts should be reviewed and components from composites have the capability to be made in a single piece.

Part integration See part consolidation.

Particulate reinforcement It is a type of reinforcement in composites used in many different composite systems and it is very practical especially for CMC.

Pigment Pigment is a material that is used to change the color of other materials and it is used for coloring polymer in polymer composites.

Ply See lamina.

Polymer matrix composites Composites in which the reinforcing phase is enveloped in a polymeric material.

Powder metallurgy It is an important solid state processing for MMCs. The metal matrix is in the form of particles and it is in contact with fiber reinforcement and after consolidation the MMC is made.

Preimpregnation See pregreg.

Prepreg A semicured fiber reinforced composites which is ready to be fabricated into composite component using molding processes.

Prepreg molding A composite fabrication process carried out by molding a prepreg as intermediate material, and the composite is either cured in autoclave or vacuum bag. For the former the process is called autoclave molding while the latter is vacuum bag molding.

Problem formulation An important step in design, carried out by the design engineers. It is necessary for them to discuss with the clients or customers at the very early stage of design process. The design need, goal, objectives, and constraints should be agreed upon during this stage.

Product design specification (PDS) A PDS is a document prepared by design engineers based on some design requirements and it is a form of control document in the design. It specifies the design, and provides guidelines for the design.

Pugh selection method Pugh selection method is a concept selection method that can be used by a design engineer to compare several design concepts. In comparing the concepts, one concept is normally an existing concept called a datum. All other concepts are compared with the datum in fulfiling design requirements. If a concept is better than datum in fulfiling one particular requirement, a "+" score is given but if it is inferior, then a "-" score is given. If the performance of that concept is the same, "S" is assigned, which means the same with datum. The system works for the remaining design requirements. The concept with the highest positive difference between "+" and "−," among all the concepts is selected as the most suitable concept.

Pultrusion A continuous manufacturing process of composites, consisting of pulling a resin impregnated fiber into a certain shape of die before the resin is cured.

Quality function deployment (QFD) It is a method that enables a development team to transform the customer's wants and needs (voice of customer), into engineering characteristics for a product or service and then evaluate each proposed product or service. It also sets their development targets.

Random reinforcement In particulate reinforced composites, it is practical to have the arrangement of the particle reinforcement to be in random orientation.

Rapid prototyping Rapid prototyping comprises several techniques used to quickly fabricate a model of a product using 3D computer aided design (CAD) data.

Reinforced plastics See polymer matrix composites.

Reinforcement It is a component, which provides a composite material with higher elastic modulus and strength compared to the base material.

Resin The term used to designate any polymers in which the fiber reinforcement is added.

Resin transfer molding (RTM) A form of liquid molding whereby positive pressure is used to drive a liquid resin into a mold cavity containing dry fibers.

Rheocasting See compocasting.

Rib structures Rib structures are normally added to polymer composite parts for structure reinforcement, and to allow greater strength and stiffness in molded polymer composite parts without the need to increase the wall thickness.

Roller Roller is a type of equipment used in the hand lay-up or spray-up to consolidate the layers of composites and to remove air bubbles. In addition to hand roller, squeegee is also used for this purpose.

Roving Roving is a collection of yarns or strands made into a parallel twisted or untwisted bundle.

Rule-based system Rule-based system is an important tool in expert system. The rule-based system employs "if–then" facility to make deductions or choices.

Sandwich structures Sandwich-structured composites are structural composites made by placing stiff skins (top and bottom) to a lightweight but thick core (such as balsa wood and honeycomb materials). These materials demonstrate exceptional strength and stiffness properties and are found in advanced applications such as in aircraft components.

Sequential engineering It is a traditional approach in product development where all activities in the design process are carried out separately. The next stage cannot start until the previous stage is completed.

Serial engineering See sequential engineering.

Sheet molding compound (SMC) SMC is an intermediate material (uncured composite) in the form of a sheet to be molded in compression molding. The polymer used is mainly thermosetting polymer and the fibers are short glass fibers.

Simultaneous engineering See concurrent engineering.

Spray co-deposition It is fabrication process for MMCs. It is a process of atomizing a melt and adding the particles (reinforcement phase) into the spray of metal droplets. Codeposition of metal matrix and particles are done on to a substrate to produce MMC.

Spray-up This fabrication process of composite material involves spraying semicured chopped GRP composites using a spray gun onto the mold. Consolidation and elimination of bubbles are done using a hand roller.

Squeeze casting Squeeze casting is a liquid state processing to produce MMCs. In this process, the metal preform (reinforcement) is inserted into die cavity and molten metal (matrix) is poured into die cavity and the die is closed. As the metal matrix starts solidifying, the pressure is applied. The ram that closes the die is removed and MMC is extracted.

Stereolithography It is a rapid prototyping technique to produce 3D rapid prototyping object. CAD model is used in the process and the process involves slicing of the CAD model into a series of plan view image and a computer-controlled moving laser beam is used to develop a 3D object from polymer resin layer by layer until the whole object is developed.

Stiffness-to-weight ratio It is also known as specific modulus or specific stiffness. It is an elastic modulus divided by the material density.

Strand It is an untwisted bundle of fibers.

Strategy canvas Strategy canvas is a tool used to obtain value creation in BOS. It provides information on the current technology and the new technology to be proposed. Strategy canvas is depicted graphically in the value curve to compare the existing market and the offering of a new technology or system.

Strength Resistance to stress. Yield strength is the stress to initiate plastic deformation. Ultimate tensile strength is the maximum stress that can be borne by the material, calculated on the basis of the actual area.

Strength-to-weight ratio It is also known as specific strength. It is a strength of material (force per unit area) divided by density.

Tensile modulus See Young's modulus.

Tensile strength A measure of the force, per unit area, required to break a test specimen when it is placed between two clamps then drawn.

Thermoplastic polymer A polymer which is able to soften with the increase of temperature, thus becoming moldable, and rehardens on cooling.

Thermosetting polymer A polymer which is in liquid or semiliquid state before curing, but may undergo a chemical reaction to form a well-bonded three-dimensional cross-linked structure and hence become more insoluble and infusible after curing.

Throw over the wall syndrome It is a practice in traditional manufacturing companies where there is no communication between design and manufacturing engineers. They work in isolation; once the design is completed design engineers throw the design over the wall to manufacturing engineer.

Total design model A design model developed by Pugh for product design process that comprises market investigation, PDS, conceptual design, detail, design, manufacture, and sale.

Toughness Resistance to fracture by impact loading and measured by the Izod or Charpy tests which give the amount of energy required to fracture a standard sample. It can also be represented by the total area under stress–strain diagram.

TRIZ TRIZ is a problem-solving method taken from the study of hundreds of patents worldwide.

TRIZ 40 inventive solutions TRIZ 40 invention principles are unique solution ideas in overcoming technical contradictions in the inventions.

Twin screw extruder It is a machine to produce molding compounds for molding process like in injection molding process. Molding compounds are made in the form of pallets with fixed formulation on the percentages of fibers, polymers, and additives.

Unidirectional fiber composite In this composite, fibers are oriented in one direction only.

Vacuum bagging See vacuum bag molding.

Vacuum bag molding It is a composite manufacturing process employing a vacuum to consolidate the intermediate material (prepreg). This prepreg is laid up in a mold and enclosed in a vacuum bag.

Wettability The ability of one phase to spread over the surface of another. In composites, wettability refers to the ability of polymer to impregnate the fibers.

Why?Why?Why? It is a concept generation technique for expanding the search space. The behavior of a persistent child is imitated in terms of the habit of asking a lot of questions. By asking "Why?Why?Why?" some design problems can be solved.

Woven roving It is high performance reinforcement in the form of a plain coarse weave fabric generally made from glass fiber tow (untwisted continuous filaments).

Yarn Yarn composed of twisted strands of fibers or filaments, to be processed into fabrics.

Yield strength Stress that causes the material to reach its elastic limit, that is, resistance to the onset of plastic deformation.

Young's modulus (E) Stress per unit of elastic strain, being measured by the slope of the elastic part of the stress–strain curve, indicating the stiffness of material. It is important property in composites.

Index

A

ABS. *See* Acrylonitrile-butadienestyrene (ABS)
Acrylonitrile-butadienestyrene (ABS), 60
AFM. *See* Atomic force microscopy (AFM)
AFP. *See* Automated fiber placement (AFP)
AHP. *See* Analytical hierarchy process (AHP)
Aluminum-graphite composites, 83
Amalgamation, 285
Analytical hierarchy process (AHP), 15, 31, 120, 198, 245, 246
  for automative brake lever, 250–253
  case study, on composite material selection, 250–253
  in concurrent engineering, 247
  consistency index (CI), 247
  consistency ratio (CR), 247
  for environmental assessment, in composite material selection, 254–255
  formulation of matrix, 246
  framework, 253
  hierarchical framework of, 255
  industrial applications, 248
  methodology, for material selection, 252
  model for selecting polymer base type, for DPF, 249
  process flow chart of, 254
  random index (RI), 247
  sensitivity analysis test, 256
ANN. *See* Artificial neural network (ANN)
Artificial neural network (ANN), 15, 259
  applications, 260
  layers of, 259
  topological structure of, 259
Ashby's chart, 228–229
  sketch of, 232
A 39 system/engineering parameters, 191
Atomic force microscopy (AFM), 85
Automated fiber placement (AFP), 106
Automotive industry
  computer-aided design (CAD), use of, 67
  filament winding process, use of, 68
  hybrid fiber reinforced epoxy composites, 68
  natural fiber composites, use of, 67
Automotive pedal box system, 32, 235
  polymer composite, development from, 68

B

Biocomposites, 65, 282
  definition, 262
  environmental sustainability of, 282
  materials selection, 262–265
    materials hierarchy of NFCs, 264
  motorcycle helmet, 53
  product development, 149
Biological neuron, 259
Blue ocean strategy (BOS), 4
  actions, 4
  frameworks of, 4
BMC. *See* Bulk molding compound (BMC)
Boron fibers
  vs. carbon fibers, 63
BOS. *See* Blue ocean strategy (BOS)
Bulk molding compound (BMC), 75, 80
Bumper absorber, 172
Burden shifting, 280

C

CAD. *See* Computer aided design (CAD)
Calendaring, 213
Cambridge Materials Selector (CMS), 227
Carbon emissions, 282
Carbon footprint, 282
Car window regulators
  kenaf fiber composites, made from, 66
CE. *See* Concurrent engineering (CE)
Ceramic matrix composites (CMC), 84–85, 126
  constituents, 84
  fabrication processes, 84–85
  properties, 84
  shear-lag model, 84
Chemical vapor deposition (CVD), 10
Chemical vapor infiltration (CVI), 10
Chopped roving, 79
CIE. *See* Concurrent/integrated engineering (CIE)
Clean technology, 277
CMC. *See* Ceramic matrix composites (CMC)
Composite-based automotive clutch pedals, 149
  rapid prototyping models, 150

Composite boat fabrication, using hand lay-up process, 52
Composite characteristics (CC), 242
Composite clutch pedal, ribbing structures in injection molded, 53
Composite drain cover, 173
Composite energy absorber, 119
Composite fabrication, 97
characteristics, 97
Composite materials, 9, 29, 141
application at Airbus, evolution of, 110
automotive parking brake lever, 128
biocomposites
design, 165–167
films, 131
carbon fiber
recycling of, 104
ceramic matrix (CMC), 126
chemical vapor deposition (CVD), 10
clutch pedal, analysis of, 125, 126
conceptual design under CE environment, 16–20
fiber reinforced polymer composite, 17
generation techniques, 17
manufacturing processes, 17
morphological chart, 18
problem formulation for, 17
product design specification (PDS), 17
constituent elements, 9
design process of, 101
engineering applications, 10
fabrication technique, 10
fiber
development, 99
reinforced epoxy, 131
fully degradable biopolymer, 10
history of, 58–59
ancient history, 58
fibers, 58
matrix materials, 58
modern history, 59
hybrid, 116
development, 116
manufacturing processes, 10
materials selection, under CE environment, 20–22
importance of, 21
polybutylene terephthalate (PBT), 105
polymer
automotive bumper fascia, 119
design of, 100
pressure vessel, design of, 99
product development, 111, 127
phases, 113
properties of, 11
selection of, 211
advantages of, 211
for automotive components, 217
concurrent engineering for, 213–214
manufacturing processes, 213
need for, 212
permutation of candidate materials, 211
thermal analysis of, 101
use in high-performance applications, 10
Composite product development, 143
approaches of publication, 146
conceptual design in, 145–149
design methods, 149
importance of, 146
literature review, 145–149
properties, 151
Composite products design, preconcept generation methods, 150–152
Compression molding, 80, 213
phenol formaldehyde/bamboo composite panel, 72
Computer aided design (CAD), 5, 100
Computer-aided material preselection by uniform standards (CAMPUS), 227
Computer integrated manufacturing (CIM), 5
Computer numerical control (CNC), 5
Concept evaluation/selection methods, for composites, 195–203
direct comparison method, for concept design selection, 197
Pugh selection method for composites, 196–197
TOPSIS method, 203
using AHP for products, from composites, 198–203
AHP hierarchy framework, 201
concept designs, final eigenvector score, 202
design concept ratings, 201
natural fiber composites ironing board, case studies, 198–203
selection criteria, weight importance, 202
weighted objectives method, for design evaluation for composites, 195–196
Concept generation methods, for composites, 160–165
analysis of technical systems, 170–172
biomimicry/biomimetics, 164–165
design of composite products, 165
blue ocean strategy, 181–185
brainstorming, 161–164
for natural fiber composite product, 162
cross-industry innovation, 167–170
in design of composite products, 167
gallery method, 175–176
in composite product development, 176
mind mapping, 185–188
morphological chart method, 177–180
in design of composite products, 179
problems, 163
sessions, 163
Conceptual design, 47, 142–144
in CE environment, 144–145
Concurrent engineering (CE), 1–3, 29, 95, 102, 143, 273
aliases, 6
basic principles of, 6
benefits of, 7–8
categories, 6
composite design process
information technology (IT) tools in, use of, 97
phases, 98
in composite materials, 116–131
in automotive components, 126–130
bumper system, 116–121
in defense industry, 99
development/advances, 12–13, 106–109
components, 13
graph theory, 14

implementation of, 12
manufacturing process, 13
nature of product development, 13
previous studies of, 14–15
tools and techniques, 15
marine application, 131
pedal box system, 121–124
definitions of, 5
design for sustainability of composites under, 22–24
industrial use of composites, 111–115
advantages
high quality, 114
reduced capital cost and expense, 114
short development time, 115
influencing agents, categories, 96
life-cycle product development activities, 4
materials selection, 213–214
in product development, 4, 95, 111
early work (1989-99), 98–101
new millennium era (2000-09), 101–104
role in, 98–105
software tools, 108
CATIA, 109
FiberSIM software, 108
thermoplastic composite, design of, 98
Concurrent/integrated engineering (CIE), 106
Concurrent product development (CPD). *See* Concurrent engineering (CE)
Continuous laminated composites, 54
Corporate social responsibility, 277
Corrosion resistance, 141
Cost effective small aircraft (CESAR), 106
composite wingbox, 106, 107
Crashworthiness test, 16

D

Design engineer, 150
Design for assembly (DFA), 107. *See also* Concurrent engineering (CE)
Design for CE of composites, 49–54
Design for environment (DfE), 273
environmentally conscious design (ECD), 273
green design (GD), 273
life-cycle design (LCD), 273
Design for manufacture (DFM), 98, 106. *See also* Concurrent engineering (CE)
RTM process, implications on, 103
Design for metal inserts (DFMI), 14
Design for minimal weight (DFMW), 14, 103
Design for quality (DFQ). *See* Concurrent engineering (CE)
Design for recyclability (DFR). *See* Concurrent engineering (CE)
Design for sustainability (DFS), 5, 8–9, 273, 275–280
bioplastic for food packaging, case study, 285–291
economic impacts, 285
environmental impact, 286
manufacturability, 291
societal impacts, 287
design-for-environment (DfE) evolution, 274, 275
of natural fiber composite materials, 281–291
Design for the environment (DFE), 8. *See also* Concurrent engineering (CE)
design for, 9
green manufacturing, 9
Design of products, from composites, 156
Design tasks in composites, classification characteristics of, 49–54
branch, 51
complexity, 52
goals, 54
novelty, 50
organization, 50
origin of task, 49, 50
production, 51
Design to cost (DTC). *See* Concurrent engineering (CE)
Detail design, 49
DFA. *See* Design for assembly (DFA)
DfE. *See* Design for environment (DfE)
DFM. *See* Design for manufacture (DFM)
DFMW. *See* Design for minimal weight (DFMW)
DFS. *See* Design for sustainability (DFS)
Digital logic method (DLM), materials selection, 218
application, 218
determination, of weightage of selection criteria, 218
screening process, 218
Digital mock-up (DMU), 109
Dough molding compound (DMC), 75
D4S. *See* Design for sustainability (DFS)

E

ECD. *See* Environmentally conscious design (ECD)
Elimination and et choice translating reality (ELECTRE) method, 245
Embodiment design, 47–48, 101
Empty fruit bunch (EFB), 216
Engineering design, 54, 195
Environmental degradation, 275
Environmentally conscious design (ECD), 273
Epoxy resin, 221
Expert systems, 230–233
components of, 234
Exsys Corvid software, 238–239

F

FEA. *See* Finite element analysis (FEA)
Fiber reinforced composites, 142
Fiber-reinforced plastics, 59
Fiber reinforcement technology, 169
Filament winding process, 213
helical winding pattern in, 79
Finite element analysis (FEA), 5, 101, 213
Fourier transformed infrared spectroscopy (FTIR), 85
Friction stir welding (FSW), 107
FTIR. *See* Fourier transformed infrared spectroscopy (FTIR)

G

GD. *See* Green design (GD)
Glass
cloth, 59
fiber reinforced epoxy composite drive shafts, 68
fiber reinforced thermoplastic composites
goals, 54
fibers, 281
rovings, 62
types, 61
A-glass, 62
AR glass, 62
C-glass, 62
E-glass, 61
S-glass, 62
Glass mat thermoplastics (GMT), 75
Glass-sugar palm fiber, 180
GMT. *See* Glass mat thermoplastics (GMT)
Graphene-based composites, 9
Graphical user interfaces (GUI), 240
Graphite fiber, 62
Green-composite sandwich structure, 75
Green design (GD), 273

H

Hand lay-up process, 78, 213
HDPE. *See* High density polyethylene (HDPE)
High density polyethylene (HDPE), 152
Honeycomb structures, 166

I

Identified TRIZ solution principles, design strategy based on, 194
IMC. *See* Injection molding compound (IMC)
Information and communication technology (ICT) tools, 213
Information system for concurrent/integrated engineering (ISCIE), 101
Injection molded carbon fiber reinforced nylon composite, 75
Injection molding, 291
machine, 76
Injection molding compound (IMC), 76
Input module, defined, 213
In situ chemical reaction techniques
chemical vapor deposition (CVD), 85
chemical vapor impregnation (CVI), 85
Integrated morphological chart, 180
Integrated product development (IPD), 109. *See also* Concurrent engineering (CE)
ISCIE. *See* Information system for concurrent/integrated engineering (ISCIE)

J

Java-based materials selection, 239–241
login screen, for user authentication, 240
mechanism of Java language, 240
user interface, of module screen, 241
for web-based application development, 239

K

KBE. *See* Knowledge-based engineering (KBE)
KBS. *See* Knowledge-based system (KBS)
Kenaf fiber composites
Crashworthiness test, 16
parking brake lever design, contraction matrix, 193
Kenaf plant
fiber extracted from, 12
Kevlar, 62
Knowledge-based engineering (KBE), 13
Knowledge-based system (KBS), 229–233
automotive pedal box system, 235
product design specification of, 237
selection of material for, 236
case study, 234
components, 231
frame-based system, 234
hierarchical graph, of polymeric-based composite materials, 237
KBS tool-kit (KEE), 236

L

Laminated composites, 51
LCA. *See* Life-cycle analysis (LCA)
LCD. *See* Life-cycle design (LCD)
Life-cycle analysis (LCA), 274, 280, 281
Life-cycle design (LCD), 273
Liquid crystal olymers (LCPs), 60

M

Manufacturing function, 49
Materials selection, 209
attributes of, 214–215
biocomposites, 262–265
publications (2010-2015) on material selection for, 263
composite materials
Ashby's chart, 228–229
case study of, 234–238
expert systems, 230–233
knowledge-based system, 229–233
materials databases, 215, 226–227
multicriteria decision making, 241–245
neural network, 258–261
computer-based systems, 211
constraint satisfaction and constraint violation, 234
design stage, 209
digital logic method, 218–219
determination, of weightage of selection criteria, 220
integrated modified digital logic method, 219
integrated weighted property method, 219
knowledge required in, 210
procedure-based vs computer-based, 215
quality function deployment (eQFD), 226
range of materials, 209
rule-based system for, 233–234
scaled property value, 221
structure, of prototype system for, 235
use, of Exsys Corvid, 238–239
use, of material handbooks, 209
weighted property method (WPM), 218–219

ranking alternatives, with respect to selection criteria, 221
weighting factor for, 220
Matrix-fiber interface, 59
MCDM. *See* Multicriteria decision making (MCDM)
Melting pot, 100
Metal matrix composites (MMCs), 82–84, 129, 141
applications, 83
brake disk, 130
constituents, 82
fabrication processes, 82
self-healing, 83–84
Microelectronics, 88
MMCs. *See* Metal matrix composites (MMCs)
Mold release agent, 78
Monolithic alloys, 82
Monte Carlo simulation, 105
Multicriteria decision making (MCDM), 203, 218
for automotive dashboard panel, 242
comparison, of different natural fiber types, 244
combined specific strength, to cost ratio criterion, 245
specific modulus, 244
specific strength, 244
criteria of natural fibers for NFC materials, 243
evaluations, of natural fibers, 242
materials selection, 241–245
multiple attribute decision making (MADM) methods, 245
properties, of used fibers, 243
screening methods, 245
Multiple attribute decision making (MADM) methods, 245

N

Nanocomposites, 85–88
advantage, 88
characterization techniques, 85
classification, 86
National Blue Ocean Strategy (NBOS), 5
Natural fiber composites (NFCs), 18, 216
Natural fiber properties (NFP), 242
Natural fibers, 281
advantages, 63
composites, 281, 287
drawbacks, 65
plant-based, 281
processing methods, 283
sugar palm fibers, 287, 288
NBOS. *See* National Blue Ocean Strategy (NBOS)
Neural network, 258–261
architecture of, 260
definition, 259
structure of composite materials selection system, 261
NFCs. *See* Natural fiber composites (NFCs)
Nonlinear optics, 88

O

OEMs. *See* Original equipment manufacturers (OEMs)
Old Stone age, 58
Original equipment manufacturers (OEMs), 46
Orthotropic materials, 141
Out-of-autoclave (OOA) methods, 10, 213

P

PACKS. *See* Parametric Composite Knowledge System (PACKS)
Pahl and Beitz model, 35
Paleolithic age, 58
Parametric Composite Knowledge System (PACKS), 102
Parking brake lever component, 193
Particle swarm optimization (PSO), 33
Pattern of rib in composite clutch pedal, 147
PBT. *See* Polybutylene terephthalate (PBT); Polybutylene terepthalate (PBT)
PDM. *See* Product data management (PDM)
PDSs. *See* Product design specifications (PDSs)
PEEK. *See* Polyether etherketone (PEEK)
Plasticizers, 61
Plastics Computer Aided Material Selector (Plascams), 227
PMCs. *See* Polymer matrix composites (PMCs)
Polybutylene terephthalate (PBT), 105
Polybutylene terepthalate (PBT), 60
Polyester resin, 221
Polyether etherketone (PEEK), 60
Polyether sulfone (PES), 61
Polyethylene terephthalate, 165
Poly(lactic acid) (PLA), 23, 61
Polymer base properties (PBP), 242
Polymer matrix composites (PMCs), 59–65, 281
advantages, 65–67
applications, 67–75
aerospace industry, 69–70
hemp fiber composites, use of, 70
high-strength fibers, use of, 69
Naca cowling, use of, 70
nonbiodegradable resins, use of, 69
automotive industry, 67–68
building industry, 70–72
edge loading test, 70
hybrid glass fiber reinforced polymer-concrete beams, use of, 72
impact indentation test, 70
jute composite door shutters, use of, 70
shock test, 70
slamming test, 70
vacuum assisted resin transfer molding (VARTM) process, use of, 72
wood-plastic composites, use of, 71
furniture products, 73–74
utilized oil palm kernel shells, use of, 74
woven banana-stem fiber reinforced epoxy composites, use of, 73, 74
leisure, 75
marine application, 72–73
fiber-reinforced polymer (FRP) composites, use of, 73
hybrid glass-sugar palm fiber, use of, 72
polyurethane foam core, use of, 73
sporting goods, 75
automotive interior applications, 284
biodegradable polymers, 61
disadvantages, 67
fabrication techniques, 75

Polymer matrix composites (PMCs) (*cont.*)
fibers, 61–63
aramid fibers, 62
boron fibers, 63
carbon fibers, 62
manufacturing methods, 75–82
autoclave molding, 82
compression molding, 80
filament winding, 79
hand lay-up, 78
injection molding, 76
pultrusion, 81
resin transfer molding (RTM), 77
spray-up, 79
vacuum bag molding, 82
materials classes, 75
matrix materials, 60–61
natural fibers, 63–65
composites, 65
reinforced socio-economic sustainability of, 282
polymers used, 60
thermoplastic polymers, 60
thermosetting polymers, 60
Polymer matrix nanocomposites, preparation method, 87
Polymer pyrolysis techniques, 85
Polyparaphenylene terephthalamide, 62
Poly(vinyl chloride) (PVC), 58
Preconcept generation
problem definition, 150–152
problem formulation, 150
product design specification, 152–155
Preference selection index (PSI) method, 245
Product architecture, 48
Product data management (PDM), 109
Product design, 29
definitions, 31
design wheel, 45
process model, 37, 39–41, 43
consideration of materials, 43
developed by design experts, 32–42
engineering design, 38, 42
total design model, 32, 48
Product design specifications (PDSs), 32, 46–47, 131, 142, 152, 209
attributes, 154
elements, 154
information gathering, 156
based on expert knowledge and experiences, 160
books, 156
company brochures, and product catalogs, 159
composite material databases, 157
conference/seminar papers and proceedings, 158
indexes and abstracts, 157
intellectual property rights, 159
internet sources, 159
journal papers, 158
standards, 158
technical reports, 158
theses, 160
Product development, 29
definitions, 31
economic impact on, 280
and economic sustainability, 279
environmental impacts, classification, 277, 278
ecological damages, 277
human health damages, 277
resource depletion, 277
and environmental sustainability, 277
innovative design approach, 44
integration of TRIZ and morphological chart, 48
market investigation, 46
money spent and labor needed in, 31
social impact on, 278, 279
and social sustainability, 277
Product innovation, 275
Prototype, 148
PSO. *See* Particle swarm optimization (PSO)
Pugh's selection method, 216
case study of, 216–218
composite material selection for automotive components, 217
four-step procedure, 216
graphical method, 216
matrix, 216
Pultruded composite drain cover, 71
Pultruded composite rods, standard components, 51
Pultrusion process, 213
continuous fibers impregnated with polymer, 81
PVC. *See* Poly(vinyl chloride) (PVC)

Q

Quality function deployment (eQFD), 226
Quality function deployment for environment (QFDE), 226

R

Rapid prototyping, 121
Reinforcement fibers, 9
Resin, 60
Resin transfer molding (RTM), 10, 77, 100, 107, 213
RTM. *See* Resin transfer molding (RTM)

S

SAN. *See* Styrene acrylonitrile (SAN)
Scanning electron micrograph, of leaf, 166
Scanning tunnelling microscopy (STM), 85
Self-propagating high temperature synthesis (SHS), 85
Sequential engineering, 3, 101
Serial engineering, 2
Sheet molding compound (SMC), 80, 116
SHS. *See* Self-propagating high temperature synthesis (SHS)
SMA. *See* Styrene maleic anhydride (SMA)
SMC. *See* Sheet molding compound (SMC)
Spray-up processing, 213
Steel drain cover, 173
Stereolithography, 149
STM. *See* Scanning tunnelling microscopy (STM)
Structural component, from kenaf fiber reinforced composites, 142
Styrene acrylonitrile (SAN), 60
Styrene maleic anhydride (SMA), 60
Sugar palm fibers, in woven form, 23, 24
Sugar palm starch, 23

Sustainability
development, 274, 276, 277
elements of, 276
for pollution prevention, 277
product development and, 277, 279

T

TBL. See Triple bottom line (TBL)
Technique for order preference by similarity to ideal solution (TOPSIS) method, 245
Thermoforming, 213
Thermoplastic toughening, 118
Thermoset resin materials selection
case study on, 221
challenge, 221
criteria for, 222
properties, 223
ranking process, 223, 225
types of, 221
weighting process, 223, 224
Throw over the wall syndrome, 3
TOPSIS method, 256–258
principle of, 256
steps of, 257
trend in alternative ranking, in AHP and, 258
Total design model, 209
clarifying need, 44–45
description of stages, 43–49
Triple bottom line (TBL), 276

V

Vacuum assisted resin transfer molding (VARTM) process, 72
Vacuum bagging, 76, 213
VARTM process. *See* Vacuum assisted resin transfer molding (VARTM) process
Vinylester resin, 221
VIse Kriterijumska Optimizacija Kompromisno Resenje (VIKOR) method, 245

W

The Waterfall model, 1, 2
Weighted property method (WPM), 218
World War II, 59
Woven aramid fiber, 64
Woven carbon fiber, 63
Woven kenaf fiber, 64

X

X ray photoelectron spectroscopy (XPS), 85

FSC
www.fsc.org
MIX
Papier | Fördert gute Waldnutzung
FSC® C083411

Kolibri 360 GmbH
Ferdinand-Jühlke-Straße 7
99095 Erfurt, Deutschland
produktsicherheit@kolibri360.de